Titles in Series:

Sample Pretreatment and Separation

Analytical Chemistry by Open Learning

Author:
RICHARD ANDERSON
Wolverhampton Polytechnic

Editor:
NORMAN B. CHAPMAN

on behalf of ACOL

Published on behalf of ACOL, Thames Polytechnic, London
by
JOHN WILEY & SONS
Chichester · New York · Brisbane · Toronto · Singapore

Library of Congress Cataloging in Publication Data:

Anderson, Richard (Richard G.)
  Analytical Chemistry by Open Learning
Sample pretreatment and separation

  Bibliography: p.
  1. Chemistry, Analytic—Technique—Programmed
instruction.   2. Chemistry, Analytic—Programmed
instruction.   I. Chapman, N. B. (Norman Bellamy),
1916- .   II. ACOL (Project)   III. Title.
IV. Title: Sample pretreatment and separation.   V. Series:
Analytical Chemistry by Open Learning (Series)
QD75.4.S24A53   1987   543'.007'7   87-10655
ISBN 0 471 91360 X
ISBN 0 471 91361 8 (pbk.)

British Library Cataloguing in Publication Data:

Anderson, Richard
  Sample pretreatment and separation.—(Analytical
  chemistry).
  1. Sampling   2. Chemistry,Analytic
  I. Title   II. Chapman, Norman B.
  III. ACOL   IV. Series
  543'.02   QD75.4.S3

  ISBN 0 471 91360 X
  ISBN 0 471 91361 8 Pbk

Printed and bound in Great Britain

Analytical Chemistry

This series of texts is a result of an initiative by the Committee of Heads of Polytechnic Chemistry Departments in the United Kingdom. A project team based at Thames Polytechnic using funds available from the Manpower Services Commission 'Open Tech' Project has organised and managed the development of the material suitable for use by 'Distance Learners'. The contents of the various units have been identified, planned and written almost exclusively by groups of polytechnic staff, who are both expert in the subject area and are currently teaching in analytical chemistry.

The texts are for those interested in the basics of analytical chemistry and instrumental techniques who wish to study in a more flexible way than traditional institute attendance or to augment such attendance. A series of these units may be used by those undertaking courses leading to BTEC (levels IV and V), Royal Society of Chemistry (Certificates of Applied Chemistry) or other qualifications. The level is thus that of Senior Technician.

It is emphasised however that whilst the theoretical aspects of analytical chemistry can be studied in this way there is no substitute for the laboratory to learn the associated practical skills. In the U.K. there are nominated Polytechnics, Colleges and other Institutions who offer tutorial and practical support to achieve the practical objectives identified within each text. It is expected that many institutions worldwide will also provide such support.

The project will continue at Thames Polytechnic to support these 'Open Learning Texts', to continually refresh and update the material and to extend its coverage.

Further information about nominated support centres, the material or open learning techniques may be obtained from the project office at Thames Polytechnic, ACOL, Wellington St., Woolwich, London, SE18 6PF.

How to Use an Open Learning Text

Open learning texts are designed as a convenient and flexible way of studying for people who, for a variety of reasons cannot use conventional education courses. You will learn from this text the principles of one subject in Analytical Chemistry, but only by putting this knowledge into practice, under professional supervision, will you gain a full understanding of the analytical techniques described.

To achieve the full benefit from an open learning text you need to plan your place and time of study.

- Find the most suitable place to study where you can work without disturbance.

- If you have a tutor supervising your study discuss with him, or her, the date by which you should have completed this text.

- Some people study perfectly well in irregular bursts, however most students find that setting aside a certain number of hours each day is the most satisfactory method. It is for you to decide which pattern of study suits you best.

- If you decide to study for several hours at once, take short breaks of five or ten minutes every half hour or so. You will find that this method maintains a higher overall level of concentration.

Before you begin a detailed reading of the text, familiarise yourself with the general layout of the material. Have a look at the course contents list at the front of the book and flip through the pages to get a general impression of the way the subject is dealt with. You will find that there is space on the pages to make comments alongside the text as you study—your own notes for highlighting points that you feel are particularly important. Indicate in the margin the points you would like to discuss further with a tutor or fellow student. When you come to revise, these personal study notes will be very useful.

∏   When you find a paragraph in the text marked with a symbol such as is shown here, this is where you get involved. At this point you are directed to do things: draw graphs, answer questions, perform calculations, etc. Do make an attempt at these activities. If necessary cover the succeeding response with a piece of paper until you are ready to read on. This is an opportunity for you to learn by participating in the subject and although the text continues by discussing your response, there is no better way to learn than by working things out for yourself.

We have introduced self assessment questions (SAQ) at appropriate places in the text. These SAQs provide for you a way of finding out if you understand what you have just been studying. There is space on the page for your answer and for any comments you want to add after reading the author's response. You will find the author's response to each SAQ at the end of the text. Compare what you have written with the response provided and read the discussion and advice.

At intervals in the text you will find a List of Objectives which are based on the material you have just read. The Objectives will give you a checklist of tasks you should then be able to achieve.

You can revise the Unit, perhaps for a formal examination, by re-reading the Objectives, and by working through some of the SAQs. This should quickly alert you to areas of the text that need further study.

At the end of the book you will find for reference lists of commonly used scientific symbols and values, units of measurement and also a periodic table.

Contents

Study Guide

Welcome to this Unit on sample pretreatment, separation, and preconcentration! In the Unit we shall study ways in which raw samples received for analysis are converted into a form suitable for making useful measurements. If you have had any previous analytical experience, I am sure you have already found that it is most unlikely that you could analyse a sample in *exactly* the form in which you received it. Virtually every analytical procedure calls for some degree of sample pretreatment. In the simplest situation, this might merely involve, say, the filling of a sample cell. On the other hand, a complex and tricky sequence of operations might be called for, perhaps involving grinding a solid sample to give a fine powder, leaching this with acid, fusing the residue, taking the product up in acid, then complex separations and preconcentrations on the resulting solution, together with derivatisation steps at one or more stages.

In many analyses, possibly in most, the sample pretreatment stage is the longest part of the procedure, and the one requiring the highest degree of manipulative skill from the analyst. It is very important, therefore, that you appreciate the need to develop appropriate skills and knowledge in the field of sample pretreatment, separation, and preconcentration. The purpose of this Unit is to help you to acquire such skills.

The basic structure of the Unit is as follows. Part 1 reviews the major stages of an analysis and the position of pretreatment and separation in the overall analysis. Part 2 will be concerned with a general survey of a number of preliminary sample pretreatments which might precede other processes to be discussed later. We then go on to look at methods for dissolution and opening-out of samples, ie methods of obtaining a solution of the analyte (probably an aqueous solution of a species of interest in an appropriate chemical form) from either inorganic or organic matrices (Parts 3 and 4). In Parts 5 and 6 we study ways in which the analyte (now assumed to be in solution) is beneficially converted into a different chemical form, either to make its determination possible or more efficient, or to aid its separation, preconcentration, or further pretreatment. In Part 7 we introduce the topics of separation and preconcentration, initially in a very general way, and then review very briefly a wide selection of separation and preconcentration techniques. In Parts 8 to 10 we consider in more detail a number of the more important separation techniques including solvent extraction, ion exchange, adsorption and desorption, precipitation and co-precipitation, and use of vapour-phase equilibria. Perhaps I should point out at this stage that the most widely used group of separation techniques, namely chromatography and electrophoresis, although touched upon from time to time, will not formally be covered in this Unit. This subject is so large that five individual ACOL Units are devoted entirely to it.

Inevitably, I must assume that you, the reader, will have a certain level of preknowledge of chemical principles before you embark on the Unit. Most of these will be familiar to you if you have reached GCE 'A' level standard in chemistry. Where there is any doubt, I have provided exercises with remedial instruction in the responses, together with suitable references where necessary. In this way, you should be able to make good any deficiencies in your chemical background. On this point, you will find that the contents of *ACOL; Classical Methods*, especially those parts concerned with equilibria, complex formation, and precipitation, will be relevant to this Unit. Since pretreatment implies that an analysis will follow, we also refer from time to time to various common analytical measurements. No detailed knowledge of these will be assumed, but I shall assume that you have an interest in chemical analysis, and will therefore have a current awareness knowledge of at least some such techniques.

In studying this Unit, you will, no doubt, wish to consult textbooks and other references. Although there are some excellent coverages of such topics as solvent extraction and ion-exchange, some of the less common separation techniques, as well as much of the material on sample pretreatment, is rather scattered in the literature and is not very well covered in general textbooks on chemical analysis, where the emphasis is generally on more specific analytical (ie measurement) techniques. References are provided, but for some general textbooks that cover most (but not all) of this Unit, perhaps I could direct your attention to texts by Skoog and West and by Fifield and Kealey (see Bibliography).

A multiple choice test on the contents of the Unit itself follows this Study Guide and Bibliography. I want you to have a go at this test before starting on the main Parts of the Unit, but I do want you to understand that the material being tested is that which is to be covered in the Unit. *Therefore do not be surprised or worried if you score a low mark or even zero, as you may well do* (indeed, if you were to get a particularly high mark, I should wonder whether you need to study the Unit at all!) However, record your mark, whatever it is, in the place indicated, and then after you have completed your study of the Unit, I shall ask you to come back and do the test again. I hope that on your second attempt your mark will have substantially improved, the extent of the improvement being a measure of the amount that you have learned from the Unit.

Good luck with the Unit!

Preliminary Multiple-choice Pre- and Post-test

Work through this test before you start on the Unit and then repeat the test on completion of the Unit. Record your marks for both attempts in the boxes provided at the end of the test. Your first mark is likely to be low, but I hope that the mark for your second attempt will be significantly higher. The improvement in the mark obtained will serve as a rough guide to the extent to which you have benefited from your study of the Unit.

For each question simply select the most suitable response and note it on a *separate* sheet of paper. To mark the test, compare your selection of responses with those at the end of the test. Where they agree, ie where your response is correct, *score +1 mark*; where they disagree, ie where your response is incorrect, *score −1/3 mark*. If for any question you have not selected any response, *score 0 mark*. Finally, count up your marks, remembering to subtract the 1/3 mark for wrong responses. Your total mark will be out of 20.

1. The term 'leaching' means:

   (*a*) extraction of lipophilic compounds from an aqueous solution into an organic solvent;

   (*b*) heating a solid sample with an electrolyte to form a melt, which is then dissolved in dilute acid;

   (*c*) extraction of a soluble component from a solid sample by using a specific solvent or reagent mixture;

(*d*) exhaustive oxidation of an organic matrix to release the elements therein into an aqueous solution for inorganic analysis.

2. A simple procedure for obtaining an aqueous solution for determination of Al and Si in a finely powdered aluminosilicate would involve:

(*a*) dissolution of the powder in boiling concentrated HF to which NaF had been added (to raise the boiling point);

(*b*) exhaustive treatment by boiling the sample with the acids, HNO_3, H_2SO_4 and $HClO_4$ in that order;

(*c*) dissolution in a hot concentrated solution of NaOH;

(*d*) fusion with Na_2CO_3 and dissolution of the clear melt in dilute HCl containing a little H_2O_2.

3. Metallic gold is soluble in:

(*a*) concentrated H_2SO_4 plus concentrated H_3PO_4 (1 : 4 v/v);

(*b*) concentrated HCl plus concentrated HNO_3 (3 : 1 v/v);

(*c*) concentrated $HClO_4$;

(*d*) concentrated H_2SO_4 plus concentrated HF (1 : 1 v/v).

4. A Kjeldahl flask is:

(*a*) a sealed conical flask containing oxygen and used for dry-ashing small amounts of organic material held in a platinum basket;

(*b*) a sealed flask used inside a bomb for reductive fusions with sodium metal for halide determinations;

(*c*) a special flask that can be mounted in a radiofrequency field for low temperature dry ashings;

(*d*) a round-bottomed, long-necked pyrex flask used for wet-ashing organic material.

5. In dry-ashing a sample of salted fish for a trace determination of lead, the following precaution is particularly important:

(*a*) a platinum or nickel crucible must be used to avoid adsorption of Pb on, say, a glass surface;

(*b*) the temperature should not be too high or else the Pb is volatilised;

(*c*) the sample must be ashed in an oxygen-rich atmosphere to increase the oxidisability of the material in the presence of NaCl;

(*d*) the sample must be heated to at least 700 °C to ensure complete oxidation.

6. A common way of improving the volatility of fatty acids for analysis by gas–liquid chromatography is:

(*a*) to reduce them to the corresponding hydrocarbons;

(*b*) to convert them into their trimethylsilyl esters;

(*c*) to convert them into their methyl esters;

(*d*) to pyrolyse them.

7. The term *chelate effect* is used to describe:

(*a*) the increase in stability of a metal complex due to the formation of a ring structure on complex formation, when compared with that of a similar metal complex where no ring is formed;

(*b*) the increase in stability of the resulting metal complex observed by ensuring that a 'hard' metal ion is complexed with a 'hard' ligand, and likewise for 'soft' metal ions and ligands;

(*c*) the different stabilities observed for metal complexes differing only in ring sizes formed on complexation;

(*d*) the different stabilities of metal complexes, which differ only with respect to the presence or otherwise of large groups adjacent to the complexing atoms in the organic ligand.

8. The true stability constant of a metal complex is, to a first approximation:

(*a*) dependent on pH, with a sigmoid relationship;

(*b*) linearly dependent on pH, the slope being determined by the stoichiometry of the complex;

(*c*) independent of pH;

(*d*) dependent on pH only if the ligand is a weak acid.

9. A batch separation technique is most likely to be encountered when:

(*a*) a preconcentration is to be performed involving a particularly large preconcentration factor;

(*b*) the species to be separated have similar, but not identical, distribution ratios between the two phases involved;

(*c*) extreme differences exist in the separability of the species to be separated;

(*d*) the components of large amounts of material are to be separated.

10. Dialysis is a separation technique involving:

(*a*) differential diffusion, on the basis of size, of molecules in solution, through a semi-permeable membrane, separating two liquid phases;

(*b*) differential rates of migration of molecules in solution under the influence of an intense centrifugal force;

(*c*) differential rates of migration of charged particles under the influence of an electric field;

(*d*) fractionation of molecules in solution on the basis of molecular size by means of a molecular sieve (a solid material containing pores of controlled size).

11. Solvent extraction might be useful for:

(*a*) preconcentrating traces of F⁻ in water (the pH is adjusted to produce non-ionic HF, which is then extracted into a polar organic solvent);

(*b*) resolution of a mixture of amino acids on the basis of their differing solubilities in organic solvents at different values of pH;

(*c*) separation of the alkali metals on the basis of pH control and the differing stabilities of their complexes with a suitable organic ligand (eg 8-hydroxyquinoline);

(*d*) preconcentrating traces of a non-polar organic compound, present in a large volume of water, into a small volume of organic solvent.

12. Two metal ions form extractable neutral chelate complexes with the same organic ligand, although the stabilities of these complexes are very different. The metal ions are thus separable by solvent extraction of these complexes provided:

(*a*) the pH of the aqueous phase is carefully adjusted to a suitable value;

(*b*) the concentration of the ligand is carefully adjusted to a suitable value;

(*c*) the aqueous : organic-phase volume-ratio is carefully adjusted to a suitable value;

(*d*) a carefully chosen masking agent is added to the aqueous phase.

13. Ferrous and ferric ions can be separated by solvent extraction, by:

(*a*) extracting the ferric ions into diethyl ether from $6M$-HCl, leaving the ferrous ions in the aqueous phase;

(*b*) extracting the ferric ions into trichloromethane as ion-pairs formed between the *o*-phenanthroline complex of Fe^{3+} and ClO_4^-, leaving the ferrous ions in the aqueous phase;

(*c*) extracting the ferrous ions into diethyl ether from a concentrated aqueous solution of ammonium thiocyanate, leaving the ferric ions in the aqueous phase;

(*d*) extracting the ferrous ions into trichloromethane as ion-pairs formed between the neocuproine complex of Fe^{2+} and NO_3^-, leaving the ferric ions in the aqueous phase.

14. In order to fractionate a mixture of organic compounds in aqueous solution into neutral extractable, acidic extractable, basic extractable, and non-extractable compounds, by extraction into an organic solvent, you would:

(*a*) check that the aqueous phase is neutral and extract. Then extract solutes from the resulting organic phase (*i*) with an acidic, (*ii*) with an alkaline aqueous phase;

(*b*) acidify the sample and extract. Then make the aqueous phase alkaline and re-extract. Then back-extract solutes from the organic phase obtained from the first extraction with an alkaline aqueous phase;

(*c*) make the sample alkaline and extract. Then acidify the aqueous phase and re-extract. Then back-extract solutes from the organic phase obtained from the first extraction with an alkaline aqueous phase;

(*d*) acidify the sample and extract. Then make the aqueous phase alkaline and re-extract. Then back-extract solutes from the organic phase obtained from the first extraction with an acidic aqueous phase.

15. Increased cross-linking in an ion-exchange resin:

(*a*) increases its affinity towards ions of smaller solvated radius;

(*b*) increases its affinity towards ions of higher charge;

(*c*) increases it selectivity towards ions of higher charge;

(*d*) increases it selectivity towards ions of smaller solvated radius.

16. The best conditions for preconcentrating trace transition- and heavy-metal ions in sea water by ion-exchange are:

(*a*) to buffer the solution to pH 6–8, pass it through a chelating ion-exchange resin in the H^+ form and then collect the preconcentrated ions by eluting them with $2M$-HNO_3;

(*b*) to pass the solution through a strong cation-exchange resin in the Na^+ form, ash the resin and take up the residue in $2M$-HNO_3;

(*c*) to pass the solution through a strong cation-exchange resin in the H^+ form, collect the preconcentrated ions by eluting them with $2M$-HNO_3;

(*d*) to make the solution $12M$ in HCl pass it through a strong anion-exchange resin in the Cl^- form, and collect the ions by elution with water.

17. The best conditions for separating the components of a mixture of amino-acids by ion-exchange would be:

   (*a*) to retain the acids as cations on a cation-exchange resin and then elute the acids in sequence by using a series of buffers of gradually decreasing pH;

   (*b*) to retain the acids as anions on an anion-exchange resin and then elute the acids in sequence by using a series of buffers of gradually increasing pH;

   (*c*) to retain the acids as cations on a cation-exchange resin and then elute the acids in sequence by using a series of buffers of gradually increasing pH;

   (*d*) to convert the acids into their zwitterionic forms when they can be retained on either a cation- or an anion-exchange resin, then elute the acids in sequence, either by increasing the pH (cation exchange) or decreasing the pH (anion exchange).

18. Adsorption involves the reversible retention of molecules from the gaseous or solution phase:

   (*a*) within the bulk of a solid or liquid phase by mechanisms akin to solvent extraction;

   (*b*) on the surface of a solid phase by physical or chemical means;

   (*c*) in the pores of a porous solid medium permeable to small molecules, no chemical interactions being involved;

   (*d*) within the crystal structure of a solid by means of gaps in the crystal lattice of similar size and shape to the adsorbed molecules.

19. Coprecipitation:

   (*a*) is the only known method for preconcentrating anions in water;

   (*b*) can be used for the non-specific preconcentration of material in general in aqueous solution;

   (*c*) is a nuisance in gravimetry, where it often causes contamination of precipitates;

   (*d*) can be used for the gravimetric determination of trace metal-ions in sea water.

20. Reduction with $NaBH_4$ will cause the following elements, in inorganic form in aqueous solution to be converted into a volatile form:

   (*a*) Cl, Br, I;

   (*b*) Sn, Hg, Zn;

   (*c*) Ru, Os, Cr;

   (*d*) As, Se, Te.

Answers

Score $+1$ for each correct response, $-1/3$ for each incorrect response and zero for each missing response. Add up your marks, remembering to subtract any negative marks. The maximum score is 20.

Question No.	Correct Response	First Attempt		Second Attempt	
		Your Response	Mark	Your Response	Mark
1	(c)	☐	☐	☐	☐
2	(d)	☐	☐	☐	☐
3	(b)	☐	☐	☐	☐
4	(d)	☐	☐	☐	☐
5	(b)	☐	☐	☐	☐
6	(c)	☐	☐	☐	☐
7	(a)	☐	☐	☐	☐
8	(c)	☐	☐	☐	☐
9	(c)	☐	☐	☐	☐
10	(a)	☐	☐	☐	☐
11	(d)	☐	☐	☐	☐
12	(a)	☐	☐	☐	☐
13	(a)	☐	☐	☐	☐
14	(b)	☐	☐	☐	☐
15	(d)	☐	☐	☐	☐
16	(a)	☐	☐	☐	☐
17	(c)	☐	☐	☐	☐
18	(b)	☐	☐	☐	☐
19	(c)	☐	☐	☐	☐
20	(d)	☐	☐	☐	☐

Record your scores here:

First attempt – before studying the Unit.   / 20

Second attempt – after studying the Unit.   / 20

Supporting Practical Work

Many of the techniques described in this Unit are just that – techniques. In other words, they tend to be just part of an overall analytical procedure and in order to use them or to monitor their effectiveness, they must be used in conjunction with various other analytical techniques.

The following extensive list of laboratory exercises all involve sample pretreatment. A suitable selection of these exercises could be used to provide a laboratory course in support of this ACOL text.

The following key is a guide to the level of experience and the facilities required.

Key   E – elementary exercise,

       A – more advanced exercise,

       I – requires specialist instrumentation,

       S – has safety implications.

— To determine alloying elements in a standard alloy by atomic absorption, after dissolution of the alloy in an acid or acid mixture (EI).

— To ascertain the completeness of the sample dissolution and the amount of time, heating, etc required to dissolve, for a small number of alloys (including steels) in a selection of acids and acid mixtures (A).

— To determine the calcium and the aluminium content of a cement sample by dissolution of the sample in HF/HCl followed by atomic absorption (ESI).

— To determine the calcium, aluminium and silicon content of a cement sample following a fusion technique using NaOH, and to compare this procedure with the dissolution in HF/HCl, above (AI).

— To determine a trace-metal (Pb) in a sample of dried plant material by wet-ashing with HNO_3 and H_2SO_4 (or $HClO_4$), and atomic absorption (ESI).

— To use a dry-ashing technique on dried plant material, and to look for volatility losses (AI).

— To determine the nitrogen content of a protein or similar sample by using the Kjeldahl technique (E).

— To determine the phosphorus content of a plastic containing a phosphate plasticiser by colorimetry based on molybdenum blue after decomposition by:

(*a*) wet-ashing with HNO_3 and $HClO_4$,
(*b*) use of the oxygen flask,

and to compare the two methods (AS).

— To identify an unknown organic compound by using spot tests to identify the functional groups present, and then by using melting points of the compound and derivatives prepared therefrom (E).

— To determine the molar mass of a polyethyleneglycol sample by acetylation of the -OH groups and titration (E).

— To determine the distribution coefficient for the extraction of iodine from water into tetrachloromethane (E).

— To investigate qualitatively the separation of iron(III) and copper(II) by extraction from $6M$ aqueous HCl into diethyl ether (ethoxyethane) (E).

— To determine traces of lead in tap-water by colorimetry with dithizone following solvent extraction into trichloromethane and masking of interfering ions with citrate and CN^- (AS).

— To determine traces of lead in tap-water by preconcentrating the lead as its APDC complex into 2-methylpentan-4-one, and atomic absorption (EI).

— To construct a pH-extraction curve for an organic compound or a neutral chelate complex, by monitoring the extraction spectrophotometrically (A).

— To separate the Cl^- and Br^- ions in a mixture by anion exchange in a nitrate medium and by monitoring the separation titrimetrically (E).

— To separate the Co^{2+} and Ni^{2+} ions in a mixture by anion exchange in the presence of HCl, and by monitoring the separation by EDTA titration (E).

— To separate two amino-acids (eg leucine and glutamic acid) by variations in pH on an ion-exchange column and by monitoring the separation by TLC (E).

— To coprecipitate traces of lead ions onto a $BaSO_4$ precipitate and to detect the coprecipitated ions spectrographically or by XRF spectroscopy (AI).

— To monitor organic vapours in the atmosphere by adsorption onto charcoal or a porous polymer, solvent desorption, and GLC (qualitatively only) (AI).

Bibliography and References

1. F W Fifield and D Kealey, *Principles and Practice of Analytical Chemistry*, 2nd edition, International Textbook Co. Ltd., 1983.

2. D A Skoog and D M West, *Fundamentals of Analytical Chemistry*, 4th edition, Holt-Saunders International Editions, 1982.

3. G D Christian, *Analytical Chemistry*, 4th edition, Wiley, 1985.

4. P R Hesse, *A Textbook of Soil Chemical Analysis*, John Murray, 1971, p 392.

5. R Bock, *A Handbook of Decomposition Methods in Analytical Chemistry* (translated and revised by I L Marr), International Textbook Co. Ltd., 1979.

6. *CRC Handbook of Chemistry and Physics*, 65th edition (1984–1985), Chemical Rubber Publishing Company.

7. J Dolezal, P Povondra and Z Sulcek, *Decomposition Techniques in Inorganic Analysis* (English translation), Iliffe Books Ltd, 1968 (out of print).

8. T T Gorsuch, *The Destruction of Organic Matter*, Pergamon, 1970 (out of print).

9. H T Clarke, *A Handbook of Organic Analysis, Qualitative and Quantitative*, 5th edition (revised by B Haynes), Edward Arnold, 1975.

10. I Smith and J W T Seakins, *Chromatographic and Electrophoretic Techniques*, Vol I *Paper and Thin Layer Chromatography*, 4th edition, Heinemann, 1976.

11. F D Snell and C T Snell, *Colorimetric Methods of Analysis*, 3rd edition, Volume III and IV (1953–1954), plus supplements, Van Nostrand Rhinehold.

12. Z Holzbecher, L Divis, M Kral, L Sucha and F Vlacil, *Handbook of Organic Reagents in Inorganic Analysis* (translated by S Kotrly), Ellis Horwood, 1976.

13. J M Miller, *Separation Methods in Chemical Analysis*, Wiley, 1975.

14. B L Karger, L R Snyder and C Horvath, *An Introduction to Separation Science*, Wiley, 1973.

15. J A Dean, *Chemical Separation Methods*, Van Nostrand Reinhold, 1969 (out of print).

16. A K De, S M Khopkar and R A Chalmers, *Solvent Extraction of Metals*, D Van Nostrand, 1970 (out of print).

17. G H Morrison and H Freiser, *Solvent Extraction in Analytical Chemistry*, Wiley, 1957 (out of print).

18. M Marhol, *Ion Exchangers in Analytical Chemistry. Their Properties and Use in Inorganic Chemistry*, Volume XIV of Wilson and Wilson's *Comprehensive Analytical Chemistry*, Elsevier, 1982.

19. Q Riemann and H F Walton, *Ion Exchange in Analytical Chemistry*, Pergamon, 1970 (out of print).

20. L L Ciaccio (Editor), *Water and Water Pollution Handbook*, Volume 2, Marcel Dekker, 1971.

Acknowledgements

Figure 4.4a is reprinted with permission from C. E. Gleit and W. D. Holland, *Anal. Chem.*, **34**, 1454, 1962. Copyright 1962 American Chemical Society.

Figure 9.5a is redrawn from K. A. Kraus and F. Nelson, *Proceeding of the First United Nations Conference on the Peaceful Uses of Atomic Energy*, **7**, 1956. Permission has been requested.

Figure 9.5b is redrawn from K. A. Kraus and G. E. Moore, *J. Am. Chem. Soc.*, **75**, 1460, 1953. Permission has been requested.

Figure 9.5c is redrawn from S. Moore and W. H. Stein, *J. Biol. Chem.*, **192**, 663, 1951. Permission has been requested.

Figure 10.2a is copied from J. M. Miller, *Separation Methods in Chemical Analysis*, Wiley, 1975, with permission of John Wiley and Sons, Inc.

1. Introduction

1.1. THE ANALYTICAL SEQUENCE: THE MAJOR STAGES OF AN ANALYSIS

Before we start to study sample pretreatment in detail, it would be useful for you to think about *all* the stages that an analyst normally goes through when he carries out an analysis on a sample. This will help you to see how the pretreatment operations relate to the other unit operations in an analysis. I have listed these below, with notes in a fairly full form, rather as if the analyst were being called upon to solve a non-routine quantitative problem. I have chosen this type of analysis because it illustrates the full sequence rather well and is also a common situation in practice. You will have an opportunity to think about the analytical sequence in other situations afterwards.

Here is the list of stages that would be involved.

(*a*) *Definition of the Problem.* This is the first stage in planning the analysis, and can be considered as the process of deciding what analytical information is needed.

(*b*) *Choice of Method.* Having decided *what* information is required, you next decide in detail *how* you are going to obtain it.

(*c*) *Sampling.* The process of selecting and removing a small, representative part of the whole, on which you will perform the analysis.

(*d*) *Sample Pretreatment and Separation.* Only in the simplest of situations will the sample be suitable for analysis without some form of pretreatment, in order to convert it into a form for suitable analysis. This pretreatment may or may not involve some form of separation.

(*e*) *Measurement.*   Obtaining the raw analytical data from measurements on the pretreated sample.

(*f*) *Calibration.*   Obtaining raw analytical data from suitably prepared standards.

(*g*) *Evaluation.*   Working out a meaningful analytical result from the raw data obtained in (*e*) and (*f*).

(*h*) *Action.*   Here we use the result to make a decision as to what action to take in relation to the original problem.

∏   Now produce a list similar to the above, but for a *qualitative* analysis.

The list that you should have produced might look something like this:

(*i*)   definition of the problem,
(*ii*)   choice of method,
(*iii*)   sampling,
(*iv*)   sample pretreatment,
(*v*)   qualitative tests on the pretreated sample,
(*vi*)   tests on reference materials for comparison,
(*vii*)   interpretation of the tests,
(*viii*)   action.

∏   I want you now to select an analytical procedure with which you are familiar. It might be a method that you use at work, or a practical exercise, or an example of an analysis from a textbook. Then try to see if you can divide up the various unit operations involved to correspond with the analytical sequence that we have just described. Write a sentence or two under each of 8 headings which I want you to provide. Here is some space below for you to do this.

(*i*)

(*ii*)

(*iii*)

(*iv*)

(*v*)

(*vi*)

(*vii*)

(*viii*)

A few points might emerge from this exercise.

(*i*)   If the analysis was routine in nature, you may have had difficulty wondering what to write under headings (*i*) and (*ii*). However, this only means that these questions have been answered previously by the person who set up the routine analysis.

(*ii*)   The sampling may not have been carried out by the analyst. but by some other person.

(*iii*) As an analyst, you may simply *report* the result to someone else, rather than actually taking any action.

(*iv*)   In some highly automated instrumental analyses, operations (*v*), (*vi*), (*vii*) and possibly (*iv*) as well, may all be performed automatically by the instrument. They do, however, still occur. Operation (*viii*) may also be automated in an automatic process-control instrument.

(*v*)   A few absolute methods of analysis (eg gravimetry) may not need a calibration stage.

However, what should be apparent to you from all this is the fundamental importance of the *sample pretreatment stage*. Indeed, we shall see that as far as input of time and effort are concerned, it is often this stage of the analysis that is the most costly and involved. Furthermore, the total errors involved in the analysis may largely be governed by what happens at the pretreatment stage. So it is very important that we get this part of the analysis right.

| SAQ 1.1a | List, from memory, the stages of the analytical sequence as we have described them for a qualitative and for a quantitative analysis. |

SAQ 1.1a

1.2. EFFICIENCY OF AN ANALYSIS

So far, we have established that virtually every analytical method involves some sort of sample pretreatment and that often this will be a significant and major part of the overall procedure. When we are evaluating an analytical method, deciding whether its performance is suitable for our purposes, or when we are comparing two methods with a view to selecting the better one, it is important that the pretreatment stages are considered very carefully indeed. What is involved here may often be the deciding factor.

What happens at the pretreatment stage may well govern:

(*a*) the overall accuracy and reliability of the results obtained;

(*b*) the total amount of time and effort involved in performing the analysis. You would usually want to choose a method that involves the *minimum* amount of sample pretreatment, *provided* that it was capable of delivering results of acceptable accuracy and reliability. It is interesting to note that many modern instru-

mental methods of analysis require *much less* sample pretreatment than do the older classical methods, and as a result are
more attractive to the analyst for this reason alone. For example, compare the sample pretreatment procedures for analysing
a steel for the nickel present by a titrimetric method, and by X-
ray fluorescence (a classical and modern instrumental method),
respectively.

	Titrimetry		*X-ray Fluorescence*
(*a*)	Dissolve the sample by using a suitable mixture of acids.	(*a*)	Cast the sample into the shape of a flat disc.
(*b*)	Remove the excess of acid by evaporating the solution nearly to dryness.	(*b*)	Polish the surface of the disc.
(*c*)	Make up to a suitable volume, at the same time adjusting the solution to a suitable pH.	(*c*)	Mount it in the spectrometer.
(*d*)	Precipitate the nickel as its dimethylglyoxime complex.		
(*e*)	Filter off the precipitate.		
(*f*)	Redissolve the precipitate in acid.		
(*g*)	Adjust the pH and add a suitable indicator.		

Surely, there can be no doubt in your mind which procedure is the
simpler, at least from the point of view of sample pretreatment! In
fact, taken as a whole, the X-ray method involves more operations at
the calibration stage, but nevertheless it is still faster overall, because
of the simpler pretreatment operations.

SAQ 1.2a

You are required to choose between an atomic absorption and a solution spectrophotometric method for the routine determination of lead ions in tap water samples. The concentrations of the lead ions in the samples are likely to be in the region of 0.01 to 0.1 mg dm^{-3}, and we need to know the concentration to the nearest 0.005 mg dm^{-3}. Here are some facts about the two methods.

	Atomic Absorption	Solution Spectrophotometry
Accuracy at 0.1 mg dm^{-3}	± 5%	± 0.2%
Limit of detection	0.005 mg dm^{-3}	0.002 mg dm^{-3}
Sample Pretreatment	Addition of Reagents	Buffering, Addition of Reagents, Solvent Extraction, Back Extraction, Re-extraction (different conditions each time).

Which method would you recommend? Explain your choice.

SAQ 1.2a

Objectives

After having studied this Part of the Unit, you should:

● be aware of the stages of the analytical sequence, be able to name these, and be able to describe an analysis with which you are familiar in terms of them, and see the relationship of sample pretreatment to the other stages of the sequence;

● be aware that a method with a simple sample pretreatment stage would be more attractive to an analyst than one with a complex sample pretreatment stage, provided that its performance in every other respect was satisfactory.

2. Preliminary Sample Pretreatment

2.1. INTRODUCTION

This Part of the Unit is concerned with a fairly broad selection of preliminary sample pretreatments. These, in general, are procedures that are applied to the sample in the state in which it is received and before anything else is done to it. The implication is that often (but not always, as we shall see) other pretreatment procedures will follow these preliminary operations. By and large, physical, rather than chemical, operations are involved here.

The operations that we shall be concerned with include:

(*a*) grinding of solids,
(*b*) preparation of samples for solid-state analysis,
(*c*) drying of samples,
(*d*) leaching and extraction of components from solid samples,
(*e*) filtering operations.

Let us now start our considerations of these procedures without further ado.

2.2. GRINDING OF SOLIDS

Very often, when you have a solid sample to analyse, the first operation that you might wish to carry out is to grind it to produce a fine powder. There are a number of reasons why you might want to do this. The most important of these are:

(*a*) finely powdered samples are more homogeneous and can be subsampled with greater precision and accuracy if carefully mixed;

(*b*) finely powdered samples are easier to dissolve because they present a large surface-area : volume ratio to any potential solvent or reagent used in dissolution;

(*c*) samples must be finely powdered before they can be compressed into discs, or fused and made into beads for such solid-state techniques as infra-red and X-ray fluorescence spectroscopy.

The commonest practical technique for grinding solids is to use the time-honoured *pestle and mortar* and plenty of 'elbow-grease'. This is quite satisfactory for many purposes, but there are some important alternative techniques as follows.

(*a*) *Use of the Diamond Mortar*

This is useful for shattering very hard and brittle substances. The device, which is made out of hardened steel, is of cylindrical construction. Its operation is best demonstrated by a series of pictures. See Fig. 2.2a, 2.2b and 2.2c.

Fig. 2.2a.   *Diamond mortar (front view) containing uncrushed sample*

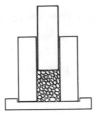

Fig. 2.2b. *Reasonably close fitting steel rod placed on the sample*

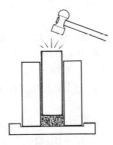

Fig. 2.2c. *Sample pulverised by hammering on the rod*

(*b*) Use of the Ball Mill

For soft materials, a ball mill may be more suitable. This consists of a porcelain crock in which you place the sample to be ground, together with a number of hard flint or porcelain balls. The crock is then sealed and mounted in a firm clamp and the whole is then shaken, rotated or vibrated mechanically, normally by a small but powerful electric motor.

After you have ground the sample, you may well wish to screen it for its maximum particle size. This can be done by using a set of sieves of known but differing mesh number. The mesh number of a sieve defines the diameter of the holes in the sieve, a *higher* mesh number indicating a *smaller* diameter or *finer* sieve. If all your sample passes through a sieve with a particular mesh number, then you know for certain that you have no particles larger than the diameter defined by that mesh number. Any particles that on a first attempt do not pass through the sieve can then be reground until they do.

∏   What kinds of error might be introduced into the analysis by the grinding operation?

The grinding operation involves a lot of friction, which can generate quite intense local heating. This may cause loss of volatile components from the sample, as well as thermal decomposition of a thermally unstable sample.

Also grinding will bring the sample into very close contact with other surfaces, and there is a distinct possibility of contamination, either from the material from which the mortar, or other components, is constructed, or, perhaps more likely, from previous samples.

∏   How could you grind materials, such as rubber or plastics, which are malleable or elastic, but not brittle?

The implication here is that we mean by 'not brittle', not brittle at room temperature. However, most substances including rubbers and plastics, are brittle at liquid-nitrogen temperatures. So the trick is to cool the sample and the grinding apparatus down to liquid-nitrogen temperatures, and then grind the cold sample while it is in fact brittle.

| SAQ 2.2a | Which of the following statements are true (T) and which are false (F)? Explain your choice of answer in each case.

(*i*) It is advantageous to grind solid samples to a small particle size, not only to reduce subsampling errors, but also to aid attack by solvents or reagents used to decompose the samples.
<div align="right">(T / F)</div>
<div align="right">⟶</div> |

SAQ 2.2a (cont.)

(*ii*) A diamond mortar is a device for pulverising especially hard samples. It is so-named because the tool used to shatter the sample is diamond-tipped.

(T / F)

(*iii*) Samples which are too hard to grind with a pestle and mortar can be more effectively ground with a ball-mill.

(T / F)

(*iv*) After a sample has been ground, its maximum particle size can be ascertained by passing the sample through sieves of known mesh number. The higher the mesh number of the sieve through which the sample completely passes the smaller is its maximum particle size.

(T / F)

2.3. PREPARATION OF SAMPLES FOR SOLID-STATE ANALYSIS

Certain fairly specialised techniques for examining solid samples require the sample to have a smooth flat surface, so that a beam of radiation, perhaps, can be reflected without scatter from this surface. Other techniques may require a smooth flat solid disc for transmission spectroscopy. In particular, *microscopy, X-ray fluorescence spectroscopy, emission spectroscopy*, and *infra-red spectroscopy* are some such techniques. Some sample preparation techniques for solid-state analysis are:

(*a*) *cutting* (eg with a *microtome* which is in effect a very fine knife used for cutting very thin sections of biological samples for microscopy);

(*b*) *polishing*, needed, say, to produce an optically flat surface for the most accurate emission or X-ray fluorescence spectroscopy;

(*c*) *compression into disc form* of powdered solids for infra-red spectroscopy (where the solids are diluted with KBr) or for X-ray fluorescence spectroscopy (the powdering is a good example of the use of a ball-mill for producing powdered solids);

(*d*) *casting of thin films*. Plastics and polymers that soften and melt on being heated (so-called 'thermoplastic' polymers) can be cast into films for X-ray fluorescence or infra-red spectroscopy, by placing the molten sample in a dish with a very smooth flat bottom, allowing it to cool and then peeling off the film. There is a similar technique for soluble polymers. A solution of the polymer is poured into the dish, the solvent is evaporated off and the film peeled off as before.

(*e*) *chill-casting*, in which a sample of alloy (eg a steel) is melted in a small furnace, cast into a suitable shape and then cooled very rapidly. This is useful both for getting the sample into a suitable shape, say for emission spectroscopy, and also for reducing its heterogeneity, since segregation of individual components will not have time to occur because of the rapid cooling.

I do not wish to go any further with these pretreatment techniques in this Unit, because here we are looking only for a general current-awareness of the types of operation that might be required for these very specialised analytical techniques. You should find more practical details on solid-state sample preparation if you study the Units on infra-red, X-ray fluorescence, and emission spectroscopy, as well as microscopy.

∏   For each of the following, suggest *why* the sample should show the property indicated (achieved through a pretreatment procedure described above).

    (*i*)   In any transmission spectroscopic or microscopic technique involving solid samples, the sample, if in a relatively pure state, should be in the form of a *thin* film.

    (*ii*)   In transmission infra-red spectroscopy, where the solid sample is compressed into a flat disc, the sample should be finely powdered and mixed with a large excess of KBr (transparent in the relevant ir region), also finely powdered (two points to be considered here).

    (*iii*)   In spark emission spectroscopy the sample (an alloy) should have a homogeneous structure with any segregation of components by crystallisation kept to a microscale.

These are the reasons for needing the above properties.

(*i*)   Since the beam of radiation in spectroscopy, or illuminating beam in microscopy is to be transmitted (ie passed through the sample) we must have a thin section, otherwise the sample will be *too opaque*, and we shall not observe any radiation at all. (Rather like diluting a solution in other forms of spectroscopy).

(*ii*)   The first point is that the KBr (which itself does not absorb infra-red radiation in the normal wavelength range) *dilutes* the sample, rather like a solvent does in solution spectroscopy, enabling you therefore to make a thicker and thus more easily

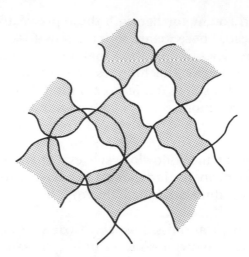

Fig. 2.3a. *Sampling by a spark of a surface with large individual crystals. The area within the circle is the area sampled by the spark. In this particular example, the white component is over represented*

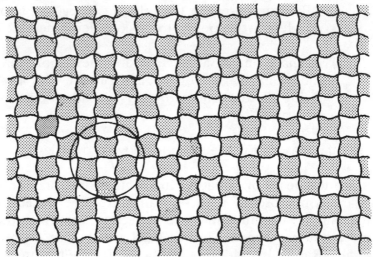

Fig. 2.3b. *Sampling by a spark of a surface with small individual crystals. The area within the circle is the area sampled by the spark. Here sampling is much more representative*

handlable disc. Secondly, the sample and KBr must be finely powdered so that a clear, homogeneous disc can be produced. Large particles will result in the production of a poor quality disc that scatters radiation, leading in turn to poor quality spectra.

(*iii*) The spark is quite small in size and therefore samples only a small area on the surface of the electrode (made out of the material that you are analysing). If this surface consists of large crystals of different components, the spark is likely to sample in a non-representative way. On the other hand, if the individual crystals are small, then the sampling by the spark will tend to be more representative. Have a look at Fig. 2.3a and Fig. 2.3b to see what I mean.

SAQ 2.3a

Name *three* analytical techniques that can be used directly on solid samples, and write one sentence per technique describing the form in which the sample might be presented in the technique.

2.4. DRYING OF SAMPLES

Many solid (and liquid) samples are received for analysis in a damp or wet state. From an analytical point of view, this means that the samples contain a potentially large, variable, and indeterminate amount of water associated with them. Furthermore, this water content may vary on storage either through evaporation or through absorption of water vapour from the atmosphere. This is particularly important for biological samples such as foodstuffs, and plant and animal material in general. It would be impossible to analyse such a sample quantitatively with the results having any real meaning unless, in some way, the water content was standardised between samples. This is normally done by *drying the sample* before analysis by some standard procedure. For biological samples a common standard procedure is to dry the sample in an oven at a specified temperature, normally close to 100 °C, and to record any weight loss. When the sample no longer loses weight, it is assumed to be dry and is then suitable for analysis. This procedure is known as *drying to constant weight*.

Alternatives to the above include drying in a vacuum desiccator, and *freeze drying*, where the sample is first frozen and then dried in the frozen state under vacuum. This latter procedure is useful for heat-sensitive samples, or where loss of volatiles is a problem. Minerals such as aluminates and silicates, on the other hand, may need temperatures of up to 1000 °C to dry them.

∏   Suppose that after you had dried your samples in the manner described above, it was not convenient to carry out the analysis immediately. What precautions should you adopt with regard to storing your samples after drying them, but before analysis?

Keep them in a desiccator to prevent them from picking up water from the atmosphere. Before analysis check that the weight has not altered.

∏   Suggest two potential sources of error when oven-drying samples, and suggest a drying procedure that might be less prone to such errors.

(*i*) Loss of volatiles other than water.

(*ii*) Thermal degradation (decomposition) of the sample.

Freeze-drying was suggested as a drying procedure less prone to these problems. Degradation of samples is likely to be especially severe in the presence of atmospheric oxygen. This can be eliminated by the use of a vacuum desiccator or by drying the samples under nitrogen.

∏   Suppose you had some samples for analysis containing differing amounts of water, but it was not convenient to dry them by one means or another. What alternative approach could you adopt for accounting for the water content?

Determine the water content itself by a separate analysis. There are a number of ways of determining water in various types of sample which it would be inappropriate to discuss in detail here. However, if you are interested in this subject, then you could make a little project of finding out about methods of water analysis. Reference 2, Chapter 24, would provide an excellent start for such a project.

SAQ 2.4a   Select from the following the procedure most likely to be adopted in practice for the preliminary pretreatment of samples of plant material, such as grasses or mosses before trace-metal determination.

(*i*) Store in a sealed container over water until the water content of the sample has been equilibrated with that of the air above it. Check that this has been achieved by weighing the sample and finding when it has come to constant weight.

(*ii*) Freeze the samples and then dry them to constant weight under vacuum, and keep them in the frozen state.   ⟶

SAQ 2.4a
(cont.)

(*iii*) Dry the sample to constant weight in a furnace at 1000 °C. Store in a desiccator and
check the weight before analysis.

(*iv*) Dry the sample to constant weight in an
oven at 105 °C. Store in a desiccator. Check
the weight before analysis.

2.5. LEACHING AND EXTRACTION OF SOLUBLE COMPONENTS

A very important aspect of many analyses is the need to obtain our
analyte in the form of a *solution*, either in water or in an organic
solvent. Indeed, this is so important that the next two Parts of this
Unit will be very largely concerned with this subject alone.

However, there are many situations where we don't want to dissolve the sample completely, but rather we try to dissolve out or extract only certain soluble components of the sample, leaving the rest of the sample (the insoluble fraction) in the solid state. The term 'leaching' is used here to describe a procedure whereby one or more components of the sample are taken into solution, perhaps accompanied by a chemical reaction. The solution is then separated from the solid phase, perhaps by filtration. Let us look at a couple of examples of the use of this type of technique in practice.

2.5.1. Leaching of Soluble Metallic Compounds from the Soil

The soil sample, after appropriate preliminary pretreatment, (see next exercise) is treated with one of the following:

(*a*) water,
(*b*) dilute acid,
(*c*) a buffered aqueous solution (eg aqueous ammonium acetate),
(*d*) an aqueous solution of an organic metal-complexing agent.

These solutions will dissolve such components of the soil as are soluble under the given conditions, but none of them will dissolve the soil completely. You can then filter off the solid residue, wash it with fresh reagent solution, and thereby separate the soluble and insoluble components of the soil. This is often used for trace metal analysis of soils. It is not, however, used for determining the total amount of any trace metal ion in a soil sample, only that proportion of the metal ion that is extracted under these conditions. Leaching is not recommended for use when a quantitative total recovery of a particular analyte is required. Under these conditions it is necessary to get the whole sample into solution.

∏   What preliminary pretreatments might be appropriate before leaching of the soil sample?

Drying the sample to constant weight and grinding it to pass a given mesh-number sieve.

Π   It is quite probable that any given metal ion is not com-
         pletely recovered in solution by this technique, since it may
         be present in the soil in different chemical forms, some of
         which are soluble and some of which are not. However, de-
         spite this fact, often useful analytical information may be
         obtained by this technique. How so?

If the purpose of the soil analysis is to determine the amount of
metal ion *available to plants growing in the soil*, which is often the
case, then it is reasonable to assume that this will correspond to
the soluble forms of the metal ion only, which is what we obtain
by leaching. It is important to make sure that the conditions under
which we leach our soil sample correspond as closely as possible
with the plant's ability to extract metal ions. This is why our list
of leaching agents includes neutral ammonium acetate buffers, and
solutions of weak complexing agents and the like.

SAQ 2.5a   Explain fully how you would pretreat a soil sam-
                   ple in order to determine the cobalt content
                   available to plants growing in the soil. The pre-
                   treatment should result in a clear, aqueous solu-
                   tion suitable for a final analysis by, say, atomic
                   absorption spectrometry.

2.5.2.   Extraction of Soluble Organic Components from an Insoluble Matrix (eg a Plastic or Rubber)

Plastics and rubbers are liable to contain a number of simple organic compounds such as:

(*a*) residual monomers,
(*b*) plasticisers,
(*c*) antioxidants,
(*d*) accelerators.

These compounds may have important effects on the properties of the plastic or rubber, and many of them also are toxic. Their determination is thus important, and if the plastic or rubber is insoluble, then the analytes must be leached out or extracted into a suitable organic solvent. The sample to be dealt with should first be ground or cut into fairly small pieces to present to the extracting solvent a large surface area per unit mass of sample. If necessary, this can be done by first freezing the sample with liquid nitrogen as we noted previously.

The extraction itself can be carried out simply by shaking the sample with the solvent and filtering, or by refluxing the sample with hot solvent (to improve solubility). The best way of extracting difficultly soluble components, however, is to carry out a *Soxhlet Extraction*. Fig. 2.5a shows a Soxhlet extraction apparatus.

The sample is placed in the Soxhlet thimble, a disposable porous container made out of stiffened filter paper. You should assume that solvent and extracted solute molecules, but not solid sample particles, can flow freely through the thimble. Refluxing organic solvent flows through the thimble and leaches out soluble components (remember, the sample will be at the boiling-point of the solvent). The solvent with extracted components flows through the thimble and the inner tube gradually fills up with the solution. When the level reaches the side-arm, however, all the solvent plus extracted components siphon off into the flask, and the thimble starts to fill with solvent again. The advantages of this procedure are as follows:

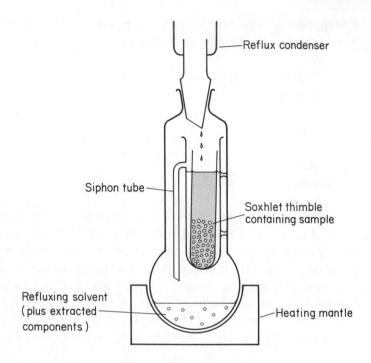

Fig. 2.5a. *Soxhlet apparatus for the continuous extraction of solutes from solids. In this diagram, the solvent is just about to siphon from the Soxhlet thimble into the heated reservoir*

(*a*) extraction of the sample always occurs with pure solvent (because it distils into the sample);

(*b*) extraction occurs at an elevated temperature (essentially that of the boiling-point of the solvent);

(*c*) extraction may be continued unattended until complete.

One disadvantage is that the extracted solutes are being continually heated in the solution-containing vessel. This might cause thermal degradation if the solutes are thermally labile. If this is so, use a solvent with a low boiling-point, such as dichloromethane or diethyl ether (ethoxyethane).

SAQ 2.5b   You are provided with a sample of rubber which
            is to be analysed for residual accelerators and
            antidegradants. The analytical technique to be
            used is high-performance liquid chromatogra-
            phy which requires a solution of the analytes in a
            non-viscous, non-uv absorbing solvent. The ac-
            celerators and antidegradants that are possibly
            present appear to be soluble in a range of halo-
            genated solvents as well as esters and ethers, but
            not in alcohols or water. You have some doubts
            about the thermal stability of some of these com-
            pounds.

            How would you set about obtaining a solution
            for analysis in this situation?

2.6. FILTERING OF LIQUIDS AND GASES

If you are analysing a sample that is a liquid, a solution, or a gas, then it may well contain suspended solid particles. If this is so, it may be necessary to separate the suspended solids from the bulk of the sample. This would normally be done by conventional filtration techniques with which I assume you are familiar, although centrifuges may also be used.

The need to separate suspended solids, say, from a solution may be because:

(*a*) the suspended particles might interfere in the subsequent analysis (eg they could cause scattering of light in solution spectrophotometry or block nebulisers in flame spectroscopy);

(*b*) the suspended particles are to be the subject of a separate analysis in their own right.

A very simple example of the latter occurs in water analysis. The water sample is filtered through a glass-fibre filter paper, previously dried to constant weight and its weight recorded. After the sample has been filtered, the paper is once again dried to constant weight and its new, increased, weight recorded. The increase in weight is then taken to represent the *total suspended solids* in the water sample, an important analytical parameter of any potable water.

You can now continue the analysis of the water sample, since the filtrate is now particle-free and can be analysed for various species in solution. Furthermore, the residue can be leached off the filter paper and analysed in solution. Note that you could find a particular analyte both in the solution and in the suspended solids. So, for instance, you might find the iron content of a water sample as follows:

(*a*) soluble Fe,
(*b*) Fe in the suspended solids,
(*c*) total Fe [the sum of (*a*) and (*b*)].

Π   When might the suspended solids cause problems in filtering or centrifuging processes? (Think – have you ever had problems in filtering solutions in the laboratory?)

I am thinking in particular of when a *colloidal* suspension is formed. In a colloidal suspension, the particles are so small that they remain suspended in solution in a stable state. This means that they are also too small to be trapped by any conventional filter. They are, however, large enough to scatter light in solution spectrophotometry.

Another situation is when gelatinous precipitates are formed, such as with metal hydroxides. The lightly charged particles fail to coagulate and filtration of such substances can be very slow.

SAQ 2.6a   Explain how you could distinguish between the total iron concentration and the soluble iron concentration in a sample of tap-water.

Objectives

After studying this Part of the Unit, you should now be able:

● to understand the need for grinding solid samples into fine powders and to describe techniques for doing this;

● to identify analytical techniques for use with solids, and to describe, in outline only, how the samples must be pretreated for such analyses;

● to appreciate that certain types of sample must be dried to constant weight before analysis, and to explain how this may be done;

● to describe how soluble components can be leached or extracted from an insoluble matrix;

● to explain how filtration techniques can be used to distinguish between an analyte present in solution and in the form of suspended particles in a liquid sample.

3. Decomposition and Dissolution of Inorganic Solids

3.1. INTRODUCTION

An important fact that I would like you to appreciate right at the outset of this Part of the Unit is that if you are intending to carry out an elemental analysis on an inorganic sample, it is extremely likely that you will require the sample to be *in the form of a solution*. This is because so many analytical techniques, especially those used for quantitative inorganic elemental analysis, either work best with solutions or fail to work with anything other than a solution. Usually these solutions are aqueous, but they may also involve organic solvents.

This is one of the reasons why I have decided to devote a whole Part of the Unit just to the problem of converting inorganic solids into the form of aqueous solutions. The other reason is that it is often quite difficult actually to do this. Many inorganic materials are inherently quite insoluble in water or other solvents and can be dissolved only by means of drastic chemical decompositions. So we shall have some interesting chemical systems to consider in this Part.

Before we go any further I should like to emphasise this point about the large number of analytical techniques requiring a dissolution step a little bit further by asking you to tackle the following exercise.

∏ Make *three* lists of common techniques suitable for elemental analysis, under the following three headings, and then comment on the lists when they contain as many techniques as you can think of (I realise that at this stage you may not be familiar with all the possible techniques. Just list as many as you can).

(*i*) Techniques *requiring* the sample to be in *solution* form,

(*ii*) techniques applicable to samples in *solid or solution* form,

(*iii*) techniques *requiring* the sample to be in *solid* form.

These are the three lists that I wrote:

> (*i*) *Sample in solution form*
> Gravimetry
> Titrimetry
> Spectrophotometry
> Spectrofluorimetry
> Flame emission
> Atomic absorption
> Atomic fluorescence
> Plasma emission
> Use of ion-selective electrodes
> Polarography
> Electrogravimetry
> Coulometry
> Ion chromatography
> Paper chromatography
> Thin layer chromatography
> Electrophoresis
> Isotope dilution analysis
>
> (*ii*) *Sample in solid or solution form*
> Activation analysis
> X-ray fluorescence analysis

Sol^r

(*iii*) *Sample in solid form*
DC arc emission spectroscopy
AC spark emission spectroscopy
Microprobe techniques
Combustion techniques

Comments

(*i*) The majority of inorganic elemental analytical techniques require the sample to be in solution form.

(*ii*) Those few techniques that can be applied to solids tend to be highly sophisticated, capital-intensive instrumental techniques likely to be available only in the larger laboratories.

So, I hope that you can see from this how important it is that we are able to prepare solutions from inorganic samples for analysis in an effective and efficient way. Some thrusts in analytical research in recent years have been aimed at improving the possibility of direct solid-state analysis, and perhaps X-ray fluorescence and associated techniques are particularly significant in this respect. However, there have been equally important developments in solution techniques. I can single out in particular the increasing importance of plasma emission spectroscopy, a solution technique that has to a certain extent taken over from arc and spark emission spectroscopy, essentially solid-state techniques. The most interesting recent development in anion analysis is ion chromatography, again a solution technique.

Where there is a choice between a solid-state and a solution technique, then if the total amount of analyte in the sample is to be determined, a solution technique has the advantage that no errors arise due to uneven distribution of the analyte in the solid (because the solid has been converted into a homogeneous solution). However, if you are interested in the *distribution* of the analyte in the solid, then a solid-state technique must be used.

Normally the solutions that we shall prepare from our solids will be aqueous solutions containing metals as simple cations and non-metallic elements as anions. The process of obtaining these solutions is often a tricky and critical part of the analysis, and there are many sources of potential error. For example, not all the analyte may end up in the solution, or contamination occurs. You may be lucky enough to have a sample such as impure rock salt (sodium chloride) which can simply be dissolved in water. However, in the real world, samples for analysis are much more likely to be metals or alloys, ceramics, refractories, minerals, soils, boiler deposits, ash samples, and the like, all of which are likely to present a challenge of one degree or another to the analyst trying to dissolve them quantitatively.

In practice, many of the procedures adopted are not just straightforward dissolution, but involve chemical change as well. Highly aggressive reagents are often involved, together with high temperatures and occasionally high pressures. Before we look at the procedures themselves let us think about the desirable features that an 'ideal' dissolution ought to possess.

∏   List as many desirable features as you can, that you would expect an 'ideal' dissolution procedure to possess.

       These are the most important features that we shall look for.

       (*i*)   It should be capable of dissolving the sample completely, ie no insoluble residue should be left behind.

       (*ii*)   It should be reasonably quick in action.

       (*iii*) If aggressive reagents are involved, they should not interfere in any subsequent analysis, or alternatively it should be possible to remove them from the sample.

       (*iv*)   The reagents used should be available in a high state of purity so as not to contaminate the sample.

(*v*)   Losses through volatility, through formation of dust and spray, through adsorption onto or absorbtion into the walls of the vessel, or indeed for any other reason should be negligible.

(*vi*) Neither the reagents nor the sample should attack the vessel in which the dissolution is being carried out.

(*vii*) The procedure should be safe.

The techniques that we shall study fall conveniently under four headings:

— direct dissolution in water or an aqueous medium without chemical change;

— dissolution in acid with chemical change;

— fusion techniques involving chemical change;

— other techniques.

Also at this stage, I want to clarify the use of two terms whose meanings are rather specific to this subject area.

Dissolution. This means taking the sample *directly into solution* either with or without chemical change.

Opening Out. This means converting the sample into a different chemical form but still in the solid state. However, the solid sample will have been altered in such a way that it can now be dissolved easily.

∏   Apply the terms *dissolution* or *opening out* of the sample as appropriate to the following:

(*i*)   a sample of brass wire is heated with $1:1$ aqueous HNO_3 until a clear solution is obtained;

(*ii*) a sample of an insoluble silicate is heated strongly with an excess of Na_2CO_3 until a clear melt is obtained, which is allowed to cool and solidify, and this is then broken up ready for dissolution in dilute HCl;

(*iii*) a sample of ferrosilicon is treated with a mixture of aqueous HNO_3 and HF; the solution is then taken to dryness and treated with aqueous HNO_3 plus H_2O_2 to give a clear solution.

The answers are as follows:

(*i*)   Dissolution.

(*ii*) Opening-out.

(*iii*) Dissolution. (The sample dissolves in the acid. The process of taking to dryness is to eliminate F^- from the matrix. Also the result of the operation is a solution).

Before we attempt to dissolve our sample, we need to ensure that it is in a suitable physical state.

Π   What pretreatments might you consider applying to a solid sample before attempting to dissolve it?

These were considered in Part 2 of the Unit and include the following.

— Drying of the sample to constant weight (perhaps more important in practice for organic materials, but may be relevant to soils, clays, and certain minerals).

— Grinding the sample to a fine powder, or in other ways reducing the particle size. This is very important, since increasing the surface-area : volume ratio will significantly increase the dissolution or opening-out rate by making more of the sample amenable to attack by reagents.

— Metal samples may need to have their surfaces cleaned and degreased before a sample is taken (eg as drillings).

Assuming that our sample is now in a suitable form for dissolution or opening-out, ie dried, ground into a fairly fine powder (or if a metal or alloy, more probably cleansed and in the form of turnings or drillings, or the like), we are ready to proceed with the dissolution.

SAQ 3.1a

Which of the following statements are *true* and which are *false*? Give reasons for your answers.

(*i*) With the advent of modern instrumental elemental analytical techniques, the need to obtain the sample in solution form has been very considerably lessened.

(*ii*) Broadly speaking, most solid inorganic samples that one encounters for analysis require rather drastic chemical decomposition techniques to bring them into solution.

(*iii*) A desirable feature of any dissolution procedure used in the elemental analysis of inorganic samples is that the sample should dissolve without undergoing any chemical change.

(*iv*) Of the procedures used for obtaining solutions from intractible solid inorganic materials, far and away the majority involve either the use of hot concentrated mineral acids or fused acidic or basic inorganic electrolytes.

SAQ 3.1a

Before we start our considerations of the actual dissolution procedures, I should like to draw your attention to some extra reading
references. Chapter 25 in reference 2 contains a fairly simple and
general account of the subject, whereas reference 5 and 7 treat the
subject much more thoroughly.

3.2. SIMPLE DISSOLUTION WITHOUT CHEMICAL REACTION

The idea here is that we simply dissolve our sample in water or other
solvent, perhaps in the presence of a buffer, but that no chemical
reaction or decomposition is involved. This is clearly the simplest
way of obtaining a solution of our sample and may come close to
our ideal in that there will be no aggressive reagents present to interfere in subsequent analyses, the solvent (most probably water) is
available in a high state of purity, losses of analyte will be minimal,
attack of the vessel will not occur, and the operation will be quite
safe.

Unfortunately, though, it is also the procedure least likely to succeed. This is because in the real world, very few of the samples that we are likely to come across will be soluble in water or other simple solvents without some form of chemical pretreatment.

∏ The list below contains some of the commoner types of solid inorganic samples submitted for analysis. Which of the types of sample listed therein are likely to be susceptible to complete dissolution in pure water and which to partial dissolution?

Metals	Building Materials
Alloys	Ores
Minerals	Soils
Ceramics	Ash
Refractories	Deposits on Filters
Glasses	Salts and Electrolytes

Really the only materials likely to be amenable to complete dissolution are simple salts and electrolytes (and of course not all of these – many inorganic salts are only sparingly soluble in water). Soil samples may contain soluble components of interest to an analyst which could be extracted, ie leached, out of the sample under mild conditions (see Section 2.5). (The filter deposits are, of course, an unknown quantity. Their solubility depends on their exact nature).

So we can conclude that the use of such mild conditions is of rather limited scope in inorganic analysis, and that most inorganic samples require the use of more or less powerful chemical reactions in order to convert them into a water-soluble form.

SAQ 3.2a

Following are five samples that are to be taken into aqueous solution. Only *one* is directly soluble in water, without any chemical change. Which is it?

(*i*) Sodium metal (for determination of traces of potassium). ⟶

**SAQ 3.2a
(cont.)**

(*ii*) Potassium perchlorate (for determination of traces of caesium).

(*iii*) Limestone ($CaCO_3$) (for determination of Fe and other impurities).

(*iv*) Sodium dihydrogen phosphate (for determination of the total phosphorus content).

(*v*) Calcium sulphate ($CaSO_4 . 2H_2O$ –gypsum) (for determination of trace metals therein).

3.3. DECOMPOSITION AND DISSOLUTION OF SAMPLES BY ACID TREATMENT

3.3.1. Use of Dilute Acids

If a sample is insoluble in water, then the simplest alternative to try is a dilute aqueous solution of mineral acid. Most metals more electropositive than hydrogen, together with many simple metal oxides, carbonates, sulphides, and the like will dissolve in dilute aqueous acidic media. The mechanism of dissolution usually involves the acid simply reacting with the sample to form a water-soluble salt of the metal, and some other product, depending on the chemical nature of the sample and, less commonly, on the acid used. Dissolution should usually occur in the cold, but gentle heating may be applied if necessary.

∏   The following substances all dissolve in hydrochloric acid to form the soluble metal chloride and some other product. Write balanced equations to show the chemical reactions that occur.

(*a*) Zinc metal,

(*b*) Magnesium oxide,

(*c*) Calcium carbonate,

(*d*) Iron(II) sulphide.

Here are the equations.

(*a*) $Zn + 2\,HCl \rightarrow ZnCl_2 + H_2 \uparrow$

(*b*) $MgO + 2\,HCl \rightarrow MgCl_2 + H_2O$

(*c*) $CaCO_3 + 2\,HCl \rightarrow CaCl_2 + H_2O + CO_2 \uparrow$

(*d*) $FeS + 2\,HCl \rightarrow FeCl_2 + H_2S \uparrow$

The implication of this type of dissolution is that normally it is the metal ion that you are interested in determining, because the anion is often converted into a volatile form and lost from the system (eg as CO_2 or H_2S in the above examples). We must also assume that the sample will, in fact, react with the acid. For *metals*, this means that they must be more electropositive than hydrogen, at least. Fig. 3.3a shows some typical standard electrode-potentials for common

Redox pair		Standard potential
Au^{3+}	$\rightarrow$ Au	+1.50
Pt^{2+}	$\rightarrow$ Pt	+1.12
Ag^+	$\rightarrow$ Ag	+0.80
Hg_2^{2+}	$\rightarrow$ Hg	+0.80
Cu^{2+}	$\rightarrow$ Cu	+0.34
H^+	$\rightarrow$ 1/2 H_2	0.00
Pb^{2+}	$\rightarrow$ Pb	−0.13
Sn^{2+}	$\rightarrow$ Sn	−0.14
Ni^{2+}	$\rightarrow$ Ni	−0.26
Co^{2+}	$\rightarrow$ Co	−0.28
Tl^+	$\rightarrow$ Tl	−0.34
Cd^{2+}	$\rightarrow$ Cd	−0.40
Fe^{2+}	$\rightarrow$ Fe	−0.45
Cr^{3+}	$\rightarrow$ Cr	−0.74
Zn^{2+}	$\rightarrow$ Zn	−0.76
Mn^{2+}	$\rightarrow$ Mn	−1.19
Al^{3+}	$\rightarrow$ Al	−1.66
Mg^{2+}	$\rightarrow$ Mg	−2.37
Na^+	$\rightarrow$ Na	−2.71
Ca^{2+}	$\rightarrow$ Ca	−2.87
K^+	$\rightarrow$ K	−2.93

Fig. 3.3a. *Standard electrode-potentials in volts at 25 °C for the reduction of some common metal ions (where these are negative, the metal should dissolve in acid with the release of hydrogen gas). (Values taken from reference 6)*

and important metal ions, in which a positive standard electrode-potential means that the metal is more electronegative than hydrogen. This is because the standard convention has been adopted of showing the electrochemical half-reactions as reductions, and using a plus sign if the reaction proceeds spontaneously.

We cannot simply assume that because a particular metal ion is more electropositive than hydrogen, it will automatically dissolve in acid with the release of hydrogen. For instance, some metals, such as aluminium and chromium become passive in the presence of certain acids, due to the formation of a thin insoluble film of oxide on the metal surface, which prevents further attack by the acid.

On the other hand, a dilute acid may react with a sample in a manner other than as a simple acid (foreshadowing the behaviour of some of the hot concentrated acids to be discussed in the next Section). For example, more metals will dissolve in 1:1 HNO_3 than would be predicted from Fig. 3.3a. Copper and silver are cases in point. Hydrogen, however, is not evolved under these conditions.

∏ In what capacity is nitric acid behaving when it dissolves copper metal to produce copper(II) nitrate and NO_2?

As an oxidising agent. The nitric acid oxidises Cu metal to Cu^{2+} ions and is itself reduced to a mixture of H_2O, NO, and NO_2.

The use of dilute acids in this manner may fail when the salt that would be produced is insoluble in water. For instance, it would be no good trying to dissolve a silver compound in HCl because AgCl is highly insoluble in water.

∏ Explain why the following attempts at dissolution of the named substance are expected to be unsuccessful:

(i) gold in dilute nitric acid,

(ii) barium carbonate in dilute sulphuric acid.

(*i*)   Gold is the most electronegative of the metals in Fig. 3.3a, and even the oxidising power of concentrated (let alone dilute) HNO_3 is insufficient to bring it into solution in an ionic state. A much more powerful oxidising agent is needed for this purpose, as we shall shortly see.

(*ii*) Barium carbonate should react with dilute H_2SO_4 to produce a barium salt of the acid, carbon dioxide, and water. Here, however, the salt is barium sulphate which is highly insoluble in water, as I would expect you to know (the simplest way of testing for sulphate ions in acid solution is to add a few drops of aqueous $BaCl_2$ solution and look for a white precipitate of $BaSO_4$). Barium carbonate should dissolve in dilute hydrochloric or nitric acid.

SAQ 3.3a

> For each of the following, first of all indicate whether the named substance can be brought into solution (with chemical change if necessary) by simple treatment with dilute aqueous acid. Then, when the answer is 'yes', suggest an appropriate acid, indicate the chemical form of the species in solution and whether any component of the sample is lost in the dissolution process.
>
> (*i*)   Brass.
>
> (*ii*) Pearls.
>
> (*iii*) Silver bromide.
>
> (*iv*) Diamonds.
>
> (*v*)   A mixture of alkaline earth oxides and carbonates.

SAQ 3.3a

3.3.2. Use of Hot Concentrated Acids

The use of dilute acids described in the last Section has shortened to some extent the list of inorganic sample types that we cannot up to now get into solution, but we are still left with a very large number of types of material that are resistant to dissolution. These include the more electronegative metals, many alloys, together with many common minerals, soils, clays, etc, especially those based on aluminates and silicates. This accounts for a lot of samples.

We can shorten this list of insoluble samples quite significantly by considering the possibility of treatment of the sample with a *hot concentrated mineral acid*. The actual practical technique here might be one of the following:

— simply boil the sample in acid in a beaker covered with a clock glass;

— boil the sample with acid under reflux;

— boil the sample with acid and allow the acid to evaporate almost to dryness. Several acids decompose under these conditions to yield decomposition products even more reactive than the acid itself (eg H_2SO_4 and $HClO_4$). This is also useful when the resulting solution is required to have a low acid content, since the bulk of the acid can be evaporated, or when treatment with different acids is indicated (add one acid, take to dryness, then add the second acid);

— use a bomb technique. The commonest version of this technique for acid treatment of inorganic samples is to place the sample plus acid in a small steel bomb whose inside is lined with platinum or PTFE, both resistant to attack by acids. The bomb is simply a container which can be sealed and then heated to a temperature above the boiling point of the acid and is designed to withstand the pressures that result. This enables you to carry out dissolutions at high temperatures without fear of loss of volatiles.

The action of the acids under these more drastic conditions is more varied than for the simple dilute acids. In particular the acids may behave as:

(*a*) concentrated strong acids (high concentration of H^+ ions);

(*b*) oxidising agents (especially where oxyanions are involved): the oxidising effect may be very powerful causing metals to be oxidised to cations;

(*c*) complexing agents (a high concentration of anions may complex with metal cations, thus retaining the latter in solution).

Hot concentrated mineral acids are especially useful for dissolving various less electropositive metals, and for alloys, such as various types of stainless steel, as well as certain metal oxides, sulphides, phosphates and silicates.

∏   What properties of a mineral acid might be relevant to its effectiveness in dissolving intractible solid inorganic materials?

(*i*)   Its acid strength.

(*ii*)   Its boiling-point, since this governs the maximum temperatures to which samples can be heated without recourse to bomb techniques.

(*iii*) Its oxidising power (including that of any decomposition products).

(*iv*) Its complexing power towards particular cations (basically a property of the anion).

(*v*)   The solubility of its metal salts.

(*vi*)   Safety aspects with respect to its handling.

(*vii*) Its purity, especially in trace analysis.

∏   When we talk about dissolution of various metals and their compounds, two terms are likely to crop up whose meaning you should be clear about. Therefore, explain what is meant by the terms:

     (*i*)   'noble metal'

     (*ii*) 'refractory metal', 'refractory oxide', 'refractory compound'.

The term 'noble metal' is used to refer to the most electronegative metals in the Periodic Table, which includes a number of the precious metals used in jewellery, and which are very resistant to attack by acids. By the same token, they are thus resistant to corrosion. This and other properties makes them attractive for jewellery. The noble metals include Au, Pd, Pt, Ru, Rh and Os.

The 'refractory metals' are those that form very stable and insoluble oxides. They are difficult to get into solution on this account. The refractory metals include, Al, Si, Ti, Zr, Hf, V, Nb, Ta, Mo and W, among others.

Let us now look at a number of common mineral acids and note especially their properties relevant to their use for dissolving inorganic materials. At the same time we can note the types of material that each acid is effective in dissolving. You will find that this next Section is fairly factual in nature. I would expect you to have, after studying this Section, a good general knowledge of these properties in a qualitative way, although memorising actual figures for boiling-points and the like is not necessary. All data relating to boiling-points, etc, have been taken from reference 6, except where stated otherwise.

1. Hydrochloric Acid

Concentrated hydrochloric acid is about $12M$ in HCl. When boiled, HCl gas is given off and the boiling-point rises until an azeotropic mixture is formed ($6M$-HCl, and boiling at 109 °C). Hydrochloric acid behaves as a strong acid but shows no oxidising properties other than those associated with the H^+ ion. The Cl^- ion, however, forms fairly strong complexes with many metal ions, especially Au^{3+}, Tl^{3+} and Hg^{2+}, and, to a less extent, Fe^{3+}, Ga^{3+}, In^{3+}, and Sn^{4+} among others. Most metal chlorides are soluble in water except for Hg_2Cl_2 AgCl and TlCl, while $PbCl_2$ is only slightly soluble in cold water, but readily soluble in hot. Hydrochloric acid is useful for dissolving many of the more electropositive metals and their oxides and hydroxides. It will also dissolve many phosphates, borates, carbonates and sulphides as well as a number of silicates, although the silica component is not retained in solution in a stable state. For a number of metals and oxides, hydrochloric acid is actually a better solvent than some of the more reactive, oxidising acids.

∏   What is meant by the term 'azeotropic mixture'?

An azeotropic mixture is a mixture of liquids which, when brought to its boiling-point, produces a vapour of composition identical to that of the liquid. Thus as the liquid boils away, it remains of constant composition. By the same token, the boiling-point of an azeotropic mixture at a fixed pressure, remains constant (hence the alternative term 'constant boiling mixture'). The individual components of an azeotropic mixture thus cannot be separated by distillation.

2. Nitric Acid

Ordinary concentrated nitric acid is about 65–69% HNO_3, an azeotropic mixture with water being formed at 67% HNO_3 with a boiling-point of about 121 °C, although it is possible to obtain more concentrated acids up to the 100% acid, boiling at 83 °C. Nitric acid is a strong acid and also a strong oxidising agent. It will oxidise all metals except the noble metals, but rather fewer can be dissolved because a number of elements such as Al, Cr, Ti, Nb, Ta become passive in the presence of the acid. Almost all metal nitrates are water soluble. The nitrate ion is a very weakly complexing ion. This means that some metal ions are hydrolysed in nitric acid solution, resulting in their precipitation as hydrated oxides. Sn, W and Sb are of this type, and in fact the ions can be separated from other ions by filtering off the precipitated oxides after reaction with nitric acid.

The main use of nitric acid on its own or diluted with water is to dissolve metals and alloys within the limitations mentioned above.

∏   What is meant by a metal 'becoming passive' in the presence of an acid?

When a metal 'becomes passive' in the presence of an acid, the acid reacts with the metal, not to form metal ions in solution but to form an *insoluble oxide*. This oxide can form a protective film on the surface of the metal, preventing further attack by the acid. Thus the metal does not dissolve.

3. Sulphuric Acid

Ordinary concentrated sulphuric acid is about 98% H_2SO_4. Pure sulphuric acid has a boiling-point of 330 °C. This is the highest boiling-point of any of the common mineral acids, and one of the main uses of sulphuric acid is to enable high temperatures to be achieved in acid dissolution processes. Sulphuric acid is a strong acid and a strong oxidising agent when hot. It also possesses powerful dehydrating properties, and will destroy any organic material present. Metal sulphates are in the main water soluble except for $CaSO_4$ (sparingly soluble), $SrSO_4$, $BaSO_4$ and $PbSO_4$ (insoluble).

One other advantage of metal sulphates is that on the whole, they are of low volatility, so that you can take the sample down to dryness without much danger of loss of volatiles, provided you have a suitable amount of H_2SO_4 present. One particular use of H_2SO_4 is to remove HF from a matrix by 'fuming to dryness'. The metal fluorides form metal sulphates and HF is driven off.

The main uses of H_2SO_4 are for dissolving metals (most, except the noble metals, dissolve) and alloys as well as many oxides, hydroxides, carbonates, sulphide and arsenide ores, and other compounds. It is used for 'fuming samples to dryness', retaining the metals as metal sulphates, and for its high boiling-point.

∏   How could you raise the boiling-point of H_2SO_4 above 330 °C?

You should be aware of one way of doing this. Use a bomb technique. But it can also be done by dissolving a high concentration of inert electrolyte (Na_2SO_4, K_2SO_4 or $(NH_4)_2SO_4$) in the acid. This is a technique we shall meet again shortly and is based on the general principle that dissolved solids raise the boiling-points of the solvents in which they are dissolved.

4. Perchloric Acid

This forms an azeotropic mixture with water, containing 72% $HClO_4$, with a boiling-point of 203 °C (reference 5). The acid is used as a 60 or 72% mixture. When hot, it is an extremely powerful oxidising agent and as such will dissolve almost all metals (except the noble metals) and alloys, and convert them into ions having their highest oxidation states, in solution. All metal perchlorates are readily soluble in water except for $KClO_4$, $RbClO_4$ and $CsClO_4$. One other feature of the perchlorate ion is that it is the most weakly complexing of the anions of the common mineral acids. However, perhaps the most important point to stress about the use of $HClO_4$ is the fact that, because of its extremely powerful oxidising properties when hot and concentrated, there is a *real and often unpredictable explosion hazard* associated with its use. This is particularly acute when either acids more concentrated than 72% are used or when the hot acid is allowed to come into contact with easily oxidisable organic or inorganic material. Some rules in handling $HClO_4$ follow.

(*a*) Never use an acid more concentrated than 72%.

(*b*) Never allow hot concentrated acid to come into contact with easily oxidisable material (eg take care metal samples have no oil or grease on them).

(*c*) If you are evaporating $HClO_4$ off, you should work in a specially lined fume cupboard containing no surfaces constructed of organic material.

(*d*) Always take the routine precautions of safety glasses, protective gloves, etc, as you should with any of these acid dissolution procedures. Use of a fume cupboard is taken for granted, but an extra protective screen is advisable in case the worst should happen.

The main use of perchloric acid is to dissolve various intractable alloys, especially steels, and other samples to produce the metal ions in solution in their highest oxidation states.

Π   Sometimes samples are first treated with HNO_3, the product is evaporated nearly to dryness and then treated with $HClO_4$. Why?

The nitric acid oxidises any easily oxidisable material, leaving only the more intractible components of the sample to the $HClO_4$, thus reducing the risk of explosions.

5. Hydrofluoric Acid

Hydrogen fluoride forms an azeotropic mixture with water containing 36% HF, and boiling at 111 °C. It is a weak, non-oxidising acid. However, the F^- ion is the most powerful complexing anion of those of the acids we have studied, and forms stable fluorides and fluoro-complexes with many elements, but especially the refractory elements, which are otherwise difficult to dissolve because they form stable insoluble oxides.

The main use of hydrofluoric acid is for dissolving silicate-containing materials. Silicon is lost from the matrix as volatile SiF_4, but the rest of the matrix is brought into the solution. However, there are a number of problems connected with the use of HF.

— Glass vessels cannot be used because glass being a silicate-based material is attacked by HF. Platinum or PTFE, or other plastic-based materials must be used.

— It is often necessary to remove all traces of F^- complexed or otherwise to prevent interference in subsequent analysis. In practice this is often difficult to do (eg AlF_6^{3-} is very stable and does not behave in solution as the free Al^{3+} ion).

— Hydrofluoric acid can inflict serious burns if it gets on the skin. These often take a very long time to heal. Full protective clothing, such as face masks, plastic aprons, gloves, etc, must be worn when handling HF.

Nevertheless, HF finds extensive use, either on its own or as an adjunct to other acids, in the dissolution of samples containing refractory elements especially silicates, when silicon is not the subject of the analysis.

∏   How are F^- ions removed from the matrix after dissolution with HF?

'Fume to dryness' with H_2SO_4. Repeat a sufficient number of times until no more F^- remains in the sample. $HClO_4$ can also be used in this process.

6. Phosphoric Acid

This is the least important acid for use in the dissolution of inorganic samples. It is a weak, non-oxidising acid and many phosphates are insoluble, or else the phosphate ion interferes in many subsequent analyses. However, it does have a high boiling-point. It is normally supplied as 85% H_3PO_4, which boils at 158 °C (reference 5). However, when you heat this, you get 'condensed' phosphoric acid, corresponding to about 100% H_3PO_4 but in actual fact consisting of a mixture of dipoly-, tripoly- and tetrapoly-acids. This mixture can be used in the range 250 to 300 °C. (Pure orthophosphoric acid boils at 213 °C). The acid finds occasional use for dissolving sulphides with the loss of sulphur as H_2S. It is also used for ferrites, chromites (where the metals are retained in low oxidation states) and silicates.

SAQ 3.3b

Consider only the acids HCl, HNO_3, H_2SO_4, $HClO_4$, HF and H_3PO_4.

(i)   Which acid has the highest boiling point?

(ii) Which acid is the most powerful oxidising agent?

(iii) Which acid is likely to be the most effective for dissolving refractory oxides?

(iv) Which acid's anion is the least likely to form stable complexes with the metal cation?

(v)   Which acid forms water-insoluble salts with the heavy alkali-metal cations?

SAQ 3.3b

SAQ 3.3c

Recommend the most suitable acid from the following list for each of the applications below:

HCl, HNO_3, H_2SO_4, $HClO_4$, HF, H_3PO_4.

Each acid may be recommended once, more than once, or not at all. For one of the applications, no single acid is suitable. Which is this?

(i)   An acid for dissolving an intractable stainless steel and which oxidises many of the metals present to their highest oxidation states.

(ii)   An acid for 'fuming a sample to dryness' without loss of metal ions as volatile compounds. $\longrightarrow$

SAQ 3.3c (cont.)

(*iii*) An acid for bringing refractory metals and their compounds into solution.

(*iv*) An acid for bringing the common transition-metal ions into solution.

(*v*) An acid for bringing the noble metals into solution.

(*vi*) An acid for dissolving an acid-soluble alloy with the elimination of Sn from the matrix.

3.3.3.   Use of Mixtures of Acids

When we come to look at actual procedures for dissolving inorganic samples by using hot concentrated mineral acids, we find that often rather than just using a single acid, *a mixture of two or three acids* is used. Sometimes the recommendation is simply to treat the sample with an acid mixture containing specified proportions of different acids. On other occasions dissolution is taken as far as it will go with one acid or mixture and then different acids are added. On yet other occasions a sample is treated with one acid or mixture, then 'fumed to dryness' and treated with another acid or mixture.

Some of the reasons for using acid mixtures are as follows.

(*a*)   We use one useful property of one acid and a different useful property of another acid (eg an oxidising acid together with an acid with a complexing anion).

(*b*)   The two acids may react to form products with greater reactivity then either acid on its own.

(*c*)   We moderate an excessive property of one acid by the presence of another acid.

(*d*)   We dissolve the sample with one acid, but then have to eliminate that acid which we do by replacing it with another.

Let us now look at some practical examples.

1. A complexing acid and an oxidising acid

This is well illustrated with mixtures such as the following,

$$HF + HNO_3$$
$$HF + HClO_4$$
$$HF + H_2SO_4$$

where the HF provides the complexing anion and the other acid the oxidising power. HF plus two of the above acids, or HF plus HCl and HNO_3 (see later) may also be used. These mixtures are useful for steels and other alloys containing refractory metals. Either the sample may be treated with the mixed acids or dissolution may be taken as far as it will go with the oxidising acid and then HF is added in the final stages to complete the dissolution.

2. Two acids reacting to form reactive products

There is one outstanding example of this which is given below.

> 3 parts HCl to 1 part HNO_3

This mixture is known as *aqua regia*. When the acids are mixed, the mixture gradually turns yellow and you can smell chlorine. What is happening is that the HNO_3 is oxidising the HCl to give various reactive oxidation products such as chlorine and nitrosyl chloride. These are very powerful oxidising agents and this, together with the complexing power of the chloride ion means that we now have a sufficiently powerful system to dissolve the noble metals. This is how such metals are normally dissolved.

3. Moderating the power of one acid with another

The best example of this is:

> HNO_3 + $HClO_4$

where the high oxidising power of the $HClO_4$ is moderated by the HNO_3. Using the acids in sequence (HNO_3 first!) means that in fact the $HClO_4$ need never come into contact with any easily oxidised material, because this will already have been oxidised by the HNO_3.

4. Replacing one acid with another

We have in fact already come across this in connection with:

$$\boxed{HF + H_2SO_4}$$

where the sample is dissolved in a mixture containing HF and then 'fumed to dryness' several times with H_2SO_4.

I have given you a number of examples of the use of acid mixtures, but this list of mixtures is by no means exhaustive. Some other mixtures that have been used include:

(*a*) $HCl + HNO_3$ in proportions other than $3:1$,

(*b*) $HF + HCl$,

(*c*) $HNO_3 + H_2SO_4$,

(*d*) $H_2SO_4 + HClO_4$,

(*e*) $H_3PO_4 + HNO_3$, H_2SO_4, or $HClO_4$

(*f*) various mixtures of 3 acids.

In fact, almost any combination you care to imagine has probably been tried somewhere!

SAQ 3.3d	Here are *four* dissolution treatments involving the use of mixtures of acids or of more than one acid in sequence.
	Explain briefly why the mixed-acid treatment or the use of acids in sequence is to be preferred to the use of a single acid. $\longrightarrow$

SAQ 3.3d
(cont.)

(*i*) The sample is dissolved with heating in a mixture of 3 parts concentrated HCl and 1 part concentrated HNO_3.

(*ii*) The sample is heated with concentrated HNO_3. As the HNO_3 boils off, $HClO_4$ is added and heating continued until dissolution is complete.

(*iii*) The sample is heated with HF and H_2SO_4 in the ratio of about $1:10$. It is then taken to fumes of SO_3 and subsequently treated with a further small amount of H_2SO_4.

(*iv*) The sample is heated with concentrated HCl and then concentrated HF is added dropwise to complete the dissolution.

3.3.4. Mixtures of Acids with other Reagents

It is often beneficial to add to individual acids or acid mixtures
other chemical reagents to improve the dissolution performance of
the former. Here are some examples.

Oxidising Agents. The oxidising powers of acids can be augmented
by using auxiliary oxidising agents, or a non-oxidising acid can be
used in conjunction with an oxidising agent. Three such oxidising
agents are:

(*a*) H_2O_2 (for the dissolution of steels),

(*b*) Br_2 (for Te-containing ores),

(*c*) $KClO_3$ (with HCl for As- and S-containing ores).

Inert Electrolytes. These are added to an acid simply to raise the
latter's boiling-point, resulting in higher available temperatures for
dissolution. The best example of this is the use of Na_2SO_4, K_2SO_4,
or $(NH_4)_2SO_4$ to raise the boiling-point of H_2SO_4.

Complexing Agents. These can be used to keep a metal ion in solution as a complex when otherwise it might be precipitated (say as a
hydroxide or oxide) or not dissolve at all. I am thinking here of the
anions of organic acids such as citrate or tartrate.

Catalysts. Reactions between some metals and acids can be very
significantly speeded up by the use of suitable catalysts. Examples
are the use of Cu^{2+} and Hg^{2+} as catalysts in this context.

The following SAQ, as well as testing your understanding of the
above, contains some specific examples of the use of auxiliary substances with acids for dissolution.

SAQ 3.3e

In each of the following dissolution procedures one of the components of the dissolution medium is not a mineral acid. Identify this component and explain its function.

(*i*) *Zinc–aluminium alloy.* Dissolve in concentrated HCl and then treat with 10-volume H_2O_2.

(*ii*) *Tin–lead solder.* Dissolve in a mixture of 10 parts HCl, 5 parts HNO_3 and 50 parts of 10% aqueous tartaric acid.

(*iii*) *Oxygen-containing compounds of arsenic and antimony.* Heat strongly with concentrated H_2SO_4 to which Na_2SO_4 has been added.

(*iv*) *Aluminium metal.* Dissolve in concentrated HCl to which a small quantity of Hg^{2+} ions has been added.

3.4. FUSION TECHNIQUES

The list of inorganic materials that remain intractable to dissolution
after we have tried all the various methods based on hot, concentrated mineral acids that we have just discussed is now considerably
shorter than when we started. However, we do still have a number
of intractable materials that either resist all attack by acids or at
least dissolve only very slowly, or only partially in acids. Again, a
few other materials may give unstable solutions in acid with components (such as silica) tending to be precipitated. Fig. 3.4a lists
some of the materials that could prove troublesome to dissolution
in acids.

Cement
Aluminates
Silicates
Ti and Zr ores
Slags
Be, Si, Al mixed ores
Insoluble residues from Fe ores
Cr, Si, and Fe oxides
W, Si, and Al mixed oxides

Fig. 3.4a.   *A representative selection of materials difficult to
dissolve by using acids*

Many silicates can be dissolved by using HF, but there is one particularly important analytical situation where HF would be quite
inappropriate for dissolving silicates.

∏   When might HF be unsuitable for dissolving silicates?

When Si is to be determined, since when HF is used, Si is lost from
the matrix as volatile SiF_4. Boron is similarly lost as BF_3.

In any of the above situations, alternative methods of opening-out
of the sample exist. These are based on *fusion techniques*. The basic
principles are as follows.

The finely ground sample is mixed intimately with an acidic or basic electrolyte (to which an oxidising agent can be added if required, for the same reasons that we used oxidising acids or acids mixed with oxidising agents previously). The proportion of sample to electrolyte (known as 'flux') is important and may vary from $1:2$ to $1:50$. The mixture is then placed in a nickel or platinum crucible and heated until the flux melts. We then hope that the sample dissolves in the molten flux to give a clear melt. When we believe that the sample has been heated long enough, we let it cool. It is a good idea at this stage to swirl the melt around the sides of the crucible to aid its subsequent removal. This cooled, solidified melt is then chipped out of the crucible and broken up into small manageable pieces. If the fusion has been successful, then this solidified melt should dissolve easily either in water or in dilute acid. Even if this is not so, it is often possible to leach components of interest out of a partially soluble melt.

The reasons why fusion techniques are successful are basically three:

(*a*) molten inorganic electrolytes are known to be *extremely powerful solvents* in their own right;

(*b*) the temperatures attained in a fusion technique are potentially much higher (anything up to 1200 °C, compared with up to 300–350 °C for acids, depending on the choice of flux), and reactivities and solubilities are much increased at these *higher temperatures* (as are volatilities of analytes, though!);

(*c*) the molten electrolyte acts as a *Lewis acid or base* leading to reaction with the sample.

∏   What is a Lewis acid and what is a Lewis base?

A Lewis acid is any species (atomic, ionic, or molecular) capable of accepting electrons from another species.

A Lewis base is any species (atomic, ionic or molecular) capable of donating electrons to another species.

Thus, in the reaction of Cu^{2+} ions with NH_3 molecules to give the deep-blue cuprammonium ion, the Cu^{2+} is the Lewis acid and the NH_3 is the Lewis base, because the NH_3 donates its lone-pair electrons into empty Cu^{2+} atomic orbitals.

An example of a flux acting in this manner is Na_2CO_3, a common choice of electrolyte for the fusion of silica-containing materials. On being heated, Na_2CO_3 decomposes and acts as a source of O^{2-} ions as follows:

$$CO_3^{2-} \rightarrow CO_2 + O^{2-}$$

The O^{2-} ions react with silica as follows

$$SiO_2 + O^{2-} \rightarrow SiO_3^{2-}$$

The sodium salt of SiO_3^{2-} is water soluble, and thus the silica is solubilised.

∏   In the reaction:

$$SiO_2 + O^{2-} \rightarrow SiO_3^{2-}$$

which species is the Lewis acid and which is the Lewis base?

The SiO_2 is the Lewis acid and the O^{2-} the Lewis base, since the SiO_3^{2-} ion is formed by O^{2-} donating electrons into empty Si atomic orbitals.

Let us now look at some common choices of inorganic electrolytes used as flux materials in fusion techniques. All values for melting-points have been taken from reference 6.

(a) *Sodium carbonate* (mp 851 °C). This is a common powerful basic flux for silicates and certain other refractory compounds. Its usefulness can be increased by adding to it oxidising agents such as KNO_3, $KClO_3$, or Na_2O_2, for samples requiring an oxidising environment (such as materials containing S, As, Sb, or Cr). Potassium carbonate (mp 891 °C) is also sometimes used, as well as $NaKCO_3$ (a 50:50 mixture of Na_2CO_3 and K_2CO_3, mp 712 °C).

(*b*) *Sodium and potassium hydroxide* (mp 318 °C and 360 °C respectively). These are powerful basic fluxes for silicates, aluminosilicates, silicon carbide, and other compounds.

(*c*) *Sodium peroxide (Disodium dioxide)*. This decomposes on being heated and acts as a powerful basic *oxidising* flux for sulphides and acid-insoluble alloys of Fe, Ni, Cr, Mo, W, and La.

(*d*) *Potassium hydrogen sulphate* ($KHSO_4$) *Potassium pyrosulphate* ($K_2S_2O_7$). These are really all one and the same flux since on being heated they undergo the following changes:

$$2\,KHSO_4 \rightarrow K_2S_2O_7 + H_2O \uparrow$$

$$K_2S_2O_7 \rightarrow K_2SO_4 + SO_3$$

The SO_3 acts as a Lewis acid, thus these are acidic fluxes. They are used at about 500 °C and are particularly useful for ignited metal oxides such as Al_2O_3, BeO, Fe_2O_3, Cr_2O_3, MoO_3, TeO_2, TiO_2, ZrO_2, Nb_2O_5 and Ta_2O_5. All these oxides are converted into soluble metal sulphates.

(*e*) *Boric oxide* (mp 450 °C). This is a useful acidic flux for silicates which can be used as an alternative to the basic fluxes, when you want to determine an alkali metal. One particular advantage of this flux is that any excess of B_2O_3 can be removed from the matrix by distilling the residue with methanol.

∏   What is the volatile product formed between B_2O_3 and methanol? Write an equation for the reaction.

The volatile product is trimethyl borate [$B(OCH_3)_3$] a volatile, inorganic ester. The reaction is below.

$$B_2O_3 + 6\,CH_3OH \rightarrow 2\,B(OCH_3)_3 + 3\,H_2O$$

(*f*) $CaCO_3$ + NH_4Cl. This mixture on being heated gives CaO and $CaCl_2$. This is a specialised flux for the extraction of alkali metals from silicates.

(*g*) *Potassium fluoride and potassium hydrogen fluoride*. These are low-temperature fluxes (useful because many metal fluorides are quite volatile) used for resistant silicates and oxides of elements forming stable fluoro-complexes, such as Be, Nb, Ta and Zr. The melt can be 'fumed' with H_2SO_4 to remove F^-, exactly as we did with hydrofluoric acid dissolutions. Note that some metal fluorides are not very soluble (eg CaF_2). Also note that Si and B are completely lost from the matrix as volatile SiF_4 and BF_3 in these fusions.

(*h*) *Disodium tetraborate decahydrate* (Borax, $Na_2B_4O_710H_2O$). Do you remember the 'borax bead' test in inorganic qualitative analysis? This is a very simple example of a borax fusion. Borax fusions are very useful for Al_2O_3, ZrO_2, Zr ores, minerals containing rare earths, Ti, Nb or Ta, as well as materials containing Al, Fe ores, and slags. Borax is not normally used on its own but is better used as a mixture with Na_2CO_3. Fusion temperatures of between 1000 °C and 1200 °C are usual.

SAQ 3.4a	Explain how each of the following electrolytes functions as a flux in terms of the Lewis acid/base theory. Give an example of the type of material thus brought into solution. (*i*) K_2CO_3, (*ii*) $KHSO_4$, (*iii*) KHF_2, (*iv*) $CaCO_3$ + NH_4Cl.

SAQ 3.4a

Finally, let us note that acid dissolution and fusion procedures can often work hand-in-hand. One example of this is to use an acid dissolution technique to dissolve as much of a particular sample as possible, and then subsequently to treat the residue with a flux. The two fractions can either be analysed separately as *acid-soluble* and *acid-insoluble* fractions, or they can be combined. One such example is a procedure for ferromanganese slags. The sample is first treated with $2:1$ HCl–H_2O. The residue is then fused with Na_2CO_3 and the melt taken up in dilute HCl containing a little H_2O_2. The two fractions are then combined.

Also some acid dissolution techniques take on the character of fusions. For example you can use a mixture of H_2SO_4 and Na_2SO_4. When you take this to dryness, you get $NaHSO_4$, which on being heated gives $Na_2S_2O_7$, a flux material. Thus what starts as an acid dissolution, effectively ends as a fusion.

3.5.   OTHER DISSOLUTION TECHNIQUES

The techniques that we have considered in the previous sections of
this Part of the Unit really account for the vast majority of important
inorganic dissolution techniques. There are a few others that do not
fit easily under the headings that we have used, but by and large,
they are of minor importance.

Perhaps the only one worthy of specific mention here is the use of a
hot concentrated aqueous alkaline solution (typically a concentrated
solution of NaOH in water). This is used to dissolve Al and certain
of its alloys (such as Al/Mg and Al/Si alloys). Hydrogen is given off
and the Al forms a soluble sodium aluminate. Beryllium can also
be dissolved in this way.

3.6.   EVALUATION OF THE VARIOUS
DISSOLUTION/OPENING OUT PROCEDURES

In this section, we shall compare the various dissolution and
opening-out procedures that we have been studying in the previous sections, from a number of general points of view. This section
will help you to select the best method in certain situations. It will
also draw to your attention a number of potential pitfalls awaiting
the unwary, when using these types of procedure.

3.6.1.   General Range of Application

∏   Fig. 3.6a contains six dissolution or opening-out procedures,
       together with six descriptions of their general range of ap-
       plicability, but the order in which both procedures and de-
       scriptions occur has been randomised. Match each proce-
       dure with the most appropriate description of its area of ap-
       plication.

	Procedure		Area of Application
(*a*)	Fusion with acidic or alkaline electrolytes.	(*a*)	Al, Be metal, Al/Si and Al/Mg alloys.
(*b*)	Use of hot concentrated aqueous sodium hydroxide solution.	(*b*)	The more intractible metals and alloys, including the noble, as well as the refractory metals (if HF present).
(*c*)	Water or simple buffered aqueous electrolytes.	(*c*)	A wide range of minerals, ores, acid-resistant alloys, oxides, sulphides and silicates. Especially useful for refractories where Si is to be retained.
(*d*)	Boiling the sample with hot, concentrated mineral acids.	(*d*)	Simple water-soluble salts. A few very electropositive metals.
(*e*)	Mixtures of hot, concentrated mineral acids, possibly with other reagents.	(*e*)	Many of the more electropositive metals, certain metal oxides, carbonates, and sulphides.
(*f*)	Dissolution in dilute acids.	(*f*)	Most metals and alloys (but not the noble metals).

Fig. 3.6a. *General dissolution and opening-out procedures and areas of application in randomised order (Procedures and areas of application to be properly matched)*

The best match is shown in Fig. 3.6b. This exercise is really revision of Section 3.2 to 3.5.

Procedure	Area of application	Procedure	Area of application
(*a*)	(*c*)	(*d*)	(*f*)
(*b*)	(*a*)	(*e*)	(*b*)
(*c*)	(*d*)	(*f*)	(*e*)

Fig. 3.6b.   *Correct match of procedure and areas of applications in Fig. 3.6a*

3.6.2.   Speed of Action

Many of these dissolution and opening-out procedures are quite slow and take a long time to complete. This is because the chemical reactions involved are themselves often slow (which is why we like our sample to be as finely powdered as possible). In the worst situations we may have reaction times of many hours.

Reaction times for individual procedures vary tremendously, but as a general rule, the following is true.

Under the worst conditions, fusions take *less* time to complete than acid dissolutions. Fusions rarely take more than an hour or two, and more commonly are complete within minutes. Some acid dissolutions can be completed in a short time, but others may take many hours (even days!) to complete.

3.6.3.   Nature of the Final Solution and Removal of the Excess of Reagents

Inevitably the solutions arising from the use of hot concentrated acids and from fusion procedures will contain a large excess of acid or electrolyte. This may well interfere in the subsequent stages of the analysis, and thus it may be necessary to remove the excess of the reagents used. For acid dissolution techniques, provided that no inorganic electrolytes have been added, this presents no problem in principle, because all the acids can be boiled off, and the mixture can, if necessary, be taken right down to dryness. However, one must be careful in so doing that no species are lost through volatilisation. This can be a problem if the acids involved are HCl or HF. On the other hand, metal sulphates tend to be very involatile, so volatility losses are not severe with H_2SO_4. HNO_3 and $HClO_4$ can also be boiled off without serious risk of loss of most metals. Non-metallic elements, though, are much more liable to be lost since they themselves are generally converted into acids.

We have also noted that HF can be troublesome in this respect, because of the stability of some metal fluoro complexes and the fact that this can lead to subsequent interferences in the analysis.

∏ How do we get rid of the last traces of F^- (complexed or uncomplexed) from a matrix?

Evaporate the mixture to dryness repeatedly with H_2SO_4 (or $HClO_4$).

If it is only free F^- ion with which we are concerned, this can often be masked by adding boric acid, which reacts with HF to form tetrafluoroboric acid which is much less troublesome than free fluoride ion.

Fusions are potentially unattractive in this respect because the resulting solution *almost invariably* contains much dissolved electrolyte together with the sample. We did come across one flux material that could be easily removed from the matrix.

∏ Name a flux material that can be relatively easily removed from the matrix after a fusion.

Boric oxide, which can be distilled off as trimethyl borate by adding an excess of methanol to the cooled melt.

We also noted previously the replacement of F^- by SO_4^{2-} in KF and KHF_2 fusions, by treatment with H_2SO_4 just as with HF dissolutions.

Sometimes, the presence of a high concentration of electrolyte can be beneficial. For instance, high concentrations of Na^+ and K^+ can act as ionisation buffers (ie act to prevent certain interferences) in flame and plasma spectroscopic techniques. The source of these ionisation buffers could be the electrolyte from the fusion.

3.6.4. Vessel Materials for Dissolution and Opening-out Operations

Since these procedures are designed to get the most intractable of materials into solution, the choice of vessel material becomes highly critical, if we do not want to dissolve the vessel along with the sample! Even small amounts of vessel material dissolving into the sample solution could cause errors in the subsequent analysis. So we must be careful to choose a vessel material that will not be attacked by the reagents that we are using.

For *acid dissolutions*, glass vessels can usually be used, so long as HF is not present. HF attacks glass, which is a silicate, in the same way that it attacks silicate in the sample. Even if the acid does not eat through the glass beaker or flask, the resulting solution is liable to be contaminated with Si (even though much of the Si will be lost as SiF_4).

Teflon ware can be used with HF if the temperature does not exceed approximately 250 °C. For higher temperatures you need a platinum crucible (expensive!).

For fusions, glass, silica and porcelain crucibles are all unsatisfactory as a rule, because the flux is designed to attack such materials. Platinum crucibles should be used for carbonate, sulphate, fluoride, and borate fusions, while nickel crucibles are recommended for oxide, hydroxide, and peroxide fusions.

Another aspect relating to choice of vessel material is that the analyte may react with the walls of the vessel, resulting in *apparent loss of analyte.* An example is in alkaline fusions in the platinum crucible, where, if a metal ion is reduced to the free metal (this may happen), then the free metal may alloy with the platinum and be lost to the sample.

Π   How might you prevent reduction of a metal ion to the free metal in a fusion procedure?

Add an oxidising agent to the flux.

3.6.5.   Purity of Reagents

One problem with these procedures, especially in trace analysis, is that, since the sample is mixed with large excesses of other reagents, unless the latter are of a high degree of purity, the sample is liable to be contaminated by impurities in the reagent. It is important, therefore, that any acid or flux material is of the highest grade of purity, ie AnalaR or AristaR grade. Indeed sometimes the usefulness of a particular procedure is limited by the lack of availability of sufficiently pure reagents. Perhaps in this respect it is worth noting that *mineral acids* are often readily available in ultrapure form, whereas KOH and NaOH are limited in their usefulness by not being readily available in sufficient purity.

SAQ 3.6a	Pick out from the following list of reagents:
	(*i*) an acid giving an anion that is difficult to eliminate completely from the matrix by volatilisation;
	(*ii*) an acid that cannot be used in glass vessels;
	(*iii*) a flux material that can be removed from the matrix by volatilisation;
	$\longrightarrow$

**SAQ 3.6a
(cont.)**

(*iv*) a reagent that is unsuitable for trace analysis because of difficulties in obtaining it pure.

B_2O_3 H_3PO_4

HCl H_2SO_4

$HClO_4$ $K_2S_2O_7$

HF Na_2CO_3

HNO_3 NaOH

3.6.6. Losses Through Volatility

An extremely serious and important source of potential error in any dissolution or opening-out procedure, especially those involving elevated temperatures, is *complete or partial loss of components through volatility*. Under acid conditions, many elements, especially the non-metals, are converted into highly volatile compounds and can be lost, even at room temperature. As the temperature is raised, so more species become volatile and are lost. Under the conditions of alkaline fusion, perhaps fewer volatile compounds are formed, but on the other hand, temperatures are much higher. One situation where you need to be especially careful is when you are taking acid solutions to dryness. Make sure that you have plenty of H_2SO_4 present to keep the metals as involatile sulphates. The non-metals will be lost.

I have collected together some examples of the types of losses that may occur, in Fig. 3.6c. Please do not assume that if an element is not mentioned under a particular set of conditions it will not suffer losses. The figure is meant to be illustrative rather than comprehensive. Also note that I would expect you to use the figure for reference rather than to try to remember every individual entry in it!

∏   Under what circumstances would you expect a species to be more likely to be volatile:

    (*i*)   as a covalent solid compound,

    (*ii*) in an ionic form in solution?

As a covalent solid compound. Covalent compounds exist as discrete molecules which can pass into the vapour state as such. Ionic species must first become covalent before they will vaporise. Hence any process that results in the conversion of ionic species into the covalent state may create a volatilisation problem.

∏   Illustrate the above principle with respect to the loss of Cr from H_2SO_4 medium in the presence of Cl^-.

Type of Treatment	Components Wholly or Partially Lost
Acid (dilute, cold)	CO_2, SO_2, H_2S, H_2Se, H_2Te, HCN, HSCN
Basic	NH_3
Oxidising	Cl_2, Br_2, I_2
Reducing	PH_3, AsH_3, SbH_3
HF	SiF_4, BF_3(totally), other metal fluorides, eg of As, Ti, Nb, Ta partially
Hot HCl	Many metal chlorides including those of As, Sb, Sn, Ge, Hg, Se, Te, and Re
Hot H_2SO_4 or $HClO_4$ in the presence of Cl^-	Bi, Mn, Mo, Tl, V, Cr
Hot H_2SO_4 or $HClO_4$	H_3PO_4
Hot aqueous acid generally	H_3BO_3, HNO_3, HCl, HBr, HI, OsO_4, RuO_4, Re_2O_7
HF/$HClO_4$ or HF/H_2SO_4 mixtures	Si, B, As, Se, Sb, Hg, Ge, Cr, Re, Os, and Ru
KF or KHF_2 fusions	Si, B, Nb, Ta, Ti, and others
$K_2S_2O_7$ fusions	Hg, anions of volatile acids, PO_4^{3-}
$Na_2B_4O_7$ and Na_2CO_3 fusions	Tl, Hg, Se, As, halogens

Fig. 3.6c.   *Losses of components through volatility*

The H_2SO_4 oxidises the Cr to the Cr(VI) state and the Cr(VI) reacts with Cl^- ions to form chromyl chloride, a covalent compound (actually a red liquid in the pure state) whose formula is CrO_2Cl_2. This compound is volatile.

SAQ 3.6b

In each of the following dissolution or opening-out procedures an error is expected because of loss of a component or contamination of the sample. Identify this error in each procedure.

(*i*) Dissolution of a chrome steel by using a mixture of H_2SO_4 and HCl in a glass vessel.

(*ii*) Dissolution of an alumina sample by using HF and H_2SO_4 in a pyrex glass vessel. Trace refractories to be determined.

(*iii*) Dissolution of $CaCO_3$ in HCl. The sample is taken to dryness. Traces of Hg^{2+} are to be determined. A glass vessel is used for the process.

(*iv*) Dissolution of a mixture of sulphides, phosphates, and silicates by 'fuming to dryness' with H_2SO_4 and dissolving the residue in dilute acid. A platinum crucible is used. The anions are to be determined.

SAQ 3.6b

3.6.7. Safety Aspects

Since we are using highly drastic conditions to decompose chemically highly intractable material, it follows that we shall be working with highly reactive, and thus potentially hazardous, chemicals. Operations of the type we have described should always be carried out in fume cupboards and you, the analyst, should wear protective clothing (safety glass or face mask, thick protective gloves, at least a laboratory coat, but sometimes a thick plastic apron or overalls may be required). Extra safety screens are a sensible precaution when one uses $HClO_4$. An eye-wash, a shower, and a first-aid kit containing materials for the treatment of acid burns (especially HF) should be at hand to deal with any spillage or splashing of acids.

A few specific hazards are especially worth a mention.

(*a*) HF
This causes particularly serious burns if it is allowed to come into contact with the skin.

(*b*) HClO$_4$
There is always a risk of explosions with the hot concentrated acid especially in the presence of easily oxidised material.

(*c*) Acids
Release of noxious volatile acid gases, such as HCN, H$_2$S, SO$_2$, Cl$_2$, Br$_2$; acid burns.

(*d*) Fluxes
These are high-temperature liquids and solids capable of inflicting severe burns if spilt.

(*e*) Bomb techniques
Explosion hazard due to the high pressures involved.

SAQ 3.6c

Here is a list of dissolution and opening-out procedures. Select the one that you think is most suitable for use in each of the analytical situations below. Each procedure may be used once, more than once, or not at all.

The procedures are as follows.

(*i*) Dissolution in 2 : 1 HNO$_3$–HCl. The residue is then treated with HF.

(*ii*) Dissolution in 3 : 1 HCl–HNO$_3$.

(*iii*) Dissolution in 1 : 1 aqueous HNO$_3$.

(*iv*) Fusion with K$_2$S$_2$O$_7$. The residue is taken up in H$_2$O.

(*v*) Fusion with NaOH. The residue is then taken up in dilute HCl containing a little H$_2$O$_2$. $\longrightarrow$

SAQ 3.6c
(cont.)

(*vi*) Treatment with a mixture of HF and
        H_2SO_4 in a bomb.

The analytical situations are as follows.

(*A*) Dissolution of a mixture of ignited metal
       oxides, including Al_2O_3, BeO, Fe_2O_3, and
       Cr_2O_3. The metals are to be determined.

(*B*) Dissolution of a sample of cement (as-
       sumed to be basically a calcium aluminosil-
       icate) for determination of the major con-
       stitutents.

(*C*) Dissolution of a stainless steel containing
       small amounts of some refractory elements,
       for analysis of alloying elements.

(*D*) Dissolution of a brass sample for determi-
       nation of the copper : zinc ratio.

(*E*) Dissolution of dolomite (basically a mix-
       ture of $CaCO_3$ and $MgCO_3$).

(*F*) Dissolution of a gold/copper alloy.

SAQ 3.6c

Objectives

After studying this Part of the Unit, you should now:

- appreciate that most common methods of inorganic elemental analysis require the sample to be in solution;

- be aware that, for most inorganic samples, fairly drastic chemical decompositions will be required to dissolve the sample;

- be able to describe the features of a useful and effective dissolution procedure for inorganic samples;

- be aware that most dissolutions for inorganic samples involve either the use of acids or fusion with inorganic electrolytes;

- be able to describe the types of sample that can be dissolved in pure water or a simple aqueous buffered medium;

- be able to give examples of the types of material that dissolve in dilute aqueous mineral acids and describe the chemical reactions accompanying such dissolutions;

- be able to describe certain properties of the common mineral acids relevant to their use in the dissolution of inorganic substances;

- be able to explain how the usefulness of these acids is increased by using them in combination with each other or with other compounds;

- be able to explain how materials intractable to acid treatment can be dissolved by using fusion techniques;

- be able to name some common fluxes, describe how they operate in terms of Lewis acid/base properties, and give some examples of their use;

- be able to compare different dissolution procedures with regard to their scope, effectiveness and speed of operation;

- be aware of the need to remove the excess of dissolution reagent from the dissolved sample, and explain how this is done;

- be aware of the importance of correct choice of vessel material and grade of reagent in dissolution procedures;

- be aware of the errors that may arise through loss of volatile components in such procedures;

- be aware of the safety implications of such procedures.

4. Decomposition of Organic and Biological Matrices for Elemental Analysis

4.1. INTRODUCTION

The analytical problem we shall be concerned with in this Part is the determination of individual elements in samples of an organic or biological nature. In order to carry out such an analysis it is normally necessary to release the elements concerned from the carbonaceous organic or biological matrix in a simple inorganic form. This often but not always means into an aqueous solution. We need to do this so that we can apply common and well-established methods of inorganic analysis to the elements in simple inorganic forms. The same elements covalently bonded to a carbon structure might very well not respond to a simple inorganic analysis in the same way as when they are in the form of ions in solution.

Let us illustrate this with an example. Let us assume that we have a sample of a plastic containing a certain proportion of polyvinyl chloride (PVC). Our analytical problem is to determine the percentage by weight of chlorine in the sample, which in turn will give us some idea of the proportion of PVC present (if we assume that PVC is the only source of Cl). The simplest quantitative technique for determining Cl is gravimetric analysis, whereby Cl$^-$ ions in aqueous solution are precipitated as AgCl. PVC has the structure below.

$$-CH_2-CH-CH_2-CH-CH_2-CH-CH_2-$$
$$\quad\;\; | \qquad\qquad | \qquad\qquad |$$
$$\quad\;\; Cl \qquad\quad\;\; Cl \qquad\quad\;\; Cl$$

You should be able to see from this structure that the chlorine atoms are all firmly and covalently bonded to the carbon skeleton. They are not in an ionic form, and even if you could get the PVC into aqueous solution, which you could not anyway, the Cl atoms will not be precipitated in the presence of Ag^+ ions.

So we need to release the Cl atoms from their carbonaceous environment and convert them into Cl^- ions in aqueous solution. If we can to this, then the analysis can be completed without further difficulty. It is the purpose of this Part to consider ways in which this is done in practice, not only for Cl but for any element in any organic or biological matrix.

∏   You have seen that a suitable simple inorganic form for chlorine extracted from carbonaceous matrix is Cl^- ion in aqueous solution, and that a suitable method of determination is by gravimetry as AgCl. See if you can suggest simple inorganic forms for the following elements (not necessarily always in aqueous solution) and, where possible, methods for their determination.

C   Br   Pb
H   S   Cu
N   P   Fe

I have prepared the response to this question in the form of a table (Fig. 4.1a), showing suitable inorganic forms of the elements, together with suitable physical states and possible methods for their determination. But note that these are only typical common examples. Most elements, in solution especially, can be determined by a number of techniques.

Fig. 4.1a is illustrative of some of the elements that we might come across. More generally though, we try to get the non-metals (except for C,H and N) into the form of anions or oxyanions, and the metals into the form of simple cations, both in aqueous solution. If we can do this, there are many familiar techniques of inorganic analysis that can be used to complete the analysis.

Element	Inorganic Form	State	Method of Determination
C	CO_2	Gas	Absorption then weighing
H	H_2O	Vapour	Absorption then weighing
N	N_2	Gas	Measurement of volume
Br	Br^-	Aqueous solution	Gravimetry (AgBr)
S	SO_4^{2-}	Aqueous solution	Gravimentry (BaSO₄)
P	PO_4^{3-}	Aqueous solution	Colorimetry (Molybdenum Blue)
Pb	Pb^{2+}	Aqueous solution	Atomic absorption
Cu	Cu^{2+}	Aqueous solution	Atomic absorption
Fe	Fe^{2+}/Fe^{3+}	Aqueous solution	Atomic absorption

Fig. 4.1a. *Suitable inorganic forms and states for some common elements with possible methods for their determination*

Now that we understand the *need* for methods for releasing elements from a carbonaceous matrix into inorganic form, let us look at how this is done. In fact the actual procedures used form the bulk of this Part, but what I want to stress here are the underlying themes behind these various procedures. In some ways the techniques bear a relationship to the dissolution techniques we studied in Part 3 for inorganic samples, in that *drastic chemical decompositions* are almost always involved, with complete destruction of the organic matrix. Most often this involves *oxidation of the matrix* to convert carbon into CO_2, hydrogen into H_2O, nitrogen into N_2 or oxides of nitrogen (N_2 only, if this is to be determined). Other elements are then left in simple inorganic forms suitable for analysis. Most of the decomposition methods that we shall consider involve oxidation. However, for the analysis of non-metals there are just a few *reduction* techniques available, and occasionally metals (which can be associated with organic species in an ionic state, ie as salts or complexes) can be released by simple treatment with acid. Also for the sake of completeness, we ought to note that there are techniques available for determining individual elements in an organic matrix without the need for preliminary conversion into a simple inorganic form, such as activation analysis and X-ray fluorescence analysis, but these are highly specialised, and you will come across them only in the larger analytical laboratories. Furthermore they are not equally suitable for all elements and all sample types.

The decomposition techniques that we shall consider are:

(*a*) combustion-tube techniques (mainly for C, H and N analysis);

(*b*) dry-ashing techniques (of general application);

(*c*) oxygen-flask techniques (of general application);

(*d*) low-temperature ashing (especially for volatile species);

(*e*) wet-ashing techniques with hot, oxidising mineral acids (including the Kjeldahl method, but also of general applicability);

(*f*) the Carius technique (a bomb technique, especially useful for volatile species);

(*g*) sodium peroxide fusions (mainly for non-metals);

(*h*) sodium fusions and extractions (for non-metals).

All the above procedures are *oxidative* in nature except (*h*), which is a *reduction*.

SAQ 4.1a

In order to determine a specific element or elements in an organic or biological sample, you are most likely:

(*i*) to decompose the sample by hydrolysis to release the element into solution in a simple inorganic form;

(*ii*) to dissolve the sample in a suitable organic solvent, ensuring that no decomposition occurs;

(*iii*) to decompose the sample completely by oxidation, thus converting the element into a suitable simple inorganic form;

(*iv*) to convert the sample (by chemical means if necessary) into a water-soluble form and take it into aqueous solution.

Select the most appropriate response to the above.

Further reading on this subject can be found in references 2, 5 and 8.

4.2. COMBUSTION-TUBE TECHNIQUES

In this section we consider quite briefly a technique that has been in use for very nearly 100 years for the routine determination of the major elements in organic compounds ie such elements as carbon, hydrogen, nitrogen, oxygen, sulphur and the halogens.

Elemental (especially C, H and N) analysis is a specialised analytical technique that has been offered as a routine service to organic chemists for many years, in particular for characterisation of the compounds they prepare. It is normal practice for someone who wants a C,H,N analysis on a sample to send it to a specialised laboratory, where the analysis will be done for him. Unless you are employed by such a laboratory, it is unlikely that you will be called upon to perform such analyses yourself. That is why it is more important for you to have a general awareness of the principles of such techniques than a practical working knowledge of them. The only qualification that I make to the above is that in recent years, the technique has been automated to a high degree, and it is now possible to purchase an automated C,H,N analyser as a laboratory instrument. Such instruments might appear in more general analytical laboratories.

The basic idea behind the combustion-tube technique is that a small weighed portion of the sample is completely oxidised and the elements to be determined are converted into gaseous or volatile forms. This is performed in a tube through which a gas is flowing so that the volatile products of oxidation are swept out of the tube and can be collected and analysed. We start by looking at simple apparatus for the determination of *carbon and hydrogen* in an organic compound. Look at Fig. 4.2a, while I explain the operations, going from left to right.

(a) On the left-hand side of the apparatus we can see a small porcelain boat containing the weighed sample. This is placed in the combustion tube which is strongly heated by means of a furnace at this point.

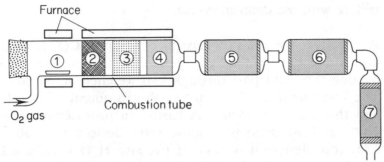

Furnace

Combustion tube

O₂ gas

(1)ʹ Sample in porcelain boat. Strong heating here.
(2) Platinum catalyst.
(3) CuO to help to complete the oxidation.
(4) A mixture of PbO_2 to retain any oxides of N and $PbCrO_4$ + Ag to retain any halogen or sulphur compounds.
(5) Desiccant to absorb H_2O produced.
(6) 'Ascarite' to absorb CO_2 produced.
(7) Guard tube containing desiccant and Ascarite.

Fig. 4.2a. *Combustion tube apparatus for carbon and hydrogen analyses*

(b) A stream of pure oxygen (or purified air, no CO_2 or H_2O vapour to be present) flows over the sample.

(c) The sample burns (or is pyrolysed) in the oxygen and the oxidation products (or pyrolysis products) are swept in the oxygen over a strongly heated platinum gauze catalyst which causes further oxidation to occur. We assume the sample is completely vaporised by this process.

(d) After they have passed over the platinum gauze, the oxidation products pass over finely ground hot CuO to complete the oxidation. We assume that after this stage oxidation is complete and that all the carbon in the sample has been converted into CO_2 and all the hydrogen into H_2O.

(e) The gases are now passed over a hot mixture of PbO_2 to retain any oxides of nitrogen, and $PbCrO_4$ and Ag to retain any sulphur or halogen compounds which would subsequently interfere with the determination.

(*f*) At this stage the gases (consisting of O_2, plus CO_2 from the carbon and H_2O from the hydrogen in the sample) leave the combustion tube and pass through two weighed absorption tubes. The first contains a desiccant to absorb quantitatively the H_2O and the second contains 'Ascarite' (a proprietary mixture of KOH and asbestos) plus some extra desiccant, to absorb the CO_2 (the desiccant is present because H_2O is released when the Ascarite absorbs the CO_2).

(*g*) Finally the excess of O_2 passes into the atmosphere through a guard tube containing desiccant and 'Ascarite' to prevent CO_2 and H_2O vapour from the atmosphere interfering with the analysis.

SAQ 4.2a

Having read through the description of the combustion tube apparatus for determining C and H in organic compounds, answer the following questions without referring to the text.

(*i*) In what chemical forms are the carbon and hydrogen when they are absorbed in absorption tubes for weighing?

(*ii*) The sample is oxidised in three stages. What are these?

(*iii*) How can you be sure that N, S, Cl, Br and I do not interfere in the analysis?

(*iv*) What is the function of the guard tube?

(*v*) What would happen if you reversed the order of the two absorption tubes?

SAQ 4.2a

Now that you understand how carbon and hydrogen can be determined by a classical combustion-tube technique, let us see how the technique can be modified to determine other elements.

Nitrogen

(a) The weighed sample is mixed with powdered CuO and placed in the porcelain boat.

(b) CO_2 gas is passed over the boat and the sample is heated. Carbon is oxidised to CO_2, hydrogen to H_2O and nitrogen to N_2 (plus some oxides of nitrogen). The oxidant here is CuO.

(*c*) The gases are swept over heated powdered Cu which reduces any oxides of nitrogen to N_2.

(*d*) The gases are then bubbled into a concentrated solution of KOH which absorbs the excess of CO_2, H_2O vapour and any other gaseous products except the N_2, which is collected in a gas burette, a device for measuring the volume of gas produced, at standard temperature and pressure. The analytical measurement is thus the volume of N_2 produced.

∏   Why is CO_2 rather than O_2 used as the gas to sweep over the sample when nitrogen is being determined?

We must be able to absorb all the gases leaving the combustion tube, except the N_2. CO_2 is conveniently absorbed in KOH solution along with all the other gases (except N_2). O_2 is not.

Oxygen

This element is less often determined than C, H, and N. If it is necessary to determine O by a combustion tube technique then it can be done, although we must reduce the sample, not oxidise it (since it is O that we are determining).

The basic idea is to mix the weighed sample with powdered carbon (the reducing agent) and heat it in a porcelain boat in a stream of hydrogen. The oxygen is released as CO, which is then passed over solid I_2O_5. The following reaction occurs.

$$5\,CO + I_2O_5 \rightarrow 5\,CO_2 + I_2(gas)$$

The gaseous I_2 is absorbed in a suitable solution and titrated.

∏   Can you suggest a suitable absorption solution, titrant and indicator for the determination of the I_2 produced above?

I_2 is absorbed in aqueous KI when the ion I_3^- is formed. It behaves as I_2 in solution. This can be titrated by using $Na_2S_2O_3$ (sodium thiosulphate) solution and a starch indicator.

Sulphur

The sample is oxidised in a stream of oxygen over a red hot platinum catalyst. The S is converted into a mixture of SO_2 and SO_3. The gases are then passed into aqueous H_2O_2 when you get the following reactions.

$$SO_3 + H_2O \rightarrow H_2SO_4$$

$$SO_2 + H_2O_2 \rightarrow H_2SO_4$$

Thus all the S is converted into H_2SO_4 which can be titrated with standard alkali or determined gravimetrically as $BaSO_4$.

Halogens (except F)

The sample is oxidised in a stream of oxygen over a red hot platinum catalyst (as for S). The halogens are converted into a mixture of the free elements and the hydrogen halides (plus perhaps a little oxyhalide). The gases are passed into a solution of Na_2SO_3 to absorb the combustion products and reduce all the halogens and oxyhalides to simple halide ions, which can then be determined conventionally (gravimetrically as silver halide).

SAQ 4.2b

Below are given block diagrams for six combustion-tube analyses. However, the individual stages of the analyses have been muddled up (ie in each column the vertical order of the blocks has been randomised). Furthermore, some inappropriate or erroneous blocks have been included.

Select one block (numbered 1 to 6) from each of the four columns A to D to produce viable procedures for the determination of the following elements by the combustion-tube technique:

$\longrightarrow$

SAQ 4.2b
(cont.)

carbon and hydrogen, nitrogen, sulphur, chlorine, oxygen.

Each block may be used once, more than once, or not at all.

For instance, if you think that element 'X' can be analysed by:

A – heating the sample in stream of O_2;

B – passing vapours over heated Cu;

C – absorbing products in aqueous KI;

D – titrating the solution with standard acid;

then your answer should appear as:

$$A = 3, \quad B = 6, \quad C = 3, \quad D = 2$$

	A	B	C	D
1	Heat sample with powdered carbon in a stream of O_2	Pass over heated Pt then over heated Cu then over heated PbO_2 and $PbCrO_4$ + Ag	Absorb in an aqueous solution of Na_2CO_3 + Na_2SO_3	Measure increase in weight of tubes
2	Heat sample with CuO in a stream of CO_2	Pass over heated CuO	Absorb all other gases in aqueous KOH	Titrate solution with standard acid
3	Heat sample in a stream of O_2	Pass over heated Pt gauze	Absorb product in aqueous KI	Titrate with standard alkali
4	Heat sample with powdered carbon in a stream of H_2	Pass over heated I_2O_5	Absorb in desiccant and Ascarite tubes in that sequence	Determine gravimetrically with $AgNO_3$
5	Heat sample in a stream of CO_2	Pass over hot Pt gauze then over hot CuO then over hot PbO_2 and $PbCrO_4$ + Ag	Absorb in tubes packed with Na_2CO_3 and Na_2SO_3	Titrate with standard $Na_2S_2O_3$ solution
6	Heat sample with powdered Cu in a stream of H_2	Pass over heated Cu	Absorb in aqueous H_2O_2	Measure volume of remaining gas in gas burette

Finally, before we leave the subject of combustion-tube techniques, let me just remind you that automated laboratory C, H and N analysers are now available in which the analysis can be completed in about 15 minutes, with essentially no intervention by the operator. The basic principles are similar to those described above (although slightly different reagents are involved) except that the evolved H_2O, CO_2 and N_2 are measured instrumentally (actually by measuring the thermal conductivity of the gas stream three times as various components are absorbed from the gases in sequence, as in the Classical method).

4.3. DRY ASHING

This is perhaps the commonest technique for ashing organic or biological samples, ie for burning off the organic part of the sample, leaving the inorganic residue as an ash for further analysis. Basically the technique amounts to no more than this. The sample is placed in a crucible (quartz, porcelain, etc.) and heated in the atmosphere (use the crucible lid to prevent any ash blowing away) until all the organic material has burnt off and you are simply left with an involatile inorganic residue. The oxygen in the atmosphere has acted as an oxidising agent and the ash residue consists of oxides of metals as well as involatile sulphates, phosphates, silicates, and the like.

The main problem with this very simple practical technique is that any element that can be converted into a volatile form is immediately wholly or partially lost. These losses through volatility become more severe the higher the temperature used for ashing. But if you don't use a high enough ashing temperature then the sample won't be completely ashed and this will also lead to errors. So the ashing procedure should be carried out in a muffle furnace so that you can control the temperature and strike a compromise between a temperature high enough to ash the sample completely in a reasonable time, and yet not to lose analyte in the form of volatile inorganic species. This compromise temperature often works out to be about 500–550 °C. Because of the problems of loss of volatiles, dry ashing in a crucible in the manner described above is useful only for the metallic elements, since most non-metals are oxidised to volatile products (think – the combustion-tube method was used for non-metals and required us to volatilise the element completely!) So to some extent the two techniques are complementary: combustion-tube method for non-metals as volatile derivatives, dry ashing for metals as non-volatile derivatives. Even some metals, though, can be lost in dry ashing. The elements (other than C, H and N) that are particularly prone to loss include the halogens, S, Se, P, As, Sb, Ge, Tl, and Hg.

Sometimes an element is more prone to loss by volatilisation if certain other species are present in the matrix. Chloride ion (eg from salt in food samples) can react with metals to produce volatile chlorides. Lead is easily lost in this way as volatile $PbCl_2$.

Another source of error is that the ash can react with the crucible material. For instance if this is porcelain or silica, reactions rather along the lines of those occurring in our fusion procedures (Section 3.4) may occur and analyte can be absorbed into the crucible walls. If this is a problem, you may need to use a platinum crucible. Nevertheless, despite these potential problems, dry ashing is used most extensively in analysis, because of its simplicity and also because of its freedom from contamination, since no other reagents are involved.

An important variant on the simple dry-ashing procedure (simply heating the sample in a muffle furnace at a fixed temperature for a fixed length of time in a covered crucible) is to use an 'ashing aid', ie to mix the sample with some other material either to aid the oxidation or to prevent losses through volatility or reaction with the crucible walls. Here are some examples.

(*a*) Sulphuric acid (forms involatile metal sulphates, thus reducing volatility losses of metallic elements);

(*b*) Nitric acid and alkali metal nitrates (act as oxidising agents, as well as decomposing to alkali metal oxides);

(*c*) Alkali metal oxides, hydroxides, or carbonates (form involatile compounds with non-metals, preventing their loss).

SAQ 4.3a

In each of the following dry-ashing procedures a loss is likely to occur. Explain why, and how the loss might be prevented.

(*i*)   Ashing of a foodstuff with a high salt content in a porcelain crucible at 600 °C. Lead to be determined.

(*ii*)   Ashing of plant material in a platinum crucible at 500 °C with H_2SO_4 as ashing aid. Gold to be determined.

(*iii*) Ashing of organic matter in a platinum crucible at 500 °C. Fluorine to be determined.

Working up the Ash

The normal result of a dry-ashing procedure is a small amount of finely powdered ash, inorganic in nature, consisting of metal oxides, sulphates, silicates, phosphates, and the like. However, if the object of the analysis is to determine the concentration of a particular element in the sample (as is most usual) then, as we saw in Part 3, we shall very likely wish to obtain a solution containing the element, and therefore of the ash. How, then, are we to set about dissolving the ash?

∏ Suggest, in general terms, how you would go about dissolving the inorganic ash resulting from the dry ashing of an organic or biological sample?

I suggest that you turn back to Part 3 of this unit to refresh your memory of the various techniques for dissolving *inorganic* samples. In practice, the actual procedure adopted will depend on the exact nature of the ash and the elements to be determined, but in outline you could consider the following:

(*i*) first try dissolving the ash in dilute HCl;

(*ii*) then try a series of more vigorous acid treatments (as described in Part 3);

(*iii*) finally consider a fusion technique.

In practice, it is not always essential that the ash be totally dissolved, if you can leach the elements of interest quantitatively from the ash, although total dissolution is usually better.

4.4. LOW TEMPERATURE ASHING

∏ What are the two main causes of error in dry-ashing of organic samples?

(*i*) Loss of volatile components in the sample.

(*ii*) Loss of analyte by reaction with the crucible.

Both of the above causes of error in dry-ashing procedures would become much less severe if only we could conduct our dry-ashing at lower temperatures than the 500–550 °C or so generally needed to burn off the organic matrix completely. What we need is a more vigorous reaction between the sample and the oxygen used to oxidise it, so that lower ashing temperatures become possible. Fortunately there is a way in which this can be done.

If you take a flowing stream of pure oxygen at a low pressure of about 1–5 torr (which is the same as 1–5 mm Hg – remember, atmospheric pressure is 760 torr or 760 mm Hg) and pass it through a high-frequency electric field (such as a radiofrequency field or a microwave electric field) then the oxygen is converted into what is known as '*excited oxygen*'. Fig. 4.4a shows the apparatus for doing this. Basically the excited oxygen is produced by allowing oxygen to flow through a silica tube at 1–4 torr, around which is a coil of wire through which an alternating electric current is flowing with a frequency of, typically, 13.5 MHz (ie a *radiofrequency*). The radiofrequency field energises the oxygen, producing 'excited oxygen'. This excited oxygen consists of a mixture of oxygen atoms, ions, and molecules in either their electronic ground or excited states. It has a life-time of about a second, after which it is reconverted into normal molecular oxygen, but for that second the 'excited oxygen' has very high reactivity.

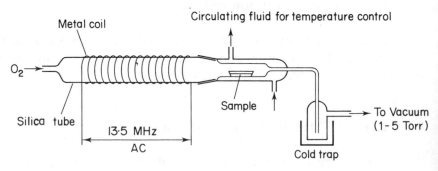

Fig. 4.4a. *Apparatus for low-temperature ashing with excited oxygen*

The 'excited oxygen' leaves the high-frequency field and flows over the sample which is oxidised at temperatures which rarely exceed 200 °C and are usually below 150 °C. Remember, the sample must be close to the high frequency field because the lifetime of the excited oxygen is only about one second.

The advantages of this technique are that *losses through volatility* are much reduced when compared with simple dry-ashing, and *reactions with the container* are also much reduced, both as a consequence of the lower temperatures involved. As, Cd, Sb, Pb, B, and Ge are just some examples of elements that can be ashed without significant loss, although they would have presented problems with simple dry-ashing. Halogens, Hg and S are still lost. One practical problem is that the ashing conditions are so gentle that samples often keep their shape. This means that the excited oxygen cannot reach the centre of the sample unless you either agitate the sample mechanically, or periodically stop the process to break up the ash. One other disadvantage is that it is a slow process, often taking hours (or even days!) for one sample. Nevertheless, the technique is useful and the necessary equipment is available commercially.

SAQ 4.4a	State two advantages and two disadvantages that low-temperature ashing has compared with conventional dry ashing.

Working up of the Ash

You would expect this to follow the same general principles as for *dry ashing*.

4.5. THE OXYGEN FLASK

(This is also sometimes known as the *Schöniger flask*, after the person who developed it to its present state).

The main difficulties associated with dry-ashing and even low-temperature ashing centre on loss of elements as volatile species. In the combustion-tube technique, we turn the problem round the other way and set out to volatilise quantitatively the elements of interest, and recover the volatile products from the gaseous phase. However, the combustion tube technique is quite tricky and the apparatus involved is complex.

It would really be rather nice if we had a technique that was relatively easy to carry out, did not involve complex apparatus and that could oxidise organic and biological samples completely so that both volatile products and involatile ash could be recovered quantitatively without loss. We come quite close to achieving these ideals with *the oxygen-flask* method for decomposition of organic and biological samples.

The oxygen flask consists of a stout walled conical flask with a ground-glass stopper. Mounted into the stopper is a thick, rigid wire with a platinum basket attached to the end. When the stopper is placed in the flask, the basket is situated inside the flask about two-thirds of the way in.

The procedure for the operation of the oxygen flask is as follows (see also Fig. 4.5a to e).

(*a*) Carefully weigh out a small amount of sample (the actual weight is fairly critical as we shall shortly see).

(*b*) Wrap this sample in a small piece of ashless filter paper and secure it in the platinum basket so that a wick of paper protrudes
(special gelatine capsules are available for liquid samples).

(*c*) Now place in the flask itself a few ml of absorbing solution.
This is normally dilute aqueous acid if you are determining a
metallic element, or dilute aqueous alkali if you are determining
a non-metal.

(*d*) Next flush out the flask with a fast stream of pure oxygen, so
that the atmosphere in the flask is of pure oxygen (just use a
lead direct from an oxygen cylinder for this purpose).

(*e*) Then immediately light the wick of paper holding the sample
and firmly place the stopper in the flask. Invert the flask so that
the absorbing solution forms a seal around the stopper. The
filter paper plus sample burns brilliantly in the pure oxygen
atmosphere and the pressure in the flask increases considei-
ably. Thus it is necessary to hold the stopper in quite firmly,
otherwise it will shoot out.

(*f*) After the sample has completely burnt away (check that no carbon or soot deposits are left behind – if there are such deposits,
the analysis is no good), leave the flask to cool, and swirl the
absorbing solution to ensure complete absorption of all volatile
oxidation products of interest as well as dissolution of any ash.

(*g*) When you are convinced that absorption or dissolution of all
oxidation products is complete (difficult to be sure how long
this takes without trial and error, but as a guide at least ensure
there is no visible mist in the flask) then open the flask (the
pressure in the flask will now be less than atmospheric) and
quantitatively wash out the absorbing solution, make it up to
volume and analyse as you will.

The oxygen-flask technique is reasonably easy to carry out practically, although a few practice runs are generally needed before you
can be sure of success. From the safety point of view, the main
hazard is an explosion in the flask during burning of the sample,
although this is, in fact, a rare occurrence. Nevertheless I would

strongly advise that, if you are using the oxygen flask, you work in a fume cupboard and wear safety glasses or a face mask and especially important, stout gloves. I have seen an oxygen flask explode, and the main danger seemed to be cuts to the hands from the broken glass.

The range of application of the oxygen flask is very wide. Below is a list of elements that have been successfully determined after the organic or biological sample in which they occur has been ashed by the oxygen-flask method. You should notice that the list contains a high proportion of non-metals and metalloidal elements. But at the same time a number of metallic elements are also represented showing the wide scope of the technique.

F	S	Ge	Co	Cd	Ca
Cl	Se	B	Ni	Mg	Ba
Br	P	Hg	Mn	Zn	Fe
I	As	Cu	Ti	Al	U

The main limitation of the oxygen-flask method is that often oxidation of the sample is incomplete. If there is any evidence that the sample has not been completely oxidised, then you must not trust any analytical results that might follow such an oxidation. You should repeat the analysis and adjust the conditions in the flask to ensure that oxidation is complete.

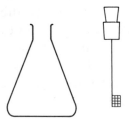

Fig. 4.5a.   *Oxygen flask and stopper and Pt basket (removed)*

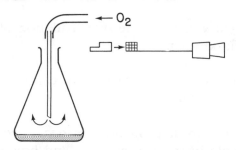

Fig. 4.5b.   *Absorbing solution in flask, now being flushed with oxygen. Sample wrapped in filter paper being inserted into basket*

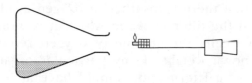

Fig. 4.5c.   *Stopper with ignited sample being inserted*

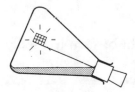

Fig. 4.5d.   *Oxidation proceeding in oxygen flask*

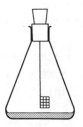

Fig. 4.5e.   *Allowing the oxidation products to be absorbed in the solution (flask also swirled regularly)*

∏   How might you conclude that oxidation of a sample in the
         oxygen flask was not complete?

Look for evidence of soot or particles of carbon left behind in the
flask after the oxidation.

∏   What is the most probable cause of incomplete oxidation
         of a sample in the oxygen flask and how would you adjust
         conditions to try to ensure complete oxidation?

The most probable cause of incomplete oxidation of a sample in the
oxygen flask is that there is insufficient oxygen available to oxidise
both sample and the filter paper in which it is wrapped. You can
work out theoretically the volume of oxygen required to oxidise
completely a known weight of any particular organic compound.
However, in practical terms you should have at least three times
this theoretical minimum of oxygen available.

If you are faced with the problem of incomplete oxidation of the
sample then you can:

(*a*) decrease the weight of sample used,

(*b*) use less filter paper,

(*c*) use an oxygen-flask of larger volume.

A common size of flask is 500 cm^3, for which the weight of sample
taken should not exceed about 50 mg. So, the oxygen flask is not
suited to oxidation of large amounts of sample.

∏   You are provided with an oxygen flask of volume 300 cm^3.
         What is the maximum weight of sample that you would con-
         sider feasible for quantitative oxidation in such a flask?

We noted in the response to the last exercise that a 500 cm^3 flask
might be used for up to 50 mg of sample, so it would be reasonable
to expect to be able to oxidise not more than 30 mg in a 300 cm^3
flask.

∏   The basket that holds the sample in the oxygen flask is made of platinum, a rare and expensive metal. Why is platinum chosen?

Platinum is a very inert metal that will not react with the sample as it is oxidised, or with the oxygen, the absorbing solution, or with any of the oxidation products. However, perhaps more important than this is that it can act as a catalyst, helping to ensure that the oxidation is rapid and quantitative.

SAQ 4.5a

Below are described some malfunctions of the oxygen flask described to you by a fellow analyst who asks for your advice. What would your advice be? Also give the reasoning behind any advice that you give.

(*i*)   'All my results are lower than expected. The oxidation proceeds smoothly without problems although I do notice some black particles in the flask after oxidation.'

(*ii*) 'I am trying to determine phosphorus in a sample. My results are low and unpredictable. I always allow the flask to stand for five minutes after the oxidation before I open the flask to allow the pressure to be reduced to about atmospheric.'

(*iii*) 'After years of use of my oxygen flask, the platinum basket eventually disintegrated and fell off. Because of the high cost of platinum, I coiled the wire (also made of platinum) connecting the basket to the stopper to make a temporary basket. My redesigned flask is shown below.

$\longrightarrow$

**SAQ 4.5a
(cont.)**

'Regrettably, however, I have had a series of explosions with flasks using this stopper.'

(*iv*) 'I have been trying to determine sulphur by using the oxygen flask. I have used a small sample and a large flask and am happy that oxidation is as complete as can be. I assumed that the sulphur is oxidised to sulphur trioxide which I have absorbed into water to give sulphuric acid which I can titrate. I have agitated the flask for up to an hour although the initial mist has dispersed after ten minutes. My results have been low. On testing the solution, I find that oxidation has produced a mixture of sulphur trioxide and sulphur dioxide.'

SAQ 4.5a

4.6. WET-ASHING TECHNIQUES

All the methods of decomposition of organic and biological samples we have looked at up to now have mainly involved oxidation in an atmosphere of oxygen (or air) by one means or another. However, when we oxidised inorganic samples before dissolution, we mainly achieved this by heating the sample with an excess of concentrated oxidising mineral acid. Now we can do exactly the same thing in oxidising organic and biological compounds.

Thus wet ashing implies *heating the sample (organic or biological) in the presence of a concentrated oxidising mineral acid or mixture of acids*. If the acids are sufficiently oxidising, if the sample is heated strongly enough, and if the heating is continued for a sufficient length of time, then it should be possible to oxidise most samples completely, leaving various elements present in the acid solution in simple inorganic forms suitable for analysis.

Π   Which of the following mineral acids have strong oxidising
      properties when hot, that might make them useful for the
      wet ashing of organic or biological samples:

      HCl,   HNO$_3$,   H$_2$SO$_4$,   HClO$_4$,   HF,   H$_3$PO$_4$

The answer to this question should be apparent to you if you remember what we learned in Part 3 of this Unit. The acids with useful oxidising properties are HNO$_3$, H$_2$SO$_4$ and HClO$_4$, and these
are the acids that are used, singly or mixed with each other, for wet
ashing organic and biological samples.

Wet ashing, by using concentrated H$_2$SO$_4$, HNO$_3$ or HClO$_4$, or mixtures of these acids, is an extremely common and important method
of decomposing organic and biological samples. It is particularly
useful for the determination of *trace metals* in various types of sample, because by and large these are converted into simple involatile
inorganic cations, which remain in the acidic medium. Wet ashing
can also be used in the determination of N, P, and S among the
non-metals, but some elements are lost wholly or partially through
volatility. These include the halogens, Hg, As, Se, B, and Sb.

The ashing procedure is normally carried out in a *Kjeldahl Flask*.

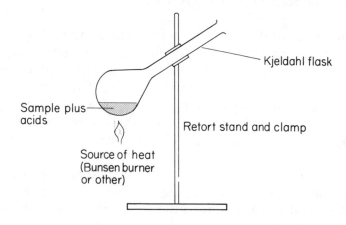

Fig. 4.6a.   *Wet ashing of organic or biological sample in a
Kjeldahl flask*

This is a round-bottomed, long-necked flask of sturdy construction and made out of Pyrex glass (see Fig. 4.6a). For use, the flask is clamped by its neck in a retort stand at an angle of about 45°. The sample is weighed and placed in the flask. The acids are then added and the flask is cautiously heated to start the reaction (which may sometimes be quite violent!). Either a single pure acid can be used or a mixture of acids or, very commonly, different acids can be added in sequence, with heating of the sample between each addition.

∏   What is the purpose of the long narrow neck of the Kjeldahl flask?

The long narrow neck acts as an air condenser, allowing you to boil the acids under reflux, thus minimising losses through volatility.

The wet ashing is considered to be complete when all signs of solid material in the flask have disappeared, although you have to be careful here:

(*a*) if you are using HNO_3, you can get soluble nitro-compounds (clear yellow solution), in which case, distil off the nitric acid and add either H_2SO_4 or $HClO_4$ until the yellow colour is gone;

(*b*) in biological samples a high silica content can leave a white insoluble residue of SiO_2.

With regard to choice of acid or acid mixture:

(*a*) it is very difficult to oxidise samples completely by using HNO_3 on its own (since HNO_3 has the lowest boiling-point and is the mildest of the three oxidising acids);

(*b*) $HClO_4$ is rarely used on its own.

∏   Why is $HClO_4$, the most powerful of the oxidising acids, not often used on its own for wet ashings?

I hope that you answered this one correctly! You remember that when we talked about $HClO_4$ for dissolving inorganic samples, we noted that, if $HClO_4$ was heated with oxidisable organic or biological materials VIOLENT AND DANGEROUS EXPLOSIONS could well be the result. These are just the conditions used in wet ashing so $HClO_4$ should be used only on its own if you know exactly what you are doing. In practice, my advice would be always to use $HClO_4$ in conjunction with another acid.

(c) H_2SO_4 is the most useful of the acids for use on its own, especially for use in the Kjeldahl method for determining nitrogen, which we shall consider shortly.

(d) For general purposes in wet ashing, though, a mixture of acids is to be preferred, or the acids can be used in sequence (the latter perhaps being the more common procedure). Two important examples of such procedures are as follows.

— H_2SO_4 and HNO_3. Heat the sample initially with H_2SO_4 until the acid begins to fume. Then continue, adding HNO_3 dropwise to begin with and then in larger portions to complete the oxidation.

— HNO_3 and $HClO_4$. Take the oxidation as far as it will go with pure HNO_3, by which time only difficultly oxidisable material will remain. Then cool the mixture and cautiously add $HClO_4$ and reheat to complete the oxidation. In this way the use of $HClO_4$ is relatively safe (I would hesitate to say that the use of $HClO_4$ is ever completely safe).

Safety Aspects

Always carry out wet ashings in a fume cupboard and wear safety glasses or a face mask, as well as acid-resistant gloves and a laboratory coat or plastic apron. If $HClO_4$ is being used routinely, then the fume cupboard should be constructed of or lined with inorganic material (since over the years, small pockets of $HClO_4$ can accumulate, due to $HClO_4$ escaping from the flask, and then one day suddenly and without warning the whole fume cupboard could go up!). A protective screen is also a wise precaution when wet ashing samples with $HClO_4$.

∏   Sometimes other substances can be added to the flask to aid the wet ashing procedure. Can you suggest two reasons why this might be useful? Hint – think back to Part 3 and the additions made to acids for dissolving inorganic substances therein.

(*i*)   Addition of inorganic electrolytes to raise the boiling-point of the acid.

(*ii*) Addition of catalysts to hasten the reaction.

(You might also have suggested the use of auxiliary oxidising agents, but this does not seem to be important in practice. The acid mixtures provide all the oxidising power needed).

I mentioned earlier the Kjeldahl method for determining nitrogen. This is an important and long-established procedure that is based on a wet ashing and for which the Kjeldahl flask was originally developed. Recently the method has been fully automated. Note that the Kjeldahl method works only for *amino and amide nitrogen*. The method is used particularly for determining protein nitrogen in various biological samples.

The principles of the method are as follows.

(*a*)   The dried and weighed sample is placed in the Kjeldahl flask and wet ashed with H_2SO_4 to which K_2SO_4 (to raise the boiling-point) and a catalyst (such as $CuSO_4$, $HgSO_4$ or SeO_2) have been added.

(*b*)   The nitrogen, so long as it is in the form of an amine or amide is converted exclusively into the NH_4^+ ion under these conditions. Since this is ionic, there are no problems over losses through volatility.

(*c*)   As soon as the oxidation is complete (the solution goes clear), the mixture is cooled and, *very carefully*, made alkaline with aqueous NaOH. The NH_4^+ is converted into NH_3.

(*d*) It is then immediately steam-distilled in a closed distillation apparatus and the NH_3 is collected in a measured volume of standard acid solution, which can then be back-titrated with standard alkali. Some practical skill is needed at this stage to make sure that you don't lose any of the NH_3 to the atmosphere.

∏   The Kjeldahl method can be applied to non-amine or -amide nitrogen, but the method must be modified. What might this modification be?

The nitrogen, say, in a nitro or nitroso group, must first be *reduced* to that in an amino-group by preliminary treatment with a reducing agent (eg Sn and HCl). It is also possible to include a reducing agent in the wet ashing procedure itself, in which case you get a strange mixture of reductions and oxidations occurring simultaneously, but under favourable conditions giving good results!

The Carius decomposition

Some elements cannot be determined by wet-ashing procedures because they are lost in the form of volatile derivatives.

∏   Name some elements that are lost through volatility in wet-ashing procedures.

We mentioned some of these earlier on in this section but some of the more important examples of volatility losses are as follows.

The halogens (except F)	As	Te
S	Sn	P
Hg	B	

These elements can be determined by using the Carius tube. The basic idea is that you place your sample in a thick-walled glass tube, together with HNO_3 (67% $\rightarrow$ 96% depending on the sample) and then seal the tube. The tube is then heated at 250–300 °C for 3 to

6 hours. The sample is completely wet-ashed (the HNO_3 is effective under these conditions because of the higher temperatures and pressures involved). The tube should be encased in a protective steel tube because there is always a risk of explosion with this technique.

After the wet-ashing is complete, the tube is allowed to cool and then, *very carefully*, opened (it could still be under pressure) and the contents recovered. You have to be very careful at this stage, not only from the point of view of accidents when the tube is opened but also you can lose volatiles in the short period when the contents are being transferred. For example, halogens can be lost this way (as the element or the hydrogen halide).

∏   Suggest an additive that can be sealed in the tube together with the sample and the HNO_3, especially to prevent loss of halogens when the tube is opened.

Silver nitrate (or metallic silver, which dissolves in the HNO_3). This results in the halogens being retained as insoluble silver halide salts.

∏   Why cannot fluorine be determined in the Carius tube?

Because the hydrogen fluoride produced would react with the walls of the glass tube.

SAQ 4.6a   Which of the following statements are *true* and which are *false*?

(*i*)   The following acids are those most commonly used, either on their own or in mixtures, for wet-ashing organic or biological samples:

HCl, HNO_3, H_2SO_4, $HClO_4$, HF, H_3PO_4.

$\longrightarrow$

SAQ 4.6a
(cont.)

(*ii*) HClO$_4$ is extremely dangerous to use on its own in wet-ashing procedures, although the hazards involved are much reduced when it is used in conjunction with other acids.

(*iii*) The long, narrow neck of the Kjeldahl flask not only acts as an air condenser, reducing losses through volatility, but also helps to reduce losses through spray and foaming if the wet-ashing reaction becomes vigorous.

(*iv*) Wet-ashing by using the Kjeldahl-flask technique is really best suited for non-metallic and metalloidal elements such as the halogens, P, S, As, Se, Sb, Te, B, and just a few metals such as Hg.

(*v*) Very special care must be taken when samples are being wet-ashed with H$_2$SO$_4$ as part of the Kjeldahl nitrogen determination, to avoid losses of N as NH$_3$.

(*vi*) The Carius-tube method is suitable for wet-ashing in the determination of all the halogens, S, P, Se, As, Hg, and other elements.

SAQ 4.6a

4.7. OTHER DECOMPOSITION TECHNIQUES

We have now studied the most important and commonly used techniques for the decomposition of organic and biological samples. To conclude this Part, therefore, we shall look very briefly at a couple of alternative approaches, but I would stress that in practice, you are much more likely to come across the other techniques, such as the oxygen-flask and the wet-ashing procedures.

4.7.1. Sodium Peroxide (Disodium Dioxide) Decompositions

The first of these involves heating the sample with sodium peroxide in a sealed, heavy-walled steel bomb. The Na_2O_2 should be in about a 200-fold excess, the sample size being about 5 to 500 mg. An auxiliary oxidant (such as KNO_3, $KClO_3$, or $KClO_4$) or diluent (such as Na_2CO_3) may also be added, as well as NaOH to lower the melting-point of the mixture, plus a readily combustible organic compound not containing the element to be determined (eg sucrose, benzoic acid, naphthalene, or ethane-1,2-diol) to aid the ignition of the sample. This technique is used to determine halogens (including F), phosphorus, sulphur, arsenic and boron.

The actual decomposition in the bomb is very quick. After it is complete, the bomb is opened and the contents are dissolved in water. The excess of Na_2O_2 is converted into NaOH and the other elements are converted as follows:

$$C \quad \rightarrow \quad CO_3^{2-}$$

$$S \quad \rightarrow \quad SO_4^{2-}$$

$$F \quad \rightarrow \quad F^-$$

$$Cl \quad \rightarrow \quad Cl^-$$

$$Br \quad \rightarrow \quad BrO_3^-$$

$$I \quad \rightarrow \quad IO_3^-$$

$$P \quad \rightarrow \quad PO_4^{3-}$$

Two disadvantages of the technique are that Na_2O_2 cannot be obtained very pure, leading to high blank values and that the large excess of NaOH in the final solution may interfere in the subsequent analysis.

4.7.2.   Reductions with Metallic Sodium

∏   Can you think of a common laboratory test that relies on decomposition of organic compounds by fusion with metallic sodium?

I am thinking of the Lassaigne's test, a test for the presence of N, S, Cl, Br, or I in organic compounds. Here are the details to remind you.

A small amount of the compound is added to molten sodium in

a small ignition tube, which is then heated strongly in a Bunsen flame. The tube is then plunged, red-hot, into a mortar containing cold water (in a fume cupboard, behind a safety screen, and with you wearing face mask, gloves etc!). The tube shatters and the product is ground up. The organic compound produces much soot, the excess of Na is converted into NaOH, but the following conversions will also have occurred.

$$Halogens \rightarrow halide\ ions$$

$$N \rightarrow CN^-$$

$$S \rightarrow S^{2-}$$

The mixture in the mortar is then filtered and tested for the presence of these ions as follows:

(*a*) halides by addition of $AgNO_3$ after acidification with HNO_3 (precipitate of silver halide);

(*b*) CN^- ion *via* the Prussian blue-test (addition of $FeSO_4$ plus a drop or two of $FeCl_3$, dissolution of precipitated Fe hydroxides in H_2SO_4 and observation of a blue colour);

(*c*) S^{2-} ion *via* the purple colour observed on the addition of aqueous sodium nitroprusside ($Na_2Fe(CN)_5NO$) or better *via* the black precipitate formed with aqueous sodium plumbite.

In addition to this simple qualitative test, organic compounds can be *reduced* quantitatively with metallic sodium or potassium, primarily for obtaining the halogens (including F) as alkali metal halides, and sulphur as an alkali metal sulphide. There are a number of ways in which this can be achieved.

(*a*) The sample can be heated with metallic sodium at 400 °C in a sealed tube. The excess of sodium is then decomposed with an excess of ethanol and the sodium halides can be extracted with water.

(*b*) The sample is dissolved in liquid ammonia. Lumps of metallic sodium are added which react to convert halogen or sulphur into anions. As soon as an excess of sodium is present, the solution turns blue due to the presence of solvated metal atoms. Then the excess of NH_3 is allowed to evaporate off and the excess of sodium is destroyed by treatment with ethanol. Alternatively, the sample plus sodium can be sealed in a glass tube together with liquid ammonia, and the sealed tube kept at room temperature for several hours. This is more effective when quantitative extraction fails when we use liquid ammonia not in a sealed tube.

(*c*) This final method is claimed to be the simplest and quickest method of extracting halogens from an organic compound. You simply shake a solution of the sample in toluene with a sodium-biphenyl or a sodium-naphthalene reagent for about 30 seconds. The reagent is prepared by heating metallic sodium, anhydrous toluene and 1, 2-dimethoxyethane together and adding biphenyl or naphthalene to give a green solution of the organometallic compound. This solution is stable if stored in the cold in a sealed container.

4.8.   COMPARISON OF TECHNIQUES FOR THE DECOMPOSITION OF ORGANIC AND BIOLOGICAL SAMPLES FOR ELEMENTAL ANALYSIS

First let me remind you of the techniques that we have studied. These are:

(*a*) Combustion-Tube Techniques,
(*b*) Dry-Ashing Techniques,
(*c*) Low-Temperature Ashing Techniques,
(*d*) The Oxygen-Flask Technique,
(*e*) Wet-Ashing Techniques in a Kjeldahl Flask,
(*f*) The Carius-Tube Technique,
(*g*) Sodium Peroxide Fusions,
(*h*) Reductive Extractions with Metallic Sodium or Potassium.

Now I want you to compare the techniques on a number of points, which you can do by simple statements, using the above letters for reference to the individual techniques. We will do this by means of a series of exercises. The answers can often be almost one-word answers. You will get the idea as we go along.

∏ First, comment on the types of elements best determined by using each of the ashing techniques (*a*) to (*h*) above.

(*a*) C, H, N, O, S, halogens.

(*b*) Best for metallic elements. The non-metallic (and some metallic) elements require special handling to avoid losses through volatility.

(*c*) Metallic elements and some metalloidal elements, especially some of those lost in (*b*).

(*d*) A wide range of metallic and non-metallic elements.

(*e*) Mainly for metallic elements. Nitrogen (Kjeldahl method) and just a few non-metals (such as P and S).

(*f*) Mainly for non-metals which may be lost through volatility from the Kjeldahl flask (but not F).

(*g*) Mainly for C, S, P and the halogens.

(*h*) More or less specific for S and the halogens.

∏ Perhaps one of the most serious limitations of any ashing technique is the loss of elements, either wholly or partially through volatility. Not all ashing techniques are equally affected, but I want you to comment on this point for each of the eight techniques.

(*a*) None – elements collected in volatile form.

(*b*) Many – most non-metals and many metals can be lost either wholly or partially. Great care is needed with this technique to avoid such losses.

(*c*) Less troublesome than in (*b*) although a few elements are still lost (halogens, Hg, and S).

(*d*) None – enclosed system.

(*e*) Not serious for metals except Hg, but non-metals may be lost this way.

(*f*) None – enclosed system.

(*g*) None – enclosed system.

(*h*) None.

∏   Now comment on any other potential sources of error for each technique.

(*a*) Sample volatilised in a chemical form other than that expected.

(*b*) Losses through the ash reacting with the crucible. Physical losses through dust dispersal. Incomplete ashing.

(*c*) Most probably incomplete oxidation at the low temperatures involved.

(*d*) Incomplete oxidation of the sample. Incomplete absorption of products in the absorbing solution.

(*e*) Incomplete oxidation.

(*f*) Incomplete oxidation.

(*g*) Contamination of sample from crucible material and impurities in the sodium peroxide.

(*h*) Incomplete extraction of halogen or sulphur.

∏   Now make a general comment about each technique concerning the ease and convenience of the technique and the cost and complexity of the equipment involved.

(*a*) A specialised technique involving fairly complex and expensive apparatus.

(*b*) Easy and convenient – no special apparatus needed.

(*c*) Not difficult – but specialised and expensive apparatus required.

(*d*) Requires a certain knack, but the apparatus is not expensive (except for platinum). More expensive, semi-automated versions are available.

(*e*) The technique in itself is not difficult. No expensive apparatus is involved, but you can buy expensive fully automated, Kjeldahl nitrogen analysers.

(*f*) A rather tricky and inconvenient technique: the equipment is not particularly expensive.

(*g*) Not a particularly convenient technique. Special bombs needed (not a major expense).

(*h*) Depends on the exact technique used. Bomb techniques are never particularly convenient. Other techniques present no particular problems on this score.

∏   Now compare the techniques with regard to speed of ashing.

(*a*) Neither particularly fast nor slow.

(*b*) Normally slow, several hours are required.

(*c*) Can be even slower than for (*b*) (hours to days).

(*d*) A matter of seconds. More time required to absorb products after oxidation.

(*e*) A matter of minutes to an hour (variable, but half-an-hour is not an unreasonable general guide).

(*f*) Three to six hours recommended.

(*g*) Very fast – just a matter of a few minutes.

(*h*) From 30 seconds to several hours, depending on the technique used.

∏   Finally, comment on any especially important aspects relating to the safety of the technique.

(*a*) None.

(*b*) None.

(*c*) None.

(*d*) Exploding flasks.

(*e*) Hot corrosive acids, explosions with $HClO_4$.

(*f*) Exploding bombs.

(*g*) None.

(*h*) Handling molten sodium and liquid ammonia.

SAQ 4.8a   Comment in an informed and critical way on the available ashing techniques for each of the following problems, and then recommend one particular ashing technique for each problem.

(*i*) Determination of traces of transition and heavy metals in a variety of samples of dried plant material collected from the environment.

$\longrightarrow$

SAQ 4.8a
(cont.)

(*ii*) Determination of phosphorus in a sample of polyvinyl chloride plastic (the phosphorus is present as an organic phosphate plasticiser and the amounts present are quite high).

(*iii*) Determination of fluorine in a newly prepared organic compound.

(*iv*) Determination of nitrogen in a sample of a soya-bean product (soya beans being a valuable source of protein).

(*v*) Determination of lead, cadmium and mercury in a sample of tuna fish in brine.

Objectives

After studying this part of the unit, you should:

- appreciate that in order to carry out a quantitative analysis for any element, present at major, minor, or trace level, in an organic or biological matrix, it is first necessary to break down that matrix to release the element in a simple inorganic form;

- know that this is normally achieved by exhaustive oxidation of the matrix, but that a few reduction procedures are also used;

- be able to describe (at current awareness level) procedures using combustion-tube techniques for organic C, H and N analysis;

- be familiar with techniques based on simple dry ashing of samples, the likely losses that can occur, and precautions to prevent these;

- be familiar with the Schöniger or oxygen-flask technique for oxidising organic and biological samples;

- be familiar with the oxidation of such samples with hot concentrated mineral acids (wet ashing) in a Kjeldahl flask, for trace-element determination;

- be able to describe the Kjeldahl method for nitrogen determination;

- be able to describe the use of Na_2O_2 fusions for oxidation of organic matrices;

- be able to describe methods for extracting non-metallic elements by reductive fusions and extractions with metallic Na;

- in a limited way, be able to compare and contrast the above techniques and make recommendations as to their use in suitable cases.

5. Organic Derivatives in Analysis

5.1. THE NEED FOR CHEMICAL REACTIONS IN ORGANIC ANALYSIS

In Part 1 of this Unit, we studied the individual stages of the analytical sequence. In other words we considered an analysis in general terms and the individual stages that one went through in carrying out that analysis. I would like you to revise that topic, which is covered in Section 1.1 with its associated SAQ. When you have done that, try the following SAQ.

SAQ 5.1a

Here is an outline procedure for the identification of an unknown drug in a sample of urine by using thin-layer chromatography (tlc). The individual stages of the procedure have been muddled up. Sort them out into their correct order. Which stage(s) involve(s) (a) chemical change(s) in the analyte?

(*i*)   Place the tlc plate in a chromatography tank and develop the chromatogram in a suitable solvent.

(*ii*)   Thus obtain the identity of the unknown drug. $\longrightarrow$

SAQ 5.1a
(cont.)

(*iii*) Compare these for the unknown with those for the standards.

(*iv*) Extract the drug from the sample with an organic solvent.

(*v*) Apply the extract to a tlc plate, along with standards (samples of drugs suspected of being present).

(*vi*) Measure the R_f values and note the colours of all spots on the chromatogram.

(*vii*) Buffer the sample to a pH at which the drug is in an extractable form.

(*viii*) Spray the plate with a solution of a reagent to convert the separated drugs into visible forms.

In this Part, we shall think very briefly about the analysis of organic compounds. By this, I refer, not to elemental analysis (we studied this in Part 4) but to the identification and determination of specific organic compounds or functional groups within such compounds. In particular, we shall consider the need to convert an organic compound, or a functional group within that compound, into a different chemical form for a particular analysis to succeed.

∏   Make a list of all the analytical techniques that you can think of for the qualitative or quantitative analysis of organic compounds or functional groups within such compounds. Include classical and modern instrumental techniques, but not elemental analysis.

Here is my list, which is not exhaustive, but covers a reasonable selection of analytical techniques commonly used for organic species. It will also serve as a basic list for what follows, which is a consideration of the role of chemical reactions in such analyses.

Identification of pure organic compounds by their melting-points.

Spot tests for organic functional groups.

Titrations for functional groups.

Paper and thin-layer chromatography.

Gas chromatography.

Liquid chromatography (especially the various forms of hplc).

Ultra-violet and visible spectrophotometry.

Spectrofluorimetry.

Infra-red and Raman spectroscopy.

Nuclear magnetic resonance spectroscopy.

Mass spectrometry.

Certain radiochemical methods (eg radioimmunoassay).

Remember, I have not attempted to include every possible technique in this list, but rather a representative selection of some of the more important ones. So you are not necessarily wrong if you have one or two techniques in your list that I have not included.

We shall spend the rest of this part working our way through this list and considering the role of chemical reactions and the preparation of derivatives for each technique. We shall also look at some examples of derivative preparation at each stage. I want you to think of this more as an illustrative review of the types of reaction that *can* occur, rather than as a comprehensive treatment of the subject. So, for instance, don't expect to become an expert in derivative preparation in any of these areas of analysis, merely by studying the following sections alone. Instead, on completion of this part, you should have an understanding of the scope and importance of 'chemistry' in organic chemical analysis. However, I would also point out to you that a number of the techniques on the list are the subject of other Units in the ACOL course.

5.2. IDENTIFICATION OF PURE ORGANIC COMPOUNDS BY THEIR MELTING-POINTS

If you have a pure organic compound and you wish to identify it or confirm its identity, then a very simple and useful test to perform is to determine its melting-point and then to compare the value you obtain with known values for similar compounds. It is normally assumed that you know what type of compound you are dealing with (eg: ketone, amide, etc.). Otherwise your list of known melting-points would be impossibly long!

However, there are some problems with the melting-point technique:

(*a*) two compounds, even of the same general chemical type, may have the same or very similar melting-point;

(*b*) if the sample is a liquid, its boiling or freezing point would have to be used instead (not so easy to measure accurately);

(*c*) some compounds decompose on being heated and do not give sharp, characteristic melting-points.

Π   Suggest a way in which the melting-point technique can be extended to combat the above problems.

We could prepare a *derivative* of our compound, determine the melting-point and compare this with known values of melting-points for derivatives of other members of that class of organic compound, prepared with the same reagent.

For many classes of simple organic compounds, reagents have been found that react with individual members of the class to form derivatives that can be obtained easily in a pure, crystalline form, and that have sharp, characteristic melting-points. Thus:

(*i*)   even though two compounds may have the same melting point, their derivatives, prepared from the same reagent, may have different melting-points;

(*ii*)   crystalline, solid derivatives may be prepared from liquid samples;

(*iii*) if the reagent has been chosen wisely, it is possible to obtain a derivative that has a sharp melting-point even when the original compound decomposed on being heated.

A good example of the use of such derivatives is the identification of aldehydes and ketones by means of derivatives formed between themselves and semicarbazide ($H_2NCONHNH_2$) or 2,4-dinitrophenylhydrazine. Here are the reactions.

$$R^1R^2C = O + H_2NNHCONH_2 \longrightarrow R^1R^2C = NNHCONH_2 + H_2O$$

(A semicarbazone)

(A 2,4 - dinitrophenylhydrazone)

More examples of such derivatives can be found in reference 9.

5.3. SPOT TESTS FOR ORGANIC FUNCTIONAL GROUPS

When we endeavoured to identify our unknown compound by means of its melting-point and the melting-points of derivatives prepared from it, we supposed we knew what type of organic compound it was. In other words, we supposed we knew what organic functional groups were present in the compound.

If we had only classical techniques available to us rather than the modern spectroscopic techniques (such as ir, nmr or mass spectrometry), then it is most probable that to find out what functional groups are present we would have resorted to a number of specific tests to identify these groups. These spot tests consist of chemical reactions that occur between the functional group and a specific reagent and which result in a product with clearly identifiable properties (colour, smell, etc), or in an otherwise identifiable change in the appearance of the system. I do not intend to give you a comprehensive list of all such spot tests because this is not the purpose of this unit. You can find these in any good text book on practical organic analysis, for example reference 9. The point here is merely to emphasise that each such spot test is based on a characteristic chemical reaction or reactions of the functional group.

Here are just a few examples of such spot tests.

Phenols

Treat with aqueous $FeCl_3$ to give a deep purple (or blue) colour due to the formation of a charge transfer complex between the phenol and the Fe^{3+} ions.

Compounds with double bonds

Treat with Br_2 in CCl_4. The orange-red colour of the Br_2 disappears (without evolution of HBr) due to the Br_2 being added across the double bond.

$$>C=C< \ + \ Br_2 \ \rightarrow \ \underset{\underset{Br}{|}}{-}C\underset{\underset{Br}{|}}{-}C-$$

Amines

These react in characteristic ways with, for instance, nitrous acid, to give different products depending on whether you have a primary, secondary, or tertiary amine. You can distinguish between such amines in this way.

∏   Can you think of a spot test for an aldehyde or ketone?

Treat the compound in solution if not a liquid, with a solution of 2,4-dinitrophenylhydrazine and look for a yellow crystalline precipitate of the 2,4-dinitrophenylhydrazone. In other words, one of the same reactions that were used to produce derivatives for melting-point determinations can be used as the basis for spot tests.

5.4.   TITRATIONS FOR ORGANIC FUNCTIONAL GROUPS

A titration is based on a chemical reaction and involves measuring the amount of reagent needed to react stoicheiometrically with the analyte. In this case, the analyte is a functional group in our sample.

Here are some examples.

(*a*) CO_2H groups can be titrated with standard KOH solution, with phenolphthalein as indicator.

(*b*) Carbon–carbon double bonds can be titrated by using Wijs' reagent, a solution of ICl, which adds across the double bond in the same way that Br_2 did in the qualitative test.

(*c*) OH groups are treated with acetic anhydride to form the acetyl derivative and acetic acid. The excess of acetic anhydride is then hydrolysed to acetic acid, which is titrated with standard KOH solution. A blank titration is carried out (ie an equivalent amount of hydrolysed acetic anhydride without added sample is titrated) and the difference between the two titrations is taken to represent the amount of acetic anhydride consumed in acetylating the —OH groups.

The acetylation reaction is:

$$ROH + (CH_3CO)_2O \rightarrow ROCOCH_3 + CH_3CO_2H$$

∏   Polyethylene glycols are simple continuous chain polymers of the following general formula:

$$HO-(CH_2CH_2O-)_nH$$

Explain very simply and in general outline only how you could estimate the value of n for such a polymer.

You could determine the proportion by weight of OH groups in the polymer by acylation and titration with KOH. As the value of n increases, this proportion decreases. The exact proportion of OH groups for each value of n can, in fact, be calculated.

See reference 9 for more details of these types of analysis.

5.5. PAPER AND THIN-LAYER CHROMATOGRAPHY

In these techniques we place a spot of solution containing a mixture of the substances that we wish to separate, onto either a sheet of chromatography paper or a tlc plate. We then develop the chromatogram by placing it in a chromatography tank in such a manner that solvent can flow over it, carrying the substances to be separated with it. The chromatogram is then removed and dried and the separated species are then present as spots of material at different positions on the chromatogram.

If the species that we have separated are strongly coloured, then we can see these spots clearly. We have a problem, however, if our species are colourless.

∏   How would you make a spot on a paper or thin layer chromatogram visible, if it arose from a colourless species?

We would form a coloured derivative of the species by means of a chemical reaction.

This is normally achieved by spraying the chromatogram with a solution of an appropriate reagent or reagents and then drying or heating it to complete the reaction (sometimes exposure to acid or ammonia fumes is also required). There are literally dozens of such chromatographic spray reagents, designed either to make spots visible for organic species in general, or to react more selectively or specifically with particular classes of compound.

Here are just a few examples, picked at random to show the kind of reactions that can be used. Reference 10 gives a more complete survey of this field.

Organic compounds in general on a silica-gel tlc plate

Spray with concentrated H_2SO_4 and heat. The organic compounds are charred and produce brown spots.

Amino-acids

Spray with the reagent, ninhydrin (in solution in butan-1-ol). Heat. The amino-acids produce purple spots.

Aromatic compounds (phenols, amines, etc)

Spray with a solution of a diazonium reagent (prepared from p-nitroaniline, hydrochloric acid and sodium nitrite). The phenols and aromatic amines couple with the diazonium salt to produce an orange azo-dye.

∏   Why cannot concentrated H_2SO_4 be used as a general spray reagent in paper chromatography?

The chromatography paper itself would be charred, as it is organic in nature.

5.6.   GAS CHROMATOGRAPHY

Gas chromatography is an extremely valuable technique for the separation and determination of organic compounds in general, but does make one important demand on the compounds to be separated. That is, that the compounds must be *volatile and thermally stable* at the instrument's operating temperature, which routinely can be anything from room temperature to 300–400 °C (selected by you, the operator).

In practice, it has been estimated that of all known organic compounds only about 30% satisfy this requirement. This leaves an awful lot of organic compounds that cannot be analysed by gas chromatography, directly. However, if a compound is involatile or is thermally unstable, all is not lost, because we can investigate the possibility of *preparing a volatile and thermally stable derivative* by treating it with a suitable reagent. There are many examples where this has been done successfully and all we can do here is to discuss a few of them.

(a) Carboxylic Acids

Carboxylic acids (compounds containing the $-CO_2H$ group) and even more so, dicarboxylic acids, tend to be of low volatility and, although it is possible to determine some of the very simplest of them directly by gas chromatography, it is almost always either necessary or at least preferable to convert them into a more volatile form.

∏ Can you think of a derivative of a carboxylic acid that is likely to be more volatile than the acid itself?

How about an ester of low molar mass, such as the methyl ester? Esters are always more volatile than their parent acids. The lower the molar mass of an ester of a particular acid then the higher the volatility as a general rule.

∏ Why do you suppose esters are more volatile than the free acids?

Because the acid contains an acidic hydrogen that can form strong hydrogen bonds with other molecules as shown below:

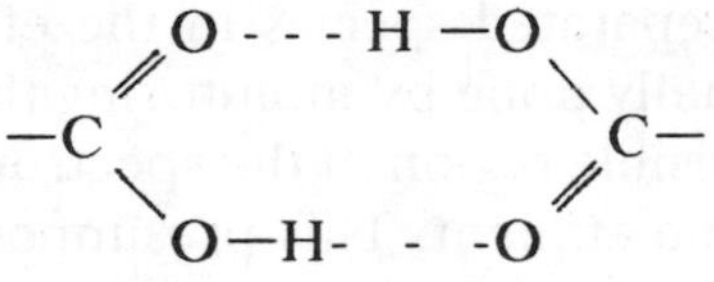

$$\begin{array}{ccc} & O\,\text{-}\,\text{-}\,\text{-}\,H-O & \\ -C\Big\langle\!\!\!\!\! & & \Big\rangle\!\!\!\! C- \\ & O-H\,\text{-}\,\text{-}\,\text{-}\,O & \end{array}$$

whereas an ester contains no such acidic hydrogen and so no such hydrogen bonds can be formed. So if you wish to analyse for carboxylic acids by gas chromatography, you can increase their volatility and improve their chromatographic properties if you first convert them into their methyl esters.

Fortunately, this is quite easy to do in practice and there are a number of reagents for the purpose, such as:

(*i*) diazomethane, CH_2N_2,

$$(RCO_2H + CH_2N_2 \rightarrow RCO_2CH_3 + N_2)$$

(*ii*) Boron trifluoride in methanol.

$$(RCO_2H + CH_3OH \xrightarrow{BF_3} RCO_2CH_3 + H_2O)$$

Simple low molecular weight carbohydrates (such as sugars)

These contain large numbers of —OH groups which can hydrogen-bond with each other. Hence sugars are not normally volatile compounds. However, if the —OH groups are converted into —$OSi(CH_3)_3$ groups, then the molecules are sufficiently volatile for gas chromatography. Reagents to effect this conversion are either chlorotrimethylsilane [$ClSi(CH_3)_3$] or hexamethyldisilazane [$(CH_3)_3Si—NH—Si(CH_3)_3$].

5.7. LIQUID CHROMATOGRAPHY

In principle, liquid chromatography (such as hplc) is applicable to any organic compound regardless of its volatility or any other property, just as long as it can be obtained in a stable solution. In practice, however, we can have problems in detecting and measuring the amounts of the separated species in the effluent from an hplc column. This is normally done by monitoring the absorbance of the effluent in the uv or visible region of the spectrum or less commonly, the fluorescence of the effluent. This presupposes that the solvents are non-uv absorbing or non-fluorescent (not a problem in practice) but that the solutes either absorb in the uv or visible, or fluoresce. Not all such solutes do.

By now, I expect that you can guess how we might tackle this problem. We could, of course, form uv-absorbing or fluorescent derivatives of our solutes and chromatograph these instead. Here is just one such example. The carbonyl group possesses only a very weak absorbance in the uv and thus carbonyl compounds give very poor sensitivity in the uv detector in hplc. Nor do they naturally fluoresce. However, they do react with the compound known colloquially as 'dansyl hydrazine' (below)

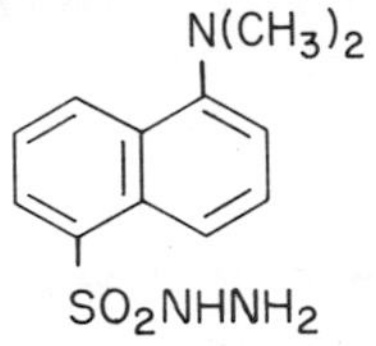

N(CH₃)₂

SO₂NHNH₂

to yield derivatives that are both uv-absorbing and fluorescent. This arises entirely from the fact that 'dansyl hydrazine' is itself uv-absorbing and fluorescent. It is as if we have 'tagged' the carbonyl compound with a uv-absorbing or fluorescent 'tag'.

∏   Write a generalised reaction for dansyl hydrazine and a ketone, $R^1R^2C=O$.

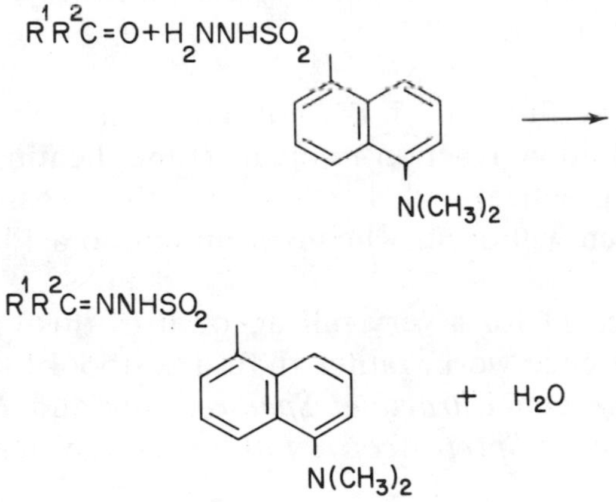

$$R^1R^2C=O + H_2NNHSO_2$$

N(CH₃)₂

$$R^1R^2C=NNHSO_2$$

+ H₂O

N(CH₃)₂

(compare with the reaction of the carbonyl group with semicarbazide or 2,4-dinitrophenyl hydrazine).

There are many other examples of such derivatisations occurring in both gas and liquid chromatography. You will come across more of these if you study the chromatography units.

5.8.   ULTRA-VIOLET AND VISIBLE SPECTROPHOTOMETRY AND SPECTROFLUORIMETRY

The situation here is exactly the same as in hplc where we are using uv-visible or fluorescence detection. In other words, the scope of the technique can be extended to include the determination of organic compounds that do not absorb in the uv or visible or do not fluoresce, by converting them into appropriate derivatives. For example, amino-acids react with the reagent, ninhydrin, under suitable conditions to give purple derivatives that can be determined spectrophotometrically.

∏   Amino-acids can be separated satisfactorily by ion-exchange chromatography but cannot be detected in a uv-visible detector. How would you suggest that they could be detected and determined?

Combine the effluent from the column with a stream of ninhydrin solution and allow reaction to occur (time, heating and buffers are required in practice) and then monitor the stream for purple coloration, which will occur whenever an amino acid is eluted.

See reference 11 for a very full account of this topic (this is definitely a reference work, rather than a textbook). Also, the ACOL Units: *Visible and Ultraviolet Spectroscopy* and *Fluorescence and Phosphorescence Spectroscopy*, will be relevant here.

5.9.   INFRA-RED AND RAMAN SPECTROSCOPY, NUCLEAR MAGNETIC RESONANCE SPECTROSCOPY, MASS SPECTROMETRY

These are the powerful spectroscopic techniques that are highly useful to the analyst for the identification and structure-elucidation of organic compounds. Each provides much useful information about the structure of an unknown molecule and, to a large extent each technique is complementary to the others. Because of this, it is not so important to derivatise molecules before analysis when we use these techniques. There may be a few specialised occasions when this statement is not true, but for our purposes, we can assume that for these techniques, there is no significant role for derivatisation to play.

∏   You have prepared a new derivative of a particular compound and you want to confirm that the structure of this derivative is what you expect it to be. In general terms only, how might you go about confirming its structure?

Here is an opportunity to use the power of ir, nmr spectroscopy and mass spectrometry in a problem concerned with the elucidation of the structure of an unknown compound.

5.10.   DERIVATISATION WITH RADIOACTIVE REAGENTS

Finally, let me draw your attention to the possibility of treating our analyte with a reagent containing a radioactive isotope. The result of this operation would be to produce a radioactive derivative whose presence could thus be detected by a radioactivity measurement. This process of producing a radioactive derivative of a non-radioactive compound has a number of uses. Perhaps you would like to suggest just one, considering the applications that we have already looked at.

∏   ·   Can you suggest one use that can be found for a radioactive derivative of an organic compound in analysis?

Perhaps the best example to quote would be in paper or thin-layer chromatography. You could chromatograph the colourless but radioactive derivatives and then determine the position of the spots either by scanning the developed chromatogram with a Geiger counter, or by autoradiography, whereby a sheet of photographic paper is placed in contact with the developed chromatogram for a suitable length of time and then developed (in the photographic sense – we are using the word 'develop' in two ways here!). Spots on the chromatogram due to radioactive species will show up on the photographic paper as black images.

Perhaps the simplest way of preparing a radioactive derivative of a compound is by an exchange reaction, whereby a non-radioactive atom in the molecule exchanges with a radioactive isotope of that element. For example, labile protons could be exchanged with tritium atoms as follows.

$$A-{}^1H + R-{}^3H \rightleftharpoons A-{}^3H + R-{}^1H$$

radioactive

analyte   reagent   analyte

There are other applications in analysis where non-radioactive analytes are treated with radioactive species such as radiometric methods of analysis and particularly important nowadays in clinical analysis, radioimmunoassay. It is beyond the scope of our simple review to delve further into these techniques. You can study a number of these techniques in *ACOL: Radiochemical Methods*.

5.11.   CONCLUSIONS

We have come to the end now, of our brief review of derivatisation techniques in organic analysis. Once again I would like to point out that I have not tried to make an 'expert' of you in the field of derivatisation in organic analysis, but simply to give you some idea of the scope of the technique and to show you how it is often possible to extend the range of application of a particular technique by preparing derivatives of compounds that otherwise could not be dealt with by that technique.

SAQ 5.11a

Which of the following statements are *true* and which are *false*? Give reasons for your answers.

(*i*)   The concept of identification of an unknown compound by comparison of melting-points with those of known compounds is applicable only to crystalline solids that are capable of being converted into

$\longrightarrow$

SAQ 5.11a
(cont.)

crystalline solid derivatives (since at least two melting-points are required to identify the compound).

(*ii*) Universal spray reagents that are capable of reacting with any organic material are invaluable in paper chromatography for the general screening of chromatograms.

(*iii*) Polymers are substances consisting of large macromolecules. They therefore have zero volatility and when heated are degraded thermally. Therefore, gas chromatography is a technique that has no relevance to their analysis.

(*iv*) hplc of non-uv absorbing compounds by using a uv-photometric detector is possible only if we can prepare uv-absorbing derivatives of the compounds that can be separated under the same conditions as for the original compounds.

SAQ 5.11a
(cont.)

SAQ 5.11b   Explain how you could distinguish between the
                  members of the following pairs of compounds
                  non-instrumentally and non-chromatographic-
                  ally:

                  (*i*) C_6H_5OH and $C_6H_5OCH_3$;

                  (*ii*) $HOCH_2CH_2OH$ and
                          $HOCH_2CH_2OCH_2CH_2OH$.

SAQ 5.11c	Consider the following list of analytical techniques and then group them in four columns under the numbers (*i*) to (*iv*) as follows:

(*i*) those techniques that are applicable to organic compounds without derivatisation (normally);

(*ii*) those techniques that are applicable to organic compounds either with or without derivatisation;

(*iii*) those techniques that are applicable to organic compounds only after derivatisation;

(*iv*) those techniques not applicable to organic compounds at all.

(*a*) Atomic absorption spectroscopy.

(*b*) Thin layer chromatography.

(*c*) Spectrofluorimetry.

(*d*) Autoradiography.

(*e*) Nuclear magnetic resonance spectroscopy.

(*f*) Spectrophotometry.

(*g*) Infra-red spectroscopy.

SAQ 5.11c

Objectives

On completing this Part of the Unit you should:

- have revised your knowledge of the stages of the analytical sequence;

- be aware that in many, but by no means all, of the available techniques for the analysis of organic compounds, a chemical conversion of the analyte just before, or as part of the actual analysis itself, is either essential or desirable;

- have listed the common and important techniques for the analysis of organic compounds (excluding elemental analysis);

- identified in which of those techniques a chemical reaction (conversion) has occurred;

- be able to quote examples of organic analyses in which chemical conversions have found use as follows:

  o derivatives in melting-point determinations;

  o spot tests for functional groups;

  o titrations for functional groups;

  o improving volatility/thermal stability in gas chromatography;

  o conversion into a coloured, UV-absorbing, or fluorescent derivative for spectrophotometry or HPLC;

  o labelling of compounds with radio-isotopes.

6. Metal-complex Formation Reactions in Analysis

6.1. INTRODUCTION AND GENERAL REVIEW OF APPLICATIONS

In Part 5 of this Unit, you learned that in many analyses of organic compounds it was either essential or beneficial at one stage or another to prepare a derivative of the analyte. In this part, we shall consider chemical reactions involved in the determination of metal ions. Far and away the most important group of such reactions is the formation of complexes, and it is with such species that we shall be almost exclusively concerned.

∏   Describe a simple metal-complex formation reaction between a metal ion M^{n+} and a ligand L in terms of the Lewis acid/base theory (Hint – we met the concept of Lewis acids and bases in Section 3.4).

We assume that the ligand has at least one lone pair of electrons available for donation into an empty atomic orbital on the metal ion. The reaction thus involves formation of a co-ordinate or dative bond between the ligand and the metal ion, the two electrons in the bond both coming from the ligand. There is thus a net transfer of electrons from the ligand to the metal ion.

$$M^{n+} + :L \rightarrow M^{(n-1)+}:L^{(+1)}$$

(lone pair)

In the Lewis acid/base theory, a Lewis acid is any species capable of accepting electrons and a Lewis base is any species capable of donating electrons. Thus the metal ion is the Lewis acid and the ligand the Lewis base.

In this Part, we shall review the way complex formation can be used in the analysis of metal ions. In fact, these applications are so important and numerous that you will come across them very extensively in other parts of this Unit and other Units of this course. In particular, in *ACOL: Classical Methods*, extensive use is made of metal-complex formation in gravimetric and titrimetric analysis. The concepts of equilibria and complex formation are also dealt with. I would strongly recommend that you study *ACOL: Classical Methods* alongside this Part, if you have not already studied it.

May I also draw your attention to reference 12, which provides more reading matter on the subject of metal complexes in analysis?

Let us now think about the different analytical techniques that can be brought to bear on metal ions in aqueous solution and those that involve metal-complex formation in their application.

∏ Make three lists of analytical techniques (the word 'analytical' here can be taken to cover separation and preconcentration as well as measurement) suitable for metal ions in aqueous solution. The lists should be under the following three headings:

(*i*) techniques normally associated with metal ions in a complexed form or that involve complex formation;

(*ii*) techniques normally associated with metal ions in a free, non-complexed form;

(*iii*) techniques applicable either to complexed or non-complexed metal ions.

Once again, as in questions of this type that you have met previously, at this stage just list the more important techniques that you can think of, rather than spend hours trying to produce a complete list.

Correspondingly, my lists, which might well be longer than yours (if you have not had the benefit yet of completing a course in analytical chemistry) are not designed to be exhaustive nevertheless!

Here are my three lists.

(*i*)	(*ii*)	(*iii*)
Techniques involving complex formation (normally)	Techniques involving free uncomplexed metal ions (normally)	Techniques involving either complexed or uncomplexed metal ions
Spectrophotometry	Redox titrimetry	Polarography
Spectrofluorimetry	Flame spectroscopy	Gravimetry
Complexometric titrimetry	Plasma spectroscopy	Ion exchange
Solvent extraction	Flameless atomic absorption	Chromatography
Masking reactions		

SAQ 6.1a   Consider the following analytical techniques. One is particularly associated with the determination of metal ions in solution in complexed form, one with the determination of metal ions in solution as free ions, and one is not particularly associated with the determination of metal ions at all. Which is which?

(*i*)   Ultra-violet and visible spectrophotometry.

(*ii*) Infra-red spectroscopy.

(*iii*) Atomic absorption spectroscopy.

SAQ 6.1a

6.1.1. General Review of Applications of Metal Complexes in Analysis

The above exercise has brought to your attention the fact that a number of common methods of analysis of metal ions in solution require the metal ion to be converted into a complex at some stage or other of the analysis. This implies that the physical or chemical properties of the metal are thus modified in some useful way for the analysis. Let us now see what these modifications are. In other words, we are answering the question 'Why do we need to convert the metal ion into a complex for these particular methods?' This will also give us a chance to review generally the applications of metal complexes in analysis.

Spectrophotometry

Whereas free metal ions are usually non-absorbing or only weakly absorbing in the UV or visible region of the spectrum, a suitably selected complex of the metal ion may absorb very strongly in this region of the spectrum. The use of strongly absorbing organic ligands whose wavelengths of maximum absorbance are significantly shifted on complex formation is particularly significant here.

Spectrofluorimetry

One can prepare fluorescent complexes from non-fluorescent metal ions and non-fluorescent ligands. Once again the use of organic ligands is particularly notable here.

Complexometric titrimetry

Here complex formation is used as the basis of the titration. Both titrant and indicator are ligands. At the end-point the indicator (up to now complexed with the metal ion) is released from the complex by the titrant, with a resulting colour change.

Solvent extraction

Whereas metal ions in the free state are typically water-soluble but insoluble in organic solvents, metal complexes, if appropriately chosen are insoluble in water but soluble in organic solvents.

Polarography

In polarography, metal ions can interfere with one another if they are reduced to the metal at similar applied potentials (ie they have similar half-wave potentials, to use the terminology that you will meet if you study this technique). These half-wave potentials can be altered by complex formation, and so by the judicious formation of certain complexes, interferences can be eliminated.

Gravimetry

One useful way of precipitating a metal ion quantitatively from aqueous solution is to convert it into a completely insoluble complex.

Ion exchange

It is possible to use complex formation reactions to manipulate the overall charge on a metal ion. This can be exploited to good effect in certain ion-exchange separations.

Chromatography

In paper and thin-layer chromatography, metal ions can be separated, either in the 'free state or by way of reversible complex formation, but the resulting spots are generally invisible. However, the paper or plate can be sprayed with a solution of a ligand that gives strongly coloured or fluorescent complexes with the metal-ion spots.

Masking reactions

These are reactions, which for metal ions normally involve complex formation, which can be used to remove the effect of an interfering ion in analysis without the need for an actual separation. What you do is to form a complex between the interfering ion and a suitable ligand, so that the complex does not interfere in the analysis, even though the free metal ion would have done so.

SAQ 6.1b

Here are *six* descriptions of the properties of certain ligands and of the resulting metal complexes, together with *six* analytical techniques. Match the descriptions to the techniques.

The descriptions

(*i*) Reacts selectively with certain types of metal ion to form water-soluble, non-absorbing complexes thus preventing the reaction of such metal ions with other ligands. $\longrightarrow$

SAQ 6.1b (cont.)

(*ii*) Reacts with metal ions to form complexes that are very much more soluble in organic solvents than in water.

(*iii*) Reacts selectively with metal ions to form water-insoluble complexes.

(*iv*) Reacts non-selectively with a wide range of metal ions to form stable, strongly coloured complexes.

(*v*) Reacts selectively with certain metal ions only (ideally under specific conditions with a single metal ion) to form stable, strongly coloured complexes.

(*vi*) Reacts with metal ions to form negatively charged complexes.

The techniques

(*a*) Spectrophotometric determination of metal ions in solution.

(*b*) Gravimetric determination of metal ions in solution.

(*c*) Detection of metal ions separated by paper or thin layer chromatography.

(*d*) Separation of metal ions by ion-exchange.

(*e*) Masking of interfering metal ions in an analysis.

(*f*) Solvent extraction of metal ions.

SAQ 6.1b

6.2. TYPES OF METAL COMPLEXES AND FACTORS AFFECTING THEIR STABILITY

In this section, we shall review the types of metal complex that are used in analysis, the types of metal ion that are most easily complexed, the types of molecule that are most commonly encountered as ligands, and a wide range of factors governing the stabilities of the complexes that are formed.

6.2.1. Types of Metal Ion Most Commonly Associated with Complex Formation

Virtually every known metal ion is capable of forming co-ordination complexes of one type or another, but the range and overall stabilities of these complexes do however, vary very greatly as we move across or down the Periodic Table. As a general rule, although, of course, the nature of the ligand has a vital part to play, we find that the most numerous and stable complexes tend to be associated with the transition-metal and 'B group' heavy-metal ions, whilst the alkali metals show the least tendency to form complexes.

6.2.2. Types of Molecules or Ions that Act as Ligands

A ligand must, by our understanding of Lewis bases (which is what a ligand is) be a molecule or ion capable of donating electrons to the metal ion. It must therefore contain at least one atom with a lone pair of electrons available for donation into empty atomic orbitals on the metal ion. (This is an over simplification, since certain other molecules which contain an excess of electrons above that needed for bonding in the molecule can form co-ordinate bonds. I am thinking about complex formation between a metal atom or ion and the π-electrons in a double bond in an organic molecule).

∏   Name some common elements whose atoms, when present in neutral molecules or anions possess lone-pair electrons available for co-ordination to metal ions.

These will be the common non-metallic elements whose atoms contain a more than half filled outer shell of electrons, so that after the normal covalent bonding has occurred to complete the atom's outer shell of electrons, at least two electrons are available to form at least one lone pair.

Thus, the list will include:

the halogens (in practice complexing only as the halide ions, F^- and Cl^- being the most important),

oxygen,

nitrogen,

sulphur.

In addition, a few other elements may be involved, such as selenium and phosphorus, but they are of less importance.

The ligands that we are most concerned with in analysis are Cl^-, F^-, certain oxygen-containing inorganic anions and a huge range of organic neutral molecules and anionic species, containing oxygen, nitrogen or sulphur atoms suitably positioned in the molecule to co-ordinate with metal ions.

SAQ 6.2a

Here are some molecular and ionic species. Which of these can act as ligands towards metal ions and which cannot?

(i) NH_3,

(ii) F^-,

(iii) CH_4,

(iv) H_2O,

(v) PO_4^{3-},

(vi) $CH_2{=}CH_2$,

(vii) $CHCl_3$,

($viii$) C_2H_5SH,

(ix) $Si(CH_3)_4$,

(x) $B(CH_3)_3$.

SAQ 6.2a

6.2.3. Chelate Complexes

A very important type of complex formed between a metal ion and an organic ligand is a chelate complex. Ligands forming such complexes are often known as chelating agents. You should in fact be familiar with chelate complexes and the chelate effect from *ACOL; Classical Methods*.

∏   What is so specifically characteristic of a chelating agent and of the chelate complexes formed from such a ligand? How do they differ in properties from similar, non-chelate complexes?

A chelating agent is a ligand containing at least two atoms capable of co-ordinating with a metal ion and thus *forming a ring structure on complex formation*, in which the co-ordinating atoms and the metal ion all form part of the ring.

Chelate complexes are invariably *more stable* than similar non-chelate structures provided no undue strain is introduced on formation of the ring (this is generally so when the ring is of the optimum size of 5 or 6 atoms).

A very simple example of a chelating agent is 1,2-diaminoethane, $H_2NCH_2CH_2NH_2$. It contains two nitrogen atoms which can both co-ordinate with the same metal ion to produce a ring structure. For example, 1,2-diaminoethane can react with copper(II) ions to form the 1:2 chelate complex shown below:

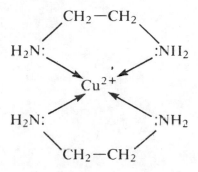

Note the formation of two 5-membered rings. Compared with, say, the complex formed between Cu(II) ions and 1-aminoethane shown below:

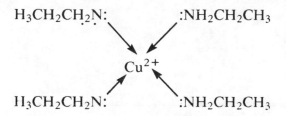

the complex with 1,2-diaminoethane is much the more stable, entirely (or almost entirely) because of the chelate effect.

SAQ 6.2l

Below are a number of organic ligands. Which are likely to be capable of forming chelate complexes and which are not? (Answer *yes* or *no* for each ligand).

(*i*) $CH_3 - \underset{\underset{O^-}{|}}{C} = CH - \underset{\underset{O}{\parallel}}{C} - CH_3$

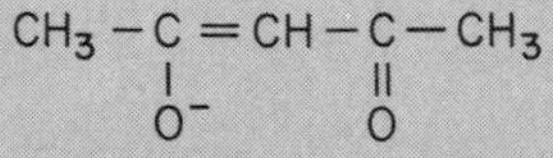

(*ii*)

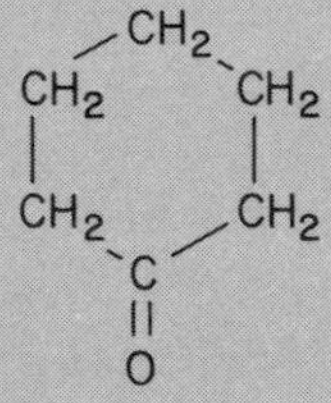

$$CH_2 - CH_2$$

(*iii*)

OH

O^-

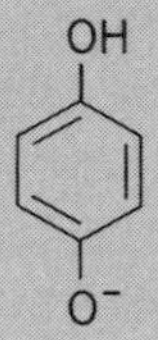

(*iv*)

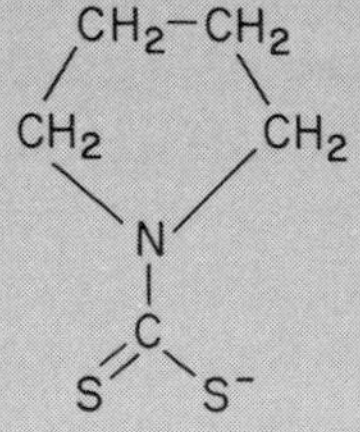

SAQ 6.2b

6.2.4. The Stabilities of Metal Complexes – Stability Constants

Metal-complex formation is a class of reversible reactions that reach thermodynamic equilibrium. The position of this equilibrium can thus be described by an equilibrium constant. This is known as the stability constant for the complex (sometimes also as the formation constant).

The expression for a stability constant is best illustrated with an example. Let us use the example that we had before, *viz* the complex formed between Cu^{2+} ions and 1-aminoethane. We saw that *four* ligand molecules co-ordinated with one Cu^{2+} ion. Now in fact it is possible to form not one complex but *four* different complexes corresponding to one, two, three and four ligand molecules respectively co-ordinating to one Cu^{2+} ion. For brevity, let us use the letter L to represent the ligand (here 1-aminoethane). Now we write a series of chemical reactions showing the stepwise formation of these complexes as follows:

$$Cu^{2+} + L \rightleftharpoons CuL^{2+}$$

$$CuL^{2+} + L \rightleftharpoons CuL_2^{2+}$$

$$CuL_2^{2+} + L \rightleftharpoons CuL_3^{2+}$$

$$CuL_3^{2+} + L \rightleftharpoons CuL_4^{2+}$$

If you read this having studied *ACOL: Classical Methods* previously to this Unit, you may well find this concept of stepwise formation of complexes familiar to you. You should also know how to write an expression for a stability constant, which is the equilibrium constant for the above reaction. You should also know that we can define *stepwise stability constants* (K_1, K_2, K_3 etc) which represent equilibrium constants for each step of the reaction (ie each equation above), and *overall stability constants* (β_1, β_2, β_3 etc) for the overall reactions to produce each complex. These are the sums of the stepwise reactions and can be written as follows:

$$Cu^{2+} + L \rightleftharpoons CuL^{2+}$$

$$Cu^{2+} + 2L \rightleftharpoons CuL_2^{2+}$$

$$Cu^{2+} + 3L \rightleftharpoons CuL_3^{2+}$$

$$Cu^{2+} + 4L \rightleftharpoons CuL_4^{2+}$$

∏   Now write down expressions for the stepwise stability constants (K_1, K_2, K_3, K_4) and overall stability constants (β_1, β_2, β_3, β_4) for the Cu^{2+}-1-aminoethane complexes. Use square brackets to denote molar concentrations and use the symbol M for Cu^{2+} to simplify the writing of the expressions. Write the expressions in terms of concentrations (ignore activity coefficients, if you are aware of these and are wondering about their relevance here). Then show the relationship between β_4 and K_1 to K_4.

If you have not come across this concept of stability constants before, then go straight to the response to see how we write these expressions.

Any equilibrium constant is defined as the ratio of the (mathematical) product of the concentrations of the (chemical) products of the reaction to the (mathematical) product of the concentrations of the starting materials of the reaction at a fixed temperature when the reaction is in equilibrium, each concentration in the expression being raised to the power of the stoicheiometric factor associated with it in the chemical equation.

Thus the equilibrium constant, K, for the reaction:

$$a\text{A} + b\text{B} \rightleftharpoons c\text{C} + d\text{D}$$

would be:

$$K = \frac{[\text{C}]^c.[\text{D}]^d}{[\text{A}]^a.[\text{B}]^b}$$

If you jumped straight to this response because you did not know how to write an expression for an equilibrium constant, then you should now be able to return to the question and at least have a go at writing an equilibrium constant expression.

The correct expressions thus are:

$$K_1 = \frac{[\text{ML}]}{[\text{M}][\text{L}]} \qquad \beta_1 = \frac{[\text{ML}]}{[\text{M}][\text{L}]}$$

$$K_2 = \frac{[\text{ML}_2]}{[\text{ML}][\text{L}]} \qquad \beta_2 = \frac{[\text{ML}_2]}{[\text{M}][\text{L}]^2}$$

$$K_3 = \frac{[\text{ML}_3]}{[\text{ML}_2][\text{L}]} \qquad \beta_3 = \frac{[\text{ML}_3]}{[\text{M}][\text{L}]^3}$$

$$K_4 = \frac{[\text{ML}_4]}{[\text{ML}_3][\text{L}]} \qquad \beta_4 = \frac{[\text{ML}_4]}{[\text{M}][\text{L}]^4}$$

The relationship between the overall stability constant, β_n and the stepwise stability constants K_1 to K_n is:

$$\beta_n = K_1.K_2.K_3 \ldots K_n$$

a fact that you can demonstrate from the expressions that I have just given. Thus for β_4

$$K_1.K_2.K_3.K_4 = \frac{[ML][ML_2][ML_3][ML_4]}{[M][L][ML][L][ML_2][L][ML_3]L]}$$

$$= \frac{[ML_4]}{[M][L]^4}$$

$$= \beta_4$$

The value of the stability constant is thus a measure of the thermo-dynamic stability of the complex, a very important property indeed when you come to consider its use in analysis. Let us look at just a few examples of this.

(a) In techniques such as spectrophotometry, spectrofluorimetry, gravimetry and solvent extraction we assume that complex formation involving the analyte ion is essentially 100% complete. This in turn implies a high value for the stability constant (for a $1:1$ complex, K_1 should be *ca* 1000 for 99.9% complex formation).

(b) On the other hand, in the same techniques as in (a), we do not want interfering ions to form complexes, so we look for small values for the stability constants of the complexes of such species.

(c) In masking reactions, we use an auxiliary (second) ligand. Here we want large values for the stability constant for analyte–ligand and interfering ion-masking agent complexes, but small values for the stability constant for the analyte-masking agent and interfering ion-ligand complexes.

(*d*) In complexometric titrimetry we have analyte-titrant and analyte–indicator stability constants to consider. We need the latter to be significantly smaller than the former so that at the end-point, the titrant will displace the analyte from the analyte–indicator complex, but not so small that this occurs too soon (ie before the end point).

(*e*) Separations of metal ions by solvent-extraction, ion-exchange and even paper or thin-layer chromatography often rely almost exclusively on the fact that stability constants for a series of different metal ions complexing with the same ligand vary greatly.

6.2.5. The Effect of pH on the Degree of Formation of a Metal–Ligand Complex

The stability constant can be assumed, at least at this stage of our studies, to be essentially a true thermodynamic constant, and as such should be independent of other factors, such as pH. However, if we plot the *degree of formation* of a metal complex (which we can define simply as the percentage of metal ion present as the complex) against pH, then at least for most organic ligands we see a very strong dependance on pH indeed. Have a look at Fig. 6.2a which shows a typical plot of the degree of formation of such a metal complex as a function of pH. The stability constant for this complex is quite high,

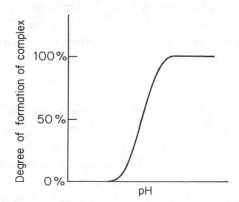

Fig. 6.2a. *Typical plot of degree of formation of a metal complex· with an organic ligand, shown as a function of pH*

which means that we can get effectively 100% complex formation, but only at certain values of pH.

∏   Why does the degree of formation of the metal complex of a typical organic ligand fall to such low values (effectively zero) at low values of pH?

The answer lies in the fact that the hydrogen ion (H^+, or H_3O^+ as it exists in water) is as much a Lewis acid as any metal cation and that organic ligands can be protonated by H^+, in competition with metal-complex formation. Thus for 1-aminoethane:

$$CH_3CH_2NH_2 + H^+ \rightleftharpoons CH_3CH_2NH_3^+$$

The protonated ligand cannot complex with a metal ion. The concentration of the protonated ligand increases as the pH decreases. A decrease in pH is equivalent to an increase in hydrogen ion concentration, since $pH = -\log[H^+]$.

The net result therefore is that as the pH decreases, the concentration of free unprotonated ligand decreases. In other words the value of [L] in the expression for the stability constant decreases, and so in turn does the value of [ML_n]. Thus less complex is formed. The shape of the curve in Fig. 6.2a is sigmoid and arises because of the negative logarithmic relationship between pH and [H^+]. The steepness of the curve is a function of the stoichiometry of the complex (ie the value of n in ML_n) and the position (say of the 50% point) along the pH axis is a function of a number of factors including the stability constant and the equilibrium constant for the protonation of the ligand.

This dependence on pH of metal-complex formation reactions is of great significance and real use in analysis, because for different metal ions complexing with the same ligand, the sigmoid curves may be well resolved on the pH axis. This means that, by careful choice of pH, we can ensure that one metal ion is complexed when another is not. This in turn leads to a useful degree of selectivity in such reactions.

We won't take this topic any further at this stage because I have an excellent application of this principle in store for you later on in this Unit, namely the selective solvent extraction of metal ions as neutral chelate complexes (Part 8).

SAQ 6.2c   The compound, 8-hydroxyquinoline has the structure below.

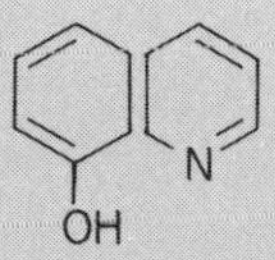

It reacts with Al^{3+} ions to form a chelate complex with a stoicheiometry of $1:3$ that is neutral (ie is not charged).

(*i*)   What is the actual molecular species that reacts with the Al^{3+} ion (ie what is L?)

(*ii*) Write expressions for the stepwise stability constants and the overall stability constant for the ML_3 complex.

(*iii*) At low pH, formation of the complex is suppressed. What other species are formed instead of ML_3?

(*iv*) Finally draw the structure of the neutral ML_3 complex.

SAQ 6.2c

6.2.6. Some Factors Affecting the Stabilities of Metal Complexes

The stabilities of metal complexes depend on a very large number of different factors, and it would be difficult to predict the magnitude of the stability constants of complexes with any degree of certainty except within a very limited series of complexes. However, having said this, let me also say that all sorts of useful trends can be observed amongst metal complex stabilities, either by making subtle changes in the structures of the ligands, or by changing the metal atom systematically across or down the Periodic Table.

Let us briefly look at two such trends.

1. The nature of the metal atom and of the co-ordinating atom

There is a useful concept here in which we can talk about 'hard' and 'soft' acids and bases (where the terms acid and base are used in the Lewis sense of electron acceptors and electron donors). The terms refer primarily to the actual atoms at either end of the co-ordination bond, although the overall structure of an organic ligand can also affect its 'hardness' or 'softness'.

The concept of what constitutes a 'hard'or 'soft' acid or base is fairly subtle and it might perhaps be more useful here simply to give examples. Note that individual acids and bases are not simply 'hard' or 'soft' but that there is a gradual and continuous range of character amongst acids and bases from 'hard' to 'soft'.

Here is a table (Fig. 6.2b) of 'hard' and 'soft' character.

Acids		Bases
	HARD CHARACTER	
High-valence transition metal ions from group IIIa, IVa and Va.	↑ Increasing in soft character / Increasing in hard character ↓	F^-
Main-group metal ions from group IIa, IIIb, IVb.		Cl^- O-containing ligands
		N-containing ligands
Transition-metal ions from later groups in the periodic table.		Ligands with the co-ordinating atom in a conjugated system.
Heavy-metal ions from groups Ib, IIb, IIIb.		S-containing ligands
	SOFT CHARACTER	

Fig. 6.2b. *'Hard' and 'soft' character of Lewis acids and bases*

The important principle behind this is that

(a) A *hard* acid + a *hard* base → *more* stable complex
(b) A *soft* acid + a *soft* base → *more* stable complex
(c) A *hard* acid + a *soft* base → *less* stable complex
(d) A *soft* acid + a *hard* base → *less* stable complex

In other words you should try to match your ligand to your metal ion with regard to its hard or soft character, for maximum stability. Some examples of stable complexes arising from this concept are:

(a) the highly stable fluoro-complexes of the more highly charged metal ions; also their refractory character (due to the high stability of their oxides) (hard character);

(b) the high stability and insolubility of the heavy metal sulphides, and the affinity shown for such metal ions by sulphur-containing ligands (soft character);

(c) the popularity of mixed O- and N-containing ligands for a wide variety of transition- and other metal ions (neither hard nor soft character predominating).

SAQ 6.2d

Which of the following statements are *true* and which are *false*? Explain your choice in each case.

(*i*) Amongst metal cations there is a tendency towards increasing 'hard' character as the charge on the ion increases.

(*ii*) Amongst both transition-metal cations and coordinating non-metallic elements there is an increasing trend towards 'soft' character as you progress to the right of the Periodic Table.   ⟶

SAQ 6.2d (cont.)

(*iii*) Amongst the coordinating non-metallic elements there is a marked increase in 'soft' character as you move down any group in the Periodic Table, but amongst metallic elements the trend is in the opposite direction.

(*iv*) Acetylacetone would be a better chelating agent for Al^{3+} than it would be for Pb^{2+}.

(*v*) Dithizone would be a good chelating agent for Pb^{2+}, whilst EDTA would be a good masking agent for interfering transition-metal ions.

Acetylacetone   is $CH_3 \overset{\text{O}}{\underset{\parallel}{C}} CH_2 \overset{\text{O}}{\underset{\parallel}{C}} CH_3$

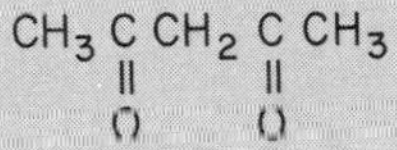

Dithizone   is

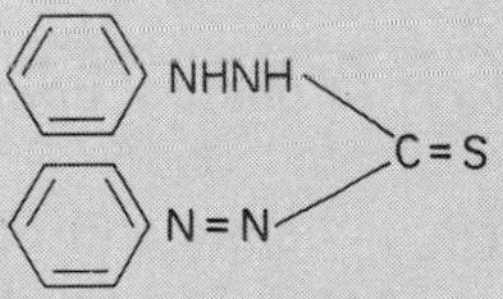

EDTA anion is

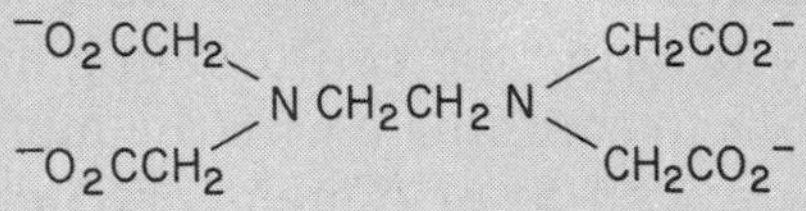

SAQ 6.2d

2. Steric effects in chelate complexes

When we form a chelate complex between a metal ion and a chelating ligand, we have to form a stable ring structure that is not unduly strained. In other words we must not have to 'bend' the molecule unduly to bring the coordinating atoms in the ligand into suitable positions to coordinate with the metal ion. This is most satisfactorily achieved with 5- or 6-membered chelate rings, since then optimal angles for eliminating ring strain of approximately $108°$ to $120°$ can exist between bonds. Chelates can be formed with ring sizes other than 5 or 6 (we have already met a 4-membered ring chelate), but the stability tends to be lower than it would be for an equivalent 5- or 6-membered ring chelate.

Another way in which we can modify the properties of a chelating agent through steric effects is to introduce a non-coordinating group into the molecule close to the coordinating atoms. This will literally 'get in the way' when an attempt is made to form the chelate.

If you look back at the structure of the Al^{3+}-8-hydroxyquinoline chelate, you will notice that things get pretty crowded around the central Al^{3+} ion (which itself is quite small). If you now try to form a chelate between Al^{3+} and 2-methyl-8-hydroxyquinoline (below),

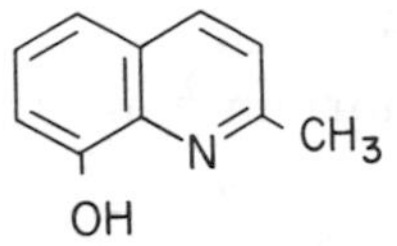

you will not succeed because the methyl groups sterically hinder the reaction ('get in the way') during chelate formation.

This molecule will form chelates with other metal ions but is *more selective* than 8-hydroxyquinoline because of the possibility of steric hindrance due to the methyl group. There are many other examples of the design of selective chelating agents by purposely building a steric effect into the molecule (see reference 12).

SAQ 6.2e   Comment on the trends in chelating properties that you might expect to observe within the following two series of chelating ligands.

(i) (a) (b) (c)

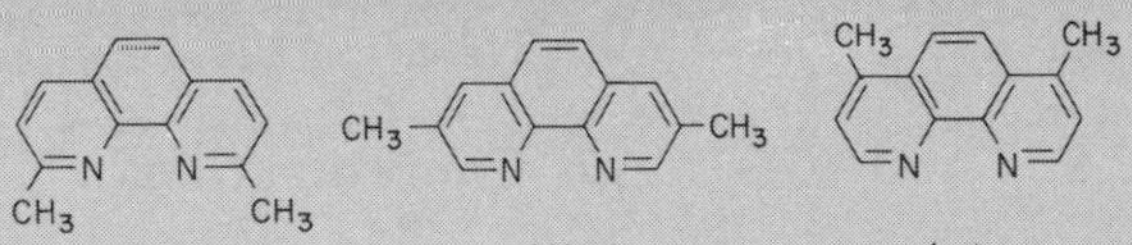

(ii) $H_2NCH_2CH_2NH_2$ $H_2NCH_2CH_2CH_2NH_2$

(a)   (b)

$H_2NCH_2CH_2CH_2CH_2NH_2$

(c)

6.2.7. Kinetic Effects in Complex Formation

All metal-complex formation reactions are reversible and come to thermodynamic equilibrium, at which point, it is possible to calculate the concentrations of the various species from appropriate expressions for the equilibrium constants.

∏   What can you say about the rates of the forward and reverse reactions when the system is in thermodynamic equilibrium?

They are exactly equal.

Normally we assume that the time taken for metal-complex formation to reach equilibrium is very short. In other words that the reactions involved are fast. However, there are circumstances when this is not so. In particular, certain metal ions with so-called 'inert' electronic configurations react very slowly in metal-complex formation, and it is quite possible in these circumstances to form a complex of another metal ion and carry out an analysis or separation before any significant reaction has occurred with the inert metal ion.

Some examples of inert metal ions are:

$$Cr^{3+}, \quad Co^{3+} \quad \text{and} \quad Pt^{2+}$$

6.2.8. Activities *versus* Concentrations

All our expressions for equilibrium constants to date have been formulated in terms of concentrations. However, a true thermodynamic constant should strictly be defined in terms of activities rather than concentrations. The activity of a species in solution can be looked upon as a measure of the total effect of that species, say, on an equilibrium. In an ideal solution, the activity of a species is the same as its concentration, but in a non-ideal solution you can imagine a proportion of the material being unavailable, due to the non-ideal behaviour of the solution. We can thus correct the concentrations to activities to account for this by multiplying the concentration by the activity coefficient, a number less than one. In general, in aqueous solutions where concentrations are $< 10^{-3}$ mol dm^{-3} we can ignore the effect of activities and that is what we shall always do, unless specific mention is made of the need to do otherwise.

SAQ 6.2f   Explain the following two observations:

(*i*) Acetylacetone appears to form stable complexes with a stoicheiometry of $1:3$ with Al^{3+} and Fe^{3+} but not with Cr^{3+}.

(*ii*) You measure the stability constant of a particular metal complex as a function of metal ion concentration. You observe the value to be independent of concentration until the concentration is of the order of 10^{-3} mol dm^{-3}. At higher concentrations you observe a dependence of the stability constant on concentration.

Objectives

On completion of this Part of the Unit, you should be able to:

- list some of the commoner and more important techniques for the determination of metal ions in solution;

- identify those which are applicable to metal ions in their free (hydrated) form, those which involve the metal ions in complex formation, and those that are applicable to either free or complexed metal ions;

- list various ligands and ligand types that complex with metal ions;

- describe some of the differences in properties between free metal ions and their complexes that are used in analysis;

- review some of the applications of metal-complex formation in analysis;

- discuss the importance of metal-complex stability;

- describe the concept of a metal-complex stability constant;

- discuss the effect of pH on apparent metal-complex stability;

- list some important properties of metals, ions and ligands that affect stability:

  (*i*)   the nature of the coordinating atoms,
  (*ii*)   the chelate effect,
  (*iii*) steric factors;

- discuss kinetic effects;

- describe (only qualitatively) the concept of activity of ions in solution.

7. Introduction to Separation and Preconcentration Techniques

7.1. THE MEANING OF SEPARATION IN ANALYSIS

I want to start this section by reminding you of a proposition that you will have met in Part 1 of this Unit, that a properly constructed analytical procedure will consist of a number of distinct stages or unit operations which are carried out in sequence. Perhaps for revision, you would like to try to write down a list of the separate operations that you might usually expect to find in an overall analytical procedure.

∏   Prepare a list of discrete operations that together make up a complete analytical procedure.

Here is my list with which you can compare yours. Yours may not be quite as extensive as mine, as I have added one or two operations at either end that in routine situations may perhaps lie outside the province of the analyst. Nevertheless, they are implicit in the overall procedure. Check especially that your list contains the asterisked items, as these are the ones we are especially concerned with in this unit.

Defining the problem in analytical terms.
Selection of the analytical method.

*Sampling.
*Pretreatment of the sample.
*Calibration (preparation and measurement of standards).
*Measurement.
*Processing of raw analytical results.
Decisions and action consequent on the final results.

Included in this list is a *sample pretreatment* step, which we have already seen includes a number of important operations.

∏   List some important operations that might have to be carried out as part of the analytical sequence, under the heading sample pretreatment.

This list is by no means exhaustive, but is designed to include some of the more important items.

Dissolving the sample.
Converting elements into simple inorganic forms.
Converting analytes into alternative chemical forms:
    eg metal ions into complexes, organic compounds into derivatives.
Separation of the analyte from other species present.
Preconcentration of the analyte.

You will notice that included in this list are *separations and preconcentrations*. It is with these two operations that we shall be concerned in this Part. Although not every analysis by any means includes a separation or a preconcentration, nevertheless this is often a vitally important part of the overall procedure. Although separations and preconcentrations have been listed as separate operations, in fact preconcentrations are special applications of separations, and in the rest of this Part we shall take it that the term separation may also cover preconcentration where appropriate.

We need now to decide exactly what we mean by the term *separation* in analysis. I want you to understand the term in this context to mean the physical transfer of a particular chemical substance from one phase or medium to another, or the actual physical separation of the components of a mixture into separate fractions. Excluded, though, is any separation of analytical signals or data (eg resolution of spectral lines in spectroscopy), or apparent 'removal' of a substance from a solution, merely by changing its chemical form into one less active ('masking').

SAQ 7.1a

Which of the following involve separations within the meaning of the term as explained in Section 7.1?

(*i*) Paper chromatography of transition-metal ions. The metal ions are resolved by development with a mixture of acetone (propanone) and hydrochloric acid, and detected and identified by reaction with various selective organic reagents.

(*ii*) Removal of the interference by iron(III) in the spectrophotometric determination of dichromate ion by complexing the former with phosphoric acid.

(*iii*) Spectrographic analysis of a powder for the metals present. Each metal is represented by a line or lines on a photographic plate spatially resolved from each other in the spectrum.

(*iv*) Removal of nickel ions from a solution containing nickel and cobalt ions by precipitation as an organic complex.

(*v*) Gas chromatography of a petroleum fraction. The chromatogram consists of a series of peaks produced as each component of the petroleum fraction passes through a detector after a length of time which is characteristic of that component.

SAQ 7.1a

7.2. APPLICATIONS OF SEPARATIONS IN ANALYSIS

In this Section we shall study a selection of the more important situations where separations can help us in analysis.

7.2.1. Separations of Analytes from Interfering Species

Whenever we set out to make a measurement on a particular analyte in a sample, we are faced with the problem as to whether there are any other substances present that can interfere in the measurement, so that an erroneous result is obtained. If this is so, then, to overcome these interferences we can either:

(*a*) look for a more selective analytical measuring technique (eg, atomic spectroscopic techniques are usually more selective than the classical techniques of gravimetry or titrimetry, or even spectrophotometry); or (*b*) modify an existing method so as to remove the interference, for example by changing the chemical form of the interfering species (a process known as 'masking'); or

(*c*) separate the analyte from the interfering species or the interfering species from the analyte.

The carrying out of separations before the measurement stage of analysis to overcome interferences represents one of the most important applications of separation in analysis.

7.2.2. Preconcentration of the Analyte in Trace Analysis

A common problem in trace analysis is to find an analytical technique that is sensitive enough to measure the very low concentrations that you might expect to encounter. If such techniques are not available and sample size is not limited, an alternative approach is to increase the concentration of the analyte to a level at which analysis is possible. This is done by transferring it from a phase of large volume to another of much smaller volume. This process, which involves separation, is known as *preconcentration* and there are many useful applications of it in analysis.

7.2.3. Transfer of the Analyte into a Phase Suitable for the Analysis

A separation may sometimes be incorporated to enable an analysis to proceed that would otherwise be impossible because of the general unsuitability of the medium. For instance, in the spectrophotometric analysis of metal ions, we often form highly coloured complexes of the ions by reaction with suitably chosen organic ligands. Unfortunately many of these complexes are extremely insoluble in water. They are, however, much more soluble in organic solvents, so, a separation in the form of a solvent extraction is incorporated and the concentration of the complex in the organic phase is measured.

7.2.4.   Isolation of the Analyte in a Pure Form

If you want to identify an unknown organic compound, you would perhaps do so *via* a number of tests, such as a melting-point determination and elemental analysis, as well as by use of its infra-red, nuclear magnetic resonance and possibly mass spectrum. Unfortunately, none of these techniques is particularly effective on impure compounds or mixtures. Therefore, a purification (ie a separation) is a prerequisite of such an analysis.

Similarly, gravimetry is a quantitative analytical technique based on obtaining the analyte (probably as a chemical derivative) in a pure form of known composition that can be accurately weighed.

7.2.5.   Identification of an Analyte *via* its Behaviour in Separations

Another way of identifying a substance, generally from a short list of known substances (eg, identifying a particular drug from a list of possible drugs) is to subject it to chromatography and to identify it *via* its chromatographic behaviour, ie by determining how far or fast it has moved through the chromatographic system in comparison with some standard substance (R_f value or retention time, etc). This is a less certain method of identification, because we rely on only one parameter, the separation or chromatographic parameter, which may not be unique and may be subject to error (it is rather like identifying a substance by melting-point alone).

7.2.6.   Simplification of a Matrix

Sometimes a reasonably selective analytical technique is to be applied to an extremely complex matrix so that general matrix effects rather than specific interferences complicate the analysis. A crude separation designed to simplify the matrix rather than to isolate individual components may be relevant here. The example that immediately comes to mind is the analysis of complex biological mixtures. Here, a preliminary solvent extraction may be used to remove all the species of interest, along with a proportion of the matrix, leaving much of the unidentifiable material behind. This may be combined, perhaps, with a preconcentration, and is sometimes known as *preliminary sample clean-up*.

7.2.7. Resolution of a Complex Mixture and Subsequent Identification/Determination of Components

This is the province of the many and varied chromatographic techniques of analysis, in which individual components of a mixture are either passed to a spectrometer where they are identified by their spectra (for example in combined gas chromatography – mass spectrometry) or in which they are quantitatively determined *via* peak areas obtained from their response in a detector. The applications of this are legion, and perhaps represent the most important group of modern techniques in quantitative organic analysis.

7.2.8. Indirect Analyses Involving Separations

An indirect analysis is one in which the actual measurement is made on a species other than the analyte (eg, a back-titration). Such a procedure may sometimes be used when direct measurements on the analyte are not possible or are unsatisfactory. Obviously, there must be some relationship between the concentration of the analyte and that of the species measured). Sometimes this relationship is established *via* a separation step. Here are some examples.

(*a*) Indirect determination of chloride ion in water by atomic absorption. Here we add a known excess of silver nitrate to the sample, precipitate the chloride as silver chloride; also therefore we precipitate a corresponding amount of silver ions. We then determine the silver ions left in solution by atomic absorption and thus we can calculate the original chloride concentration.

(*b*) Determination of phosphate ions by extraction of the phosphomolybdate complex and atomic-absorption determination of the molybdenum extracted. The extracted complex contains twelve molybdenum atoms per phosphate ion, so a useful gain in sensitivity is achieved. This is an example of an *amplification reaction* (rather like a 'pseudo-preconcentration' although no preconcentration is actually involved.)

(*c*) Isotope dilution analysis is a radiochemical technique which is useful when we can isolate the analyte in a pure form, but not quantitatively. We add a known amount of the analyte in a radioactive form (ie labelled with a radioactive isotope) to the sample and then *separate* a portion of the analyte mixed with the added labelled analyte from the rest. We measure its radioactivity as well as that of the pure labelled compound. From a simple formula involving the ratio of these two, we can calculate the extent to which the labelled added analyte has been diluted by the inactive analyte in the original sample, and hence the amount of inactive analyte.

This has been a brief review in which we have considered eight ways in which separations may be of use to the analyst. There may be one or two other ways that we haven't mentioned, although these are unlikely to be of other than rather specialised interest. Perhaps you can think of one.

Finally, before we move on from this section, there is one point that I would like you to think about. We have seen at least eight ways in which separation can be of use to the analyst, and of course there will be many situations where an analysis will be possible only because we can incorporate an effective separation step into it. However, there may be other situations where there is a choice of method.

∏   Faced with the situation where you have a choice of methods for solving a particular analytical problem, the choice in broad terms may be between:

(*i*) use of a simple, basic, perhaps low-performance technique that solves the problem only by virtue of its use in conjunction with a separation step;

(*ii*) use of a sophisticated high-performance technique that solves the problem without the need for a separation step.

What in general terms, would be the factors that would sway your decision one way or the other?

Perhaps the most significant factor would be that the need for a separation step would mean that there would be a greater number of steps in the overall analytical procedure. This in turn might imply (all other factors being equal, of course) that the method that involved separation was:

(*a*) slower,
(*b*) more complicated to carry out,
(*c*) more prone to error,

than that which did not. Of course there may be situations where this is not so, but the development in recent years of sophisticated instrumental techniques capable of greater sensitivity and selectivity has somewhat lessened the need for separation in certain areas of analysis. In other situations, such as chromatography the separation step itself has been the subject of sophisticated refinement.

SAQ 7.2a	Six analytical situations follow in which a specific problem has been identified and has yet to be solved. For each situation, explain whether incorporation of a separation stage might help to solve the problem and, if so, how, in relation to the eight areas of application of separation to analysis described in Section 7.2. (*i*) Analysis of a liver extract for a drug metabolite by liquid chromatography. Unfortunately, the chemical composition of the extract is so complex as to make identification and quantification of the peak very difficult. (*ii*) Analysis of a 5 µl sample of child's blood for lead. Unfortunately, the method available has an absolute limit of detection (ex $\longrightarrow$ ·

**SAQ 7.2a
(cont.)**

pressed in ng) that is not low enough for the analysis to be possible because of the small sample size.

(*iii*) Colorimetric determination of the copper content of tap water as its 1-(2-pyridylazo)-2-naphthol complex. Unfortunately the complex is precipitated from aqueous solution. It is however, soluble in trichloromethane.

(*iv*) Titration of cobalt in the presence of nickel by using EDTA as titrant. Unfortunately the minimum values of pH for the titration of the two ions are too close together for the titration to distinguish between the ions.

(*v*) Determination of traces of a toxic metal present in drinking water. Unfortunately we can't find a technique with a limit of detection (expressed in concentration units) that is low enough for us to carry out the analysis.

(*vi*) Total analysis of an alloy sample (ie determination of each individual metal present to give a total equal to 100% within experimental error). The trouble is, we don't know which metals are present.

SAQ 7.2a

7.3. THE INDIVIDUAL STEPS OF AN ANALYTICAL SEPARATION

There is a model that we can use to describe nearly all the analytical separations that we shall examine. Although there are one or two separations that don't fit the model very well, so many of them do, that it is worthwhile to develop this model in general terms for our purposes. The model may be basically described as follows.

> A *two-phase system* in which the species to be separated (after *chemical conversion* if necessary) are *distributed* between the phases. The phases are then *physically* (*mechanically*) *separated* from each other to complete the separation.

Put another way, a typical separation may involve up to three steps:

> *Chemical conversion* of the substances to be separated into derivatives (not always required),

Distribution of the substances between two phases in contact with one another,

Physical (mechanical) separation of the phases.

Fig. 7.3a, b and c show this pictorially. The two species, A and C are separated because A forms a derivative with X, but C does not. The derivative is then transferred to a second phase, whereas other species are not. The two phases are then separated, resulting in separation of A (as a derivative) from C. Let us look at these individual steps in a little more detail.

Chemical Conversion
Where this is required (and it isn't always by any means) then we may be talking about forming derivatives of organic compounds or

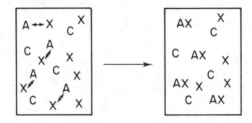

Fig. 7.3a.   *First stage of a separation, showing species A reacting with X, but species C remaining unchanged*

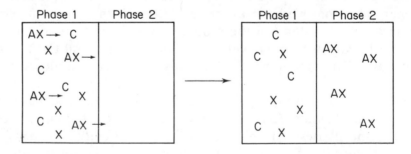

Fig. 7.3b.   *Second stage of a separation showing distribution between phases 1 and 2, so that AX is present in phase 2 and C and X in phase 1*

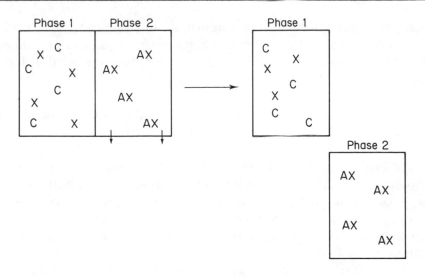

Fig. 7.3c.   *Third stage of a separation showing physical (mechanical) separation of phase 1 from phase 2*

complexes of metal ions or other such processes in order to improve the separability of the various species. When incorporated though, often the selectivity or efficiency of a separation depends entirely on this step.

Distribution
The two phases could be gas and liquid, gas and solid, liquid and liquid or liquid and solid. The distribution itself is an equilibrium process and as such is reversible. In other words species may be transferred in either direction between the phases. At equilibrium the rates of transfer in either direction are equal, and we can define an equilibrium constant to measure the position of this equilibrium (we shall do this later).

Physical separation
This could, for instance, involve separation of two immiscible liquids, filtering off a precipitate, collection of a fraction from a distillation column, or the separation that results from the flow of one phase over another, as in a chromatographic system.

Note that although up to now it might be implied from our model that these steps may occur sequentially, in practice, it may well be that two or more of the stages occur simultaneously in certain separations. This shouldn't be taken to mean that the model doesn't apply.

∏   Which two stages always occur simultaneously and continuously in any chromatographic separation?

Stage 2 (distribution) and stage 3 (physical separation). The physical separation results from the flow of one phase over another, which is a continuous process; distribution likewise occurs continuously between the two phases. Occasionally, a chemical reaction may be involved in the chromatographic system itself; this may also occur simultaneously.

The few separation techniques that do not fit this model tend to be one-phase systems in which separation is based on differential rates of migration. Superficially they may bear some resemblance to chromatography, but there is no distribution involved (eg electrophoresis).

SAQ 7.3a   Sulphate ions in water may be determined gravimetrically as barium sulphate. This analysis may be considered as consisting of:

(*i*) a separation step, involving isolation of the sulphate ions in the form of pure, insoluble barium sulphate,

(*ii*) a measurement step, involving the weighing of the separated barium sulphate.

Describe the stages of the separation part of this analysis (ie (*i*) above) under the headings of:

chemical conversion,   ⟶

| SAQ 7.3a | distribution between two phases, (name the two |
| (cont.) | phases), |

physical separation of the phases.

Which two of these stages occur simultaneously and which sequentially?

7.4. MODES OF SEPARATION

We have just seen that in general terms a separation may consist of:

(*a*) chemical conversion,
(*b*) distribution in a two-phase system,
(*c*) physical separation of the phases.

We have also seen that these individual stages may occur simultaneously or sequentially. Now, the way that the distribution and phase separation stages relate to each other leads to an important and useful means of classifying separations by what we will call *mode of separation*. Let us now look at some of these modes of separation.

7.4.1. Batch Separations

These are the simplest separations and involve a single distribution between two phases, which is first allowed to come completely to equilibrium. The two phases are subsequently separated to complete the process. Two clearly distinct operations should be discernible in a batch separation, which typically implies the use of standard laboratory glassware (eg your two phase-system may be allowed to come to equilbrium in a beaker).

Batch separations are suitable when the species that are being separated are quantitatively concentrated into a single phase at equilibrium. For example, in a preconcentration, the equilibrium constant should be such that the analytes are completely transferred to the phase of small volume. In the separation of two species A and B, A should be transferred quantitatively into one phase and B should remain in the other. In both these examples the desired separation should be achieved in a single equilibrium process. Batch separations in which such extreme transfer is not achieved *are* encountered, but here the interest is theoretical rather than analytical, and the net result is likely to be the measurement of the equilibrium constant for the distribution process.

Examples of batch separations are single-stage solvent extractions, precipitations, coprecipitations and electrodepositions. Their success frequently depends on a preliminary chemical conversion to provide derivatives with the required properties for the separation.

∏   Would a batch separation system be suitable for separating the individual components of a complex mixture?

No. You could separate the species quantitatively only into two fractions, depending on the phase into which they are concentrated, and this assumes that all the species will be quantitatively transferred one way or the other. You might be able to preconcentrate them or perhaps isolate one component *via* a selective reaction. Otherwise for isolation of each component, more sophisticated methods of separation will be required.

7.4.2. Multiple Batch Separations

These might be used when the performance of a simple batch separation is inadequate, for instance, when a simple equilibrium process is insufficient for quantitative transfer of a substance into a single phase. They involve the combination of two or more batch separations performed in sequence. In other words the sequence of operations is:

distribution → phase separation → distribution → phase separation → distribution → phase separation → distribution → phase separation, etc,

each step being still quite distinct.

They are most characteristically associated with solvent extraction, where, for instance the solutes in an aqueous phase may be repeatedly extracted with fresh portions of organic solvent to remove the last traces of a particular species. A more sophisticated example is the technique of discontinuous countercurrent extraction for the separation of species with similar solubilities. Here a special apparatus is used to produce a separation involving up to 250 or so batch separations.

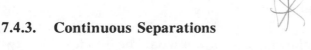

7.4.3. Continuous Separations

This is an extremely important separation technique that encompasses all forms of *Chromatography*. We shall describe it basically in terms of chromatography although in fact, fractional distillation is also a continuous separation technique.

Chromatography is used for the separation of complex mixtures of substances and is potentially effective when the properties of the substances to be separated are very similar (ie too similar for batch or multiple-batch separations). In many ways it is really a group of techniques rather than a single technique and as such its applications are very wide. In chromatography, we have two phases, one of which flows continuously over the other which is stationary. These are known as the mobile and the stationary phase respectively. Typically the mobile phase may be a gas or a liquid and the stationary phase a liquid coated onto a solid, or a solid. The stationary phase may be packed into a column as particles, the mobile phase then flowing through the column (there are arrangements other than this) to give us our continuous system.

The mixture is introduced onto the column as a narrow band of material and immediately a distribution between the two phases occurs. However, that portion of the material that remains in the mobile phase then starts to move through the column with the mobile phase and fresh distributions occur. There is a continuous process of distribution and movement (equivalent to phase separation), and these two stages of the separation occur simultaneously (as opposed to batch techniques where they occur independently). The components of the mixture then move through the column *at a rate related by a simple equation to the position of the equilibrium for their distribution between the two phases*. It is this that leads to the separation.

∏   Is a chromatographic system a system that is in thermodynamic equilibrium?

No. Because of the motion of one phase with respect to the other, the distribution only approximates to, but never actually reaches thermodynamic equilibrium.

∏   Which of the following conditions would be best for the chromatographic separation of the components of a mixture?

(*i*)   The distribution equilibria should be as different as possible for each species to be separated.

(*ii*) The distribution equilibria should be measurably different for each species, extremes being avoided (ie species should not be preconcentrated into either phase).

(*iii*) The distribution equilibria should favour preconcentration strongly into the stationary phase.

(*iv*) The distribution equilibria should favour preconcentration strongly into the mobile phase.

As a general rule (although there may be some exceptions in practice) (*ii*) would be the best choice of conditions, since these conditions would represent the situation where there are always suitable amounts of each component in each phase. If the components are concentrated entirely in the stationary phase, they will not move, and if they are concentrated entirely in the mobile phase they will not be retained. Neither situation will lead to a useful separation. Situation (*i*) could correspond more to a batch separation and in practice might lead to a crude separation rather than the high-resolution separation otherwise possible.

7.4.4. Trapping Techniques

This is rather like chromatography in that once again, we have a mobile phase flowing over a stationary phase as in a chromatographic column, but here the conditions are first of all chosen so that the species to be separated will initially be concentrated entirely into the stationary phase. This need not necessarily be quantitative as in a batch mode, because the motion of the mobile phase will have the effect of forcing equilibria in favour of retention in the stationary phase as the sample moves down the column. Nor is the sample applied in a small band. It may and generally is applied as a continuous flow of material. The sample often *is* the 'mobile phase'. The net result is that trace components are quantitatively retained in the stationary phase from a large volume of sample. Conditions are afterwards changed, so that the species are released rapidly into a mobile phase (different from the first) of small volume, ie the species are temporarily *trapped* in the stationary phase.

Π   What, therefore, do you suppose is the purpose of using a
       trapping technique in analysis?

Since we start with our species of interest present at trace concentrations in a large volume of sample and end with them in a small
volume, clearly what we have achieved is a *preconcentration* and
this is perhaps its most important use.

We shall come across applications in other areas, such as separating
two or more very different species as in batch separations, but without perhaps requiring the extreme conditions that the latter imply.
Examples of trapping techniques include the use of adsorption tubes
for preconcentrating traces of organic compounds in water and the
atmosphere, and the use of ion exchange for preconcentrating ions
in water.

Π   Look at Figs. 7.3a – c. What separation mode is implied
       therein?

A batch separation. There is clearly a single chemical conversion,
a distribution and a phase separation, all happening separately and
in sequence.

SAQ 7.4a   Here are brief descriptions of four separations.
                Classify each as one of the following:

                — a batch separation,
                — a multiple batch separation,
                — a continuous separation,
                — separation involving a trapping technique.

                (*i*)   Preparation of de-ionised water. Water is
                        continuously passed through a mixed-bed
                        ion-exchange resin. Positive and negative
                        ions are quantitatively replaced by H^+
                        and OH^- respectively. When the cartridge
                        containing the ion-exchange resin is ex-
                        hausted, ie contains an excess of ions re-

                                                              $\longrightarrow$

**SAQ 7.4a
(cont.)**

moved from the water, it is replaced with a new one.

(*ii*) Iodine is quantitatively removed from an aqueous solution by extraction with successive portions of tetrachloromethane. As soon as a portion of tetrachloromethane obtained from the extraction shows no purple colour due to extracted iodine, the extraction is assumed to be complete.

(*iii*) The simple sugars in a mixture are separated by placing spots of solution along the edge of a sheet of adsorbent paper, dipping this in a mixture of propan-1-ol, ethyl acetate (ethyl ethanoate), and water, allowing the solvent to rise by capillary action along most of the length of the paper, carrying the spots with it, drying it, spraying it with aniline phthalate (a special colour-forming reagent for sugars) and observing that each sugar has travelled a different characteristic distance along the paper.

(*iv*) Separation of copper ions from a mixture of other ions in solution. Under a particular set of chemical conditions copper ions and no other react with α-benzoin oxime to form an insoluble precipitate which can subsequently be collected by filtration.

SAQ 7.4a

7.5. QUANTITATIVE BASIS OF SEPARATION

In this section, we shall meet some of the more important quantitative parameters that are used in separation methods. What we shall do is to look at the background to each, and give a definition and a few words of explanation by way of introduction. We shan't use them in any significant way as yet. However, you will meet them all again and have opportunities to use then, either in later sections of this Unit or in the Chromatography Units.

7.5.1. Distribution Coefficient

We have already seen that an important part of most separations is a *distribution between two phases*. Now, this is an equilibrium process, and therefore can be described by an equilibrium constant. The most general form of this constant is the *distribution coefficient*, which is simply equal to the ratio of the concentrations of the substance in question in the two phases, *when the system is in equilibrium*. We can express this mathematically as follows, where K_D is the distribution coefficient.

$$K_D = \frac{\text{concentration of substance in phase 1}}{\text{concentration of substance in phase 2}}$$

at equilibrium

This in fact, is not the true thermodynamic constant, which is actually the ratio of the *activities* of the substance in the two phases. However, activities are difficult to measure, and analysts are invariably concerned with concentrations. So we use this definition based on concentration. K_D is a function of temperature and pressure, and often of concentration, although the true constant, the ratio of activities, is independent of concentration.

Distribution coefficients are used extensively in solvent extraction where we shall make great use of them. They are also used in such techniques as ion-exchange and adsorption, as well as in all forms of chromatography, in which they are fairly simply related to the rate of migration of the substance through the chromatographic system. In precipitation techniques, we use the solubility product rather than the distribution coefficient, since the concentration (or activity) of a substance in the solid phase is effectively a constant (and arbitrarily set at unity). Extremely large or extremely small values of K_D imply more or less quantitative concentration of the substance into one or the other of the two phases. Values of K_D in the region of one imply a more even distribution of the substances between the phases.

∏ Distribution of a substance between two phases involves reversible transfer of that substance across a phase boundary, ie transfer in either direction. When the system is at equilibrium, what can you say about the *rates* of transfer of substance in either direction?

The rate of transfer of material in one direction will be exactly equal to the rate of transfer of material in the reverse direction at equilibrium. However, it is important that you understand that this does not mean that transfer of material has stopped. It is still occurring, perhaps quite rapidly, even at equilibrium.

7.5.2. Distribution Ratio

A very important point to understand about the distribution coefficient when we are considering systems in which chemical equilibria are involved is that each individual chemical species has its own individual value of the distribution coefficient. For instance, if we are extracting a metal ion as a complex, then the uncomplexed metal ion and all the different complexed forms of the metal ion each have their own, different value of K_D.

However, as analysts, what we are interested in is the *total* amount of metal ion extracted in whatever chemical form it might be. This is where the distribution ratio comes in. The distribution ratio is equal to the ratio of the concentrations of the substances *in all its different chemical forms* in one phase to that in the other, at equilibrium (which now applies to chemical as well as distribution equilibrium).

Expressed mathematically, the distribution ratio, D, is as follows.

$$D = \frac{\text{concentration of analyte in all chemical forms in phase 1}}{\text{concentration of analyte in all chemical forms in phase 2}}$$

at equilibrium

The distribution ratio thus takes into consideration the fact that an analyte may partake in various chemical equilibria as well as the distribution equilibrium. The value of D varies, therefore (in practice possibly over many orders of magnitude) according to the position of any relevant chemical equilibria.

∏   Would you expect either D or K_D to be pH dependent?

K_D is independent of pH (except to a small degree due to the dependence of activities on pH which we can ignore) because K_D refers to a particular chemical species only. D refers to all chemical forms of an analyte. If the positions of any of the chemical equilibria depend on pH (which in practice they might do) then D might well be strongly pH dependent.

Π Suggest suitable orders of magnitude for the value of D (or K_D in the absence of chemical equilibria) for:

(*i*) preconcentration by solvent extraction,
(*ii*) chromatography.

(*i*) Very large values (1000 is probably the very least acceptable) would be needed for any preconcentration (this value is expressed as a ratio – concentration in phase of small volume: concentration in phase of large volume).

(*ii*) Fairly moderate values of the order of 1 (say from 0.1 to 10) would be required for chromatography.

This is because in preconcentration we want a complete and quantitative transfer into one of the phases whereas in chromatography, to obtain differential migration, we require a fairly uniform distribution between the phases. Extreme values would imply that the solutes were 'stuck' in one or other of the phases.

7.5.3. Separation Factor

If we are trying to separate two substances, than a very useful parameter is the *separation factor*. If the two substances have distribution coefficients K_{D_1}, and K_{D_2} between the same two phases (or distribution ratios D_1 and D_2), then the separation factor between the two substances is simply the ratio of the distribution coefficients (or distribution ratios) for the two substances distributed between the two phases.

$$\beta = K_{D_1}/K_{D_2} \text{ (or } D_1/D_2)$$

at equilibrium

The separation factor is used in solvent extraction and in all chromatographic techniques (where it might be known as the relative retention, but it has the same value).

Batch separation techniques (such as solvent extraction) require very large values of β, say, of the order of 10,000 or more for effectively quantitative separations. This implies very large differences in the K_D or D values for the two substances. These large differences are often obtained in practice *via* selective chemical conversions.

However in chromatography, we can separate species quantitatively with values of β much closer to unity. There is rarely any point in seeking a value of β greater than about 2; indeed the most powerful chromatographic systems can separate substances with separation factors between them as little as 1.02. Large values of β, as for batch separation, would probably mean that one component would take a very long time to separate by chromatography.

∏   How could you separate two substances with a separation factor exactly equal to one?

You couldn't! A separation factor of one means identical values of D or K_D and therefore no separation. You would have to change some property of the system so that either the D values or the K_D values were altered to give a value of β other than one. Then in principle the separation should be possible by one means or another.

7.5.4. Kinetic Aspects

In our general description of separations up to now we have assumed that all distributions and chemical reactions reach thermodynamic equilibrium before we carry out the phase-separation step. In other words we wait for the concentration ratios to reach the relevant K_D, D, or β values, after which they remain constant before the separation is completed.

Now, in practice, there are some important situations where the separation is complete before thermodynamic equilibrium is achieved. Then the rate of distribution and/or chemical reaction becomes important. Also, in principle, separations may not always be predicted quantitatively by the equilibrium constants K_D, D, or β. In other words, the kinetic aspects of the system become significant. Here are two important examples where thermodynamic equilibrium is not achieved in a separation and therefore kinetic aspects are important.

(*a*) Certain metal ions have inert electron configurations and their reactions in the formation and dissociation of complexes proceed extremely slowly, allowing one to carry out a separation effectively before any reaction has occurred. Common examples of such ions are Cr^{3+}, Co^{3+} (normally only encountered in a complexed form), and Pt^{2+}.

(*b*) In continuous systems, such as chromatography in particular, which are dynamic in nature, thermodynamic equilibrium is never established between the two phases, because of their continuous relative motion. This has a fundamental effect on the extent to which bands or peaks broaden as they pass through the chromatographic system. Basically, although there are often other important factors affecting band broadening, the less closely a chromatographic system approaches thermodynamic equilibrium, the more the bands will broaden and therefore the larger the separation factor will have to be between two species that are just separable.

Π   Explain the following observation.

> Cobalt(II) reacts with 1-nitroso-2-naphthol to form a 1:3 neutral complex in which the cobalt ion has undergone oxidation. The complex so formed shows a most unexpected and remarkable degree of 'stability'.

The oxidation step takes the cobalt(II) to the cobalt(III) state. Cobalt(III) ions have an inert electron-configuration and therefore the dissociation of the complex, even under conditions in which the complex is thermodynamically unstable, proceeds extremely slowly. This gives the complex the appearance of being highly stable, although in fact it should be described as being inert rather than stable. See Section 6.2.7., page 172.

SAQ 7.5a Here are *four* statements, only *one* of which is completely and unambiguously true. Which one is this and what are the errors in the other three statements?

(*i*)   The distribution coefficient is a constant given by the ratio of the concentrations of a substance distributed between two phases at thermodynamic equilibrium. It is a function of temperature and pressure but is independent of concentration.

(*ii*) Iron(III) reacts with acetylacetone (pentane-2,4-dione) to give a 1:3 (metal to ligand) complex. Chromium(III) apparently does not react. From this fact we can conclude that the complex formed between iron(III) and acetylacetone has high thermodynamic stability whereas that between chromium(III) and acetylacetone does not.

(*iii*) Chromatography, amongst separation techniques in general, offers the greatest amount of separating power. This is because the separation factors between various species are usually larger for chromatographic techniques than for other separation techniques.

(*iv*) The value of the distribution ratio for a particular analyte distributed between two phases is dependent, and very markedly so, on the position of any chemical equilibria in which that analyte is involved.

SAQ 7.5a

7.6. BRIEF SURVEY OF SEPARATION TECHNIQUES USED IN ANALYSIS

This Section consists basically of a list of all the more important separation techniques that you are likely to meet in analysis, plus a sentence or two of explanation about each technique, its areas of application and other useful bits of information. In later sections of this Unit you are to study some of these separation techniques in much more detail, and likewise in the chromatography Units. You will find this section fairly factual in nature, but when you move on to the next section, you ought really to know the contents of this brief survey thoroughly.

Start by reading through the survey and then tackle the exercise at the end. Consult the text as much as you need to at this stage. Then try committing its contents to memory and finally tackle the SAQ without reference to the text at least as far as you are able. Only use the text as a back-up when your memory has failed you.

7.6.1. Solvent Extraction

A solution of the substances to be separated in a particular solvent is equilibrated with a second solvent, immiscible with the first, and a distribution between the two solvents occurs. It is used in batch and multiple-batch modes for separating specific metal ions as complexes, or for their preconcentration; also in a less sophisticated way, for organic compounds. The two immiscible solvents are normally water and an organic solvent, and the most important factor governing separability is the relative solubilities of the species being separated in the two solvents. Large differences in solubility are achieved by using selective chemical reactions to form derivatives, and by pH control.

7.6.2. Precipitation

This is a simple batch technique for removing certain species from solution selectively by converting them by chemical means into highly insoluble forms. It is used largely in inorganic analysis (eg in gravimetry) as well as in other branches of analysis (eg protein precipitation in biochemical analysis).

7.6.3. Coprecipitation

Here, trace solutes (typically metal ions) are incorporated quantitatively into the precipitate formed when a substance present at high concentration (the carrier) is precipitated. Several mechanisms, of which perhaps the most important is adsorption, may account for this phenomenon which has found application for preconcentration of trace species. In gravimetric analysis, it causes erroneous analytical results by the production of impure precipitates.

7.6.4. Electrodeposition

Metal ions may be selectively removed from solution by controlled-potential electrolysis. The separations thus achieved depend on the relative position of the metal ions in the electrochemical series.

7.6.5. Ion Exchange

Here, a solution containing ions that are to be separated is brought into contact with a solid material (normally an organic resin) to which are bound ionogenic (or ionised) groups with counter-ions associated with them. The distribution involves reversible exchange between these counter-ions and ions of the same sign in solution. There are basically two types of ion-exchange resin containing either bound cationic or bound anionic groups, that are used for the separation of anions, and cations, respectively. Ion exchange is used in the trapping mode for preconcentration, for the removal, or replacement of all ions of one type (ie cations or anions) by a single specific ion, and for selective separations based on chemical reactions in the solution phase. It finds its most important applications in inorganic analysis and in chromatography (see Section 7.6.9).

7.6.6. Adsorption/Desorption

A solid/liquid or solid/gas-phase technique used in the trapping mode for the preconcentration of organic material (especially from aqueous samples or from vapours in the atmosphere). An active solid adsorbent (eg, charcoal) with a large surface-area to mass ratio, reversibly adsorbs molecules onto its surface by dipole–dipole, hydrogen-bonding or other interactions. Adsorption implies quantitative retention on the solid phase *via* the trapping technique. Desorption is the subsequent recovery of the adsorbed material, normally by use of an organic solvent, or for vapours, by heat (thermal desorption).

7.6.7. Volatilisation/Distillation/Fractional Distillation/Cold Trapping

All these involve liquid/gas (or occasionally solid/gas) equilibria. *Volatilisation* simply implies the selective removal of a substance from solution by chemical conversion into a gaseous form. *Distillation* implies simple separation *via* differences in boiling-point. *Fractional distillation* is a countercurrent technique involving continuous equilibrations between a liquid and a vapour phase in a fractionation column, and is therefore capable of separating species with much closer boiling-points than is possible with simple distillation. Cold trapping implies the collection of gases or vapours from a gaseous sample by condensing (and possibly freezing) them in a cold trap.

7.6.8. Absorption (note carefully the spelling)

The collection of gases from a gaseous sample by bubbling it through a chemically reactive solution. The chemical reaction specifically converts a particular gaseous species into an involatile, soluble form. The term is sometimes also used in connection with the collection of gases by irreversible reactions with solids (this separation technique differs from our model in that the distribution is essentially irreversible). Examples are the absorption of sulphur dioxide gas in hydrogen peroxide solution to form sulphuric acid, and the absorption of carbon dioxide by soda-lime.

7.6.9. Chromatography

The term covers a wide range of separations, all of which are based on a two-phase, continuous system in which one phase (the mobile phase) flows over the other phase (the stationary phase). The wide range of separations possible arises from the wide range of stationary and mobile phases available. Many of the separation techniques that we have already studied have been adapted to produce important chromatographic systems. Some of the more important forms of chromatography are as follows.

Liquid–Solid (Adsorption)
Based on adsorption/desorption distributions and used for organic molecules of medium-to-low polarity.

Liquid–Liquid (Partition)
Based on solvent-extraction-type distributions between two immiscible liquid phases (for example water adsorbed onto paper with an

organic solvent flowing along the paper, as in paper chromatography). Used for separating organic and inorganic species more polar than those above.

Ion-Exchange
Based on a cation- or anion-exchange resin as stationary phase and an electrolyte as mobile phase. Used for the separation of inorganic or organic ions of like sign.

Exclusion
Based on a solid material that is inert but contains pores (holes in its structure) of strictly controlled size, plus a single solvent. Solute molecules may enter the pores (ie the stationary phase) to an extent depending solely on molecular size. Used principally for separating polymers according to their relative molar mass.

Gas–Solid
Based on adsorption (plus some exclusion) but with a gaseous rather than a liquid mobile phase. Used for separating small molecules (usually above this boiling-point) and is therefore suitable for gases.

Gas–Liquid
Based on distributions between a solution in a liquid stationary phase and a vapour in a gaseous mobile phase. Used for separating volatile organic molecules, generally below their boiling-points. It is related to but, (since polarity is important), is more complex than fractional distillation, which it has largely replaced in analysis.

7.6.10. Dialysis

Here two liquid phases are separated by a *semi-permeable membrane*. The sample is on one side of the membrane and a pure solvent (water) on the other. Separation is by differential diffusion of small molecules through the membrane and is therefore based on molecular size. It is used largely in biochemical applications eg for separating smaller molecules from colloidal species, the latter being unable to permeate the membrane.

7.6.11. Electrophoresis

In this technique, charged species are separated *via* their different rates of migration in the presence of an electric field in a solid/liquid phase system. The solid acts simply as a support, the ions always remaining in the liquid phase. Therefore this separation technique does not fit our model of a two-phase system, and there is no distribution across a phase boundary. Nevertheless, the results bear some superficial resemblance to those obtained from, for instance, paper or thin-layer chromatography (spots or bands of material displaced by different distances along a sheet of paper or other flat medium). The technique is used mainly for separating proteins, sometimes in conjunction with a true chromatographic separation.

7.6.12. Ultracentrifugation

Large, possible colloidal molecules will migrate (settle) at different rates in the presence of an intense centrifugal force, resulting in separation. Another system based on differential rates of migration in a single phase, rather than a distribution and therefore falling outside our general model. Again it is mainly of biochemical application.

Some general texts for reading about these separation techniques can be found in references 13, 14 and 15.

∏   Having read through this brief factual review of the most important types of separation technique available to the analyst try to classify them in the following two ways. (Refer back to the text as much as you wish.)

(*i*)   According to the phases involved. Leaving electrophoresis and ultracentrifugation out of this classification, you should be able to describe them all as two-phase systems. Do this diagrammatically as follows by considering each chromatographic technique separately.

Phase 1 →	Solid	Liquid
Phase 2↓		
Liquid		
Gas (Vapour)		

(*ii*) Under mode of separation by using the following headings:

Batch,
Multiple Batch,
Continuous,
Trapping technique,
One-phase system based on differential rates of migration,
Other (if needed).

You might find that in this case, some techniques appear under more than one heading. On the other hand, here you should be able to treat all the chromatographic techniques together.

Here are two possible classifications for you to compare with your classifications.

(*i*)

Phase 1 →	Solid	Liquid
Phase 2↓		
	Precipitation	Solvent Extraction
	Coprecipitation	Partition Chromatography
Liquid	Electrodeposition	Exclusion Chromatography

	Ion-Exchange	Dialysis
	Ion-Exchange Chromatography	
	Adsorption/Desorption	
	Adsorption Chromatography	
	Adsorption/Desorption	Volatilisation
	Gas–solid Chromatography	Distillation
Gas (Vapour)	Absorption	Fractional Distillation Gas–Liquid Chromatography Cold Trapping Absorption

(*ii*)

Batch:	Solvent Extraction, Precipitation, Coprecipitation, Electrodeposition, Volatilisation, Distillation, Dialysis
Multiple Batch:	Solvent Extraction
Continuous:	Chromatography, Fractional Distillation
Trapping Technique:	Ion-Exchange, Adsorption/Desorption, Cold Trapping, Absorption
One Phase:	Electrophoresis, Ultracentrifugation

SAQ 7.6a

For each of the following descriptions of a separation technique, give the name of the technique to which the description most aptly applies. A different separation technique should be selected for each description. Try to do this question from memory. If you don't feel confident to do this immediately, then revise section 7.6 which contains all the information that you need for the question. Then attempt the question.

Descriptions

(*i*) A chromatographic technique used for polymers, in which molecular size is the most important factor.

(*ii*) A technique in which different rates of migration of different charged species in the presence of an electric field leads to separation.

(*iii*) A chromatographic technique which has largely supplanted fractional distillation for the separation of mixtures of volatile organic compounds.

(*iv*) A technique involving a membrane permeable to small molecules only.

(*v*) A preconcentration technique involving a solid-solution equilibrium. The solid phase is not pure, which can be a problem in a related well known classical quantitative method. $\longrightarrow$

**SAQ 7.6a
(cont.)**

(*vi*) Two liquid phases are involved in this batch technique, and solubility is the overriding factor determining the separations possible.

(*vii*) Paper chromatography is an example of this type of chromatography, which is related to (*vi*).

(*viii*) A preconcentration technique suitable for organic vapours in the atmosphere, involving reversible processes occurring on the surface of an 'active' solid with a large surface-area-to-mass ratio.

(*ix*) A technique used largely in inorganic analysis but also for organic analysis, in which charged species are preconcentrated or separated chromatographically.

(*x*) Metallic cations are separated by this technique, according to their position in the electrochemical series.

SAQ 7.6a

Objectives

When you have studied this Part you should be able to:

- define the term 'separation' in an analytical context;

- place the separation stage in context in the overall analytical sequence;

- recognise a wide variety of situations where separations are applicable in analysis;

- understand that analytical problems that can be solved by separation, may sometimes also be solved by other methods, sometimes even more satisfactorily;

- describe an idealised model of a separation in terms of chemical reaction – phase distribution – phase separation;

- understand that these steps may be combined in different ways to produce different modes of separation, and describe these;

- know some basic important quantitative parameters used to describe separations that involve thermodynamic equilibria;

- understand that some separations may be kinetically controlled;

- have a 'current-awareness' knowledge of a wide selection of separation techniques.

8.   Solvent Extraction

8.1.   GENERAL INTRODUCTION

Solvent extraction is a common, important and very useful separation technique which you may well encounter in many branches of chemical analysis. The basic idea behind the technique involves separation of the components of mixtures of substances (known as *solutes*) by *partitioning* them between two immiscible solvents. In other words, we take a solution of our mixture in a particular solvent and equilibrate this with a second solvent immiscible with the first. The solutes partition (or distribute) themselves between the two solvents according to their relative solubilities.

How would we perform such a separation in practice? A typical solvent extraction might involve the following steps.

(*a*) Prepare a solution of the solutes to be separated in a particular solvent (if the sample is not a solution to start with).

(*b*) If necessary, prepare derivatives of the solutes or make other chemical adjustments to the solution (such as complex formation and pH adjustment) in order to maximise differences in solubility between the solutes.

(*c*) Add a second immiscible solvent to produce a two-phase system.

(*d*) Shake the mixture in a sealed (stoppered) container until partitioning of the solutes between the two phases has reached equilibrium.

(*e*) Allow the two phases to separate into two layers.

(*f*) Finally, draw each layer off separately.

In a 'perfect' solvent extraction, we might expect to find that one group of solutes has dissolved entirely in the first solvent, whereas the other group of solutes is to be found only in the second solvent, resulting in a clean, quantitative separation of the two groups of solutes. Theory predicts, however, that at equilibrium, there will be finite concentrations of all solutes, however small, in both phases. In practice, we shall find situations where the separation is virtually quantitative but we must also expect to find situations where solutes are partially dissolved to a significant extent in both solvents at equilibrium. Separation will then not be quantitative. In most practical situations in solvent extraction, one of the solvents will be aqueous, such as:

— pure water,

— an aqueous buffer solution of specific pH,

— an aqueous solution of electrolytes,

— an aqueous solution containing various complexing agents,

— aqueous acid or alkali,

— a combination of two or more of the above.

We shall call this the *aqueous phase*.

The other solvent is then normally an organic solvent which must of necessity be sufficiently immiscible with water to support a two-phase system. Common choices of organic solvent might be:

— benzene (less so now because of its toxicity),

— toluene and the isomeric xylenes,

— dichloromethane, trichloromethane (chloroform), tetra-chloromethane,

— ethers, such as diethyl ether (ethoxyethane),

— esters (eg ethyl acetate),

— certain ketones (if they are immiscible with water)

— water-immiscible alcohols of higher molar mass,

— aliphatic hydrocarbons (hexane, light petroleum).

Organic solvents which are miscible with water include alcohols, ketones, aldehydes and carboxylic acids, all of lower molar mass, as well as acetonitrile, dimethyl sulphoxide and dioxan. Such solvents would be unsuitable for extraction of a solute from an aqueous phase although some could be used for extraction from other water-immiscible organic solvents.

To go with the treatment of solvent extraction that we are now about to work through, I recommend *reference 1* as providing a concise but readable coverage of the same sort of ground.

References 16 and 17 provide much fuller and more detailed treatments of the subject.

SAQ 8.1a	Consider each of the following five solvent systems. For each system decide whether or not it would be useful for solvent extraction, and write down a sentence or two of explanation for your decision. $\longrightarrow$

SAQ 8.1a
(cont.)

	Aqueous phase	Organic phase
(*i*)	Decimolar aqueous NaCl	Trichloromethane
(*ii*)	De-ionised water (unbuffered)	Acetonitrile
(*iii*)	50% aqueous methanol	Cyclohexane
(*iv*)	Water buffered to pH 3	4-Methylpentane-2-one
(*v*)	Molar aqueous ammonia	Ethanoic acid

8.2. BROAD AREAS OF APPLICATION

Solvent extraction is commonly used as follows.

(*a*) To extract metal ions from aqueous solutions into organic solvents. Such extractions normally involve a prior conversion of the metal ion into an organic complex form which is more readily extracted.

(*b*) To extract organic compounds in general, in which case, species may be transferred in either direction. Here, we often use the fact that for acidic or basic substances the uncharged form only is likely to be efficiently extracted into the organic phase. The ionised or protonated form is normally retained entirely in the aqueous phase.

There are one or two other types of compound that might be candidates for extraction such as covalent inorganic molecules (eg iodine, tin tetraiodide and osmium tetroxide) and suitable derivatives of one or two anions (eg the heteropolymolybdate complexes of anions such as phosphate, arsenate and silicate), but these examples are much rarer and we shall not be much concerned with them.

∏ Can you think of some *analytical situations* in which we would wish to carry out separations by using solvent extraction?

The text that follows describes *four* such situations (although there are some others that are more specialised, such as the use of solvent extraction in the separation stage of an isotope dilution analysis).

8.2.1. Analytical Applications of Solvent Extraction

(*a*) *Separation of one species from one or more others which might subsequently interfere in the analysis*. Here a *selective* extraction is required. In other words *only* the analyte *or* the interferents should be extracted. This is particularly useful in the analysis of metals where the required selectivity in extraction could be achieved *via* selectivity in the complex formation stage. However, solvent extraction is unlikely to be sufficiently selective for this purpose for the analysis of organic compounds, unless the analyte happens to have very different solubility or acid-base properties from all the other species present.

(*b*) *Preconcentration of trace analytes to a concentration at which their determination is possible*. This can be achieved by extracting them from a large volume into a much smaller volume of solvent. This places great demands on the efficiency of extraction into the solvent of smaller volume, as well as on the (hopefully low) miscibility of the two solvents. By using a single extraction, concentrations can be increased by factors of up to 20 to 50 in this approach. Normally, we do not worry about selectivity in this kind of application, which has proved useful for both organic compounds and metal ions.

∏   Can you think of at least two reasons why the preconcentration factor might be limited to, say 20–50 in a single-stage solvent extraction?

The limitations are: first, all solvents are miscible with each other to a certain extent, and so we cannot expect to support a two-phase system with only a very small volume of one solvent relative to that of the other; secondly, there are simple physical manipulation problems. It would be difficult to equilibrate effectively, say 10 cm^3 of organic solvent with 1 dm^3 of water and then recover the former quantitatively.

(*c*) *Preliminary sample 'clean-up'*. Complex organic and biological mixtures may be subject to extraction procedures, not so much to isolate the analyte, but perhaps to simplify the matrix by fractionating the material on a broad acid/base and solubility basis. Organic extracts of analytes often produce cleaner background traces in a subsequent chromatographic analysis.

(*d*) *To utilise a water-insoluble derivative of a metal ion in spectrophotometry*. Many complexes of metal ions with organic molecules are strongly coloured and are useful in colorimetric or spectrophotometric analysis. Some of these, though, are water insoluble but extractable into organic solvents. Dithizone complexes are good examples.

Solvent extraction is less effective for the resolution of complex mixtures into individual components, although this has been achieved for some by using complex multiple extraction schemes.

∏   What extremely important group of separation and analytical techniques might offer a means of resolving complex mixtures in solution into their individual components?

For separation of the components of complex mixtures in solution (and often of fairly simple mixtures as well) *chromatography* may well be used in one of its forms (eg gas chromatography, liquid chromatography). One of these, partition chromatography, is in many ways equivalent to a multiple solvent-extraction system. Partition chromatography basically depends on solvent extraction (solutes are partitioned between a liquid stationary and a liquid mobile phase) and the separations achieved rely on differences in extractability of solutes between the phases.

SAQ 8.2a	Which of the following separations would you expect to be achieved by a solvent extraction procedure (none, one, more than one, or all might be suitable)? (*i*) Preconcentration of traces of fluoride ion in water. (*ii*) Separation of cobalt ions from nickel ions in aqueous solution. (*iii*) Preconcentration of traces of a halogenated hydrocarbon (eg a pesticide) present in a biological fluid and removed from the bulk of the matrix. ⟶

SAQ 8.2a
(cont.)

(*iv*) Resolution of a group of amino acids into individual components on the basis of acid/base properties.

8.3. PRACTICAL PROCEDURES IN SOLVENT EXTRACTION

One of the most useful advantages of solvent extraction as a separation technique is the ease with which it can be carried out in the laboratory. The only special apparatus that we require is a separating funnel, a glass container of, say, 50–500 cm^3 capacity with a stopper at the top end and a stopcock (tap) at the bottom end. See Fig. 8.3a.

Let us now briefly go through the stages of a simple solvent-extraction procedure for the extraction of a metal ion from water into trichloromethane.

Fig. 8.3a. *Separating funnel*

(*a*) Place an aliquot of the aqueous solution of the metal ion in the separating funnel (upright position – stopcock closed!)

(*b*) Add a solution of a suitable complexing agent together with a buffer and any other reagents to ensure that the metal ions are efficiently and completely converted into an extractable complex.

(*c*) Add a measured volume of trichloromethane.

(*d*) Place the stopper in the top of the funnel and shake it moderately vigorously to mix the two solvents and allow equilibrium between the two phases to be established. Normally two to three minutes is adequate for this purpose, although more time may be required. If you shake too vigorously an emulsion of the two solvents may be formed and this leads to difficulty in separating the phases subsequently.

(*e*) Allow the separating funnel to rest in the upright position until the solutions have completely separated into two clear distinct layers with a sharp interface between them.

(*f*) Remove the stopper, carefully open the stopcock and cautiously draw off the more dense layer into another container. Here, of course, this will be the trichloromethane layer containing the complexed metal ions extracted.

SAQ 8.3a

Place the following operations in their correct sequence to produce a procedure for preconcentrating traces of metal ions from water into 4-methylpentane-2-one (MIBK). The preconcentration factor is 100 and the metal ions are extracted as their pyrrolidine-dithiocarbamate complexes. The complexes are formed in aqueous solution between pH 3 and pH 10. MIBK is less dense than water. (*N.B.* There is more than one acceptable order).

(*i*)   Add 10 cm^3 of MIBK.

(*ii*)   Evaporate the sample by gentle heating to 1/5th of its original volume.

(*iii*)   Allow the phases to separate.

(*iv*)   Shake gently for 2 to 3 minutes.

(*v*)   Add a few cm^3 of a solution of ammonium pyrrolidine dithiocarbamate.

(*vi*)   Draw off the lower phase.

(*vii*)   Draw off the upper phase.

(*viii*) Acidify this sample.

(*ix*)   Retain this phase.

(*x*)   Take a 1 dm^3 sample of water.

$\longrightarrow$

**SAQ 8.3a
(cont.)**

(*xi*) Discard this phase.

(*xii*) Transfer to a separating funnel.

(*xiii*) Add buffer to give a pH of about 5.

(*xiv*) Cool the solution.

SAQ 8.3b   Indicate which of the following statements are *true* (T) and which are *false* (F).

(*i*)   In order to carry out a solvent extraction successfully, it is necessary to shake the two solutions together as vigorously as possible to disperse the two phases into minute droplets to ensure complete mixing.

T / F

(*ii*) Equilibrium in solvent extraction is normally established quite quickly, say, after shaking the liquids for 2–3 minutes.

T / F

(*iii*) After equilibration, the different substances present will have concentrated themselves entirely into one phase or the other, depending on their relative solubilities.

T / F

(*iv*) For solvent extraction to be effective, there is no particular need for the relative volumes of the phases to be equal or even nearly equal.

T / F

The procedure just described is, in fact, a very basic, simple solvent extraction procedure corresponding to the *batch separation* procedure described in Section 7.4.1, and there are a number of fairly important refinements of technique that we may wish to consider using.

(*a*) We could add a second measured volume of trichloromethane to the aqueous phase, repeat the procedure from (*iii*) above, and combine the two trichloromethane layers. We could, in fact, repeat the extraction in this manner more than once and combine *all* the trichloromethane extracts. We would do this if we believed that a single extraction was not sufficient to remove all our metal ion (or extractable species in general). Generally *multiple extractions* are more efficient in removing a solute quantitatively than a single extraction for a given volume of solvent. This corresponds to the *multiple-batch separation* procedure described in Section 7.4.2.

(*b*) We could shake the trichloromethane layer with a second aliquot of the aqueous phase containing all the reagents, but no metal ion. This is known as 'backwashing' and may be useful for removing small traces of co-extracted impurities.

(*c*) We could shake the trichloromethane layer with an aliquot of an aqueous phase containing a *different* mixture of buffers and reagents. Typically, perhaps the pH might be altered. This is known as 'back extraction'. The altered conditions may result in some, but not all, of the extracted species being 'back extracted' into the aqueous phase. This would be used to increase the *selectivity* of an extraction procedure. A group of extractions and back extractions is sometimes known as a *'multistage' extraction scheme*.

(*d*) The aqueous phase before being shaken with the solvent or the trichloromethane phase after extraction of the solute could be evaporated down to a small volume. This is a useful adjunct to using solvent extraction for preconcentration in trace analysis.

(*e*) The trichloromethane phase could be passed through a filter paper or kept over a drying agent if it was necessary to remove traces of water from the organic phase.

An important point to understand in solvent extraction is that it is common practice to talk about retaining one phase for further treatment or analysis and *discarding* the other phase. In practice, in a well tested extraction system, all the material of interest will indeed be collected in one phase and the other phase will be of no further interest to us. However, unless we are absolutely confident about this it would be good scientific practice to interpret the words 'discard the aqueous/organic phase' as meaning to set this phase aside rather than to pour it down the sink, just in case we find it necessary to treat it further.

∏   Even if we assume 100% extraction of a particular substance from an aqueous into an organic phase, a simple single stage solvent extraction may not guarantee 100% recovery of this substance. Why is this, and what steps could you take to improve the recovery significantly?

Small amounts of organic phase may be entrained within the aqueous phase and small losses may be suffered when the last portion of solvent near the phase boundary is drawn off. So, shake the aqueous phase with a second volume of organic phase to 'scavenge' the remaining traces of solute, and combine the two organic phases.

8.4.   FACTORS AFFECTING THE EXTRACTABILITY OF SOLUTES

Now that we have some understanding of the basic procedure and areas of application of solvent extraction, we need to think about how the physical and chemical properties of a substance may influence its behaviour in a solvent-extraction system. In other words we are going to try to form some generalisations about how different solutes might distribute themselves between aqueous and organic solvents.

The two most important factors to be considered are:

(*a*) the solubilities of the solute in each of the solvents,
(*b*) any chemical equilibria that the solute might be involved in in either solvent.

Let us consider these in turn.

8.4.1. Solubility

If the solute is highly soluble in solvent A but only sparingly soluble in solvent B, then you would expect the solute, after solvent extraction, to be concentrated almost exclusively in solvent A. And, of course, *vice versa*. If the differences in solubility are less dramatic, then you would expect the solute to be distributed between the two solvents to an extent reflecting, qualitatively at least, the ratio of solubilities. This distribution between two solvents based solely on solubilities is an *equilibrium* process. We shall shortly meet the appropriate equilibrium constant (distribution coefficient) to describe this distribution.

So, what kinds of solutes are soluble in aqueous phases and what kinds are soluble in organic phases? The principle 'like dissolves like' works well here, and the ability of the solvent to *solvate** the solute is also very relevant.

We shall find it useful to think of solutes in relation to a scale of polarity which we can extend to include a distinction between covalent and ionic bonding.

* *Solvation* is a process whereby solvent molecules are attracted to solute molecules with the formation of weak bonds between them. Thus a solvated molecule can be visualised as having a sphere of solvent molecules arranged round and weakly bonded to it. Solvation has a very fundamental effect on solubility. When the solvent is water we talk about hydration and often the bonds (hydrogen bonds in this case) are relatively strong.

Π   We have just introduced the very important word 'polarity' in relation to solubilities. What does 'polarity' mean in relation to molecular properties, and how can it affect solubilities?

A covalent bond within a molecule can be considered as being *non-polar* (in which case the electronic distribution around the atoms forming the bond is symmetrical and there is no charge separation and zero bond dipole moment) or the bond may be *polar* (in which case the electronic distribution around the atoms forming the bond is not symmetrical, there is a charge separation and a finite dipole moment). For instance, carbon–carbon bonds in a hydrocarbon are to all intents non-polar, whereas the oxygen-hydrogen bond of the OH group is highly polar. Polar bonds can be represented as shown below:

$$O^{\delta -} \underline{\qquad} H^{\delta +}$$

to signify the charge separation.

A polar molecule is one which contains polar bonds so distributed geometrically that it has an overall dipole moment. It will attract other polar molecules electrostatically. These attractive forces mean that polar solutes will dissolve readily in polar solvents. However, the dipoles have a disturbing influence in a non-polar medium so non-polar solvents preferentially dissolve non-polar solutes. These dipoles arise through the bonding electrons being more closely associated with one atom than the other. In the extreme case where the electrons are associated entirely with one atom and not the other, there is no covalent bond, but *ions* are produced. Thus ionisation is the extreme case of a polar 'bond', eg in inorganic electrolytes.

Some molecules, that possess mobile electrons, may not be intrinsically polar, but in the presence of a polar molecules have dipoles induced by electrostatic forces. These dipoles are not permanent but are present only when there is an electrostatic field of force present to 'polarise' the molecule. Such molecules are said to be *polarisable* and show weak polar properties in a polar medium but non-polar properties in a non-polar medium. Many aromatic compounds are in this category.

Fig. 8.4a shows such a scale of polarity/covalent-ionic nature in which polarity or tendency to ionic nature increases as we descend the table. Examples are included.

	Bonding Characteristics	*Examples*
1.	Non-polar covalently bonded compounds	Aliphatic, alicyclic hydrocarbons
2.	Weakly polar or non-polar but polarisable covalently bonded compounds	Aromatic hydrocarbons, certain halogenated hydrocarbons
3.	Moderately polar covalently bonded compounds	Ketones, esters, amides
4.	Strongly polar covalently bonded compounds	Alcohols, polyhydric alcohols, phenols
5.	Weakly ionised compounds (in equilibrium with unionised forms)	Carboxylic acids, amines
6.	Strongly ionised compounds (ionisation complete)	Inorganic electrolytes, strong acids and bases

Fig. 8.4a. *Scale of polarity/covalent-ionic nature*

Now, to predict qualitatively how a solute might be extracted from an aqueous phase into an organic phase, we can apply the following principles. These are useful guidelines but are not infallible.

(*a*) For an uncharged species, the less polar it is the easier it will be to extract it quantitatively from an aqueous phase into a non-polar organic solvent.

(*b*) Charged (ionic) species are generally difficult to extract into any organic phase from an aqueous phase, if indeed any extraction occurs at all, unless they are able to form an ion-pair with a counter ion (when they are effectively uncharged).

(*c*) The ability of either solvent to solvate the solute strongly will cause the solute to be retained by that solvent. This applies to water and its ability to hydrate ions, and also to solvents such as diethyl ether which solvate various complex ions that are then extracted as solvated ion-pairs.

(*d*) Polar solutes are more efficiently extracted into polar organic solvents than into non-polar organic solvents. This generally makes their extraction from aqueous phases more difficult because polar solvents are more likely to be miscible with water.

SAQ 8.4a

Match the following types of compound on a one-to-one basis with the following statements which all refer to extraction from an aqueous phase.

Compounds

(*i*)   Non-polar organic compounds.

(*ii*) Organic compounds of medium to high polarity.

(*iii*) Strongly ionised inorganic electrolytes.

(*iv*) Weakly ionised organic electrolytes.

Statements

(*a*) Cannot normally be extracted.

$\longrightarrow$

SAQ 8.4a (cont.)

(*b*) Can be extracted efficiently into non-polar organic solvents of high immiscibility with water.

(*c*) Can possibly be extracted into polar organic solvents provided ionisation is suppressed.

(*d*) Can be extracted into organic solvents of medium to high polarity.

8.4.2. Effects of Equilibria

In several of the examples we have discussed so far we have already noted that the precise chemical form of the solute has a fundamental bearing on its extractability. Let us now clarify this by making two general statements.

(*a*) The extractability of any solute will depend on its precise chemical form and this in turn is in part expressed in terms of the position of any relevant chemical equilibrium in either phase.

(*b*) Selectivity in solvent extraction is most often achieved by judicious manipulations of appropriate chemical equilibria to produce either organic solvent-soluble or water-soluble species, rather than simply relying on natural differences in extractability of different solutes.

Let us illustrate this with two examples. First we consider acidic, neutral and basic organic compounds. Now neutral organic compounds do not ionise and so are potentially extractable at all values of pH. Acidic substances will be ionised at high pH, might be partially ionised at neutral pH but will tend to exist in the unionised form at low pH. Hence, acidic substances will be retained in the aqueous phase at high pH but are potentially extractable at low pH. Exactly the opposite argument applies to basic compounds by virtue of their protonation at low pH. Thus we have a simple means of fractionating complex organic mixtures on the basis of their acid/base properties. This is useful as a means of simplifying complex organic matrices before, say, more detailed chromatographic analysis.

SAQ 8.4b   Insert phrases selected from the list below to fill the numbered gaps in the following. Each phrase in the list may be used once, more than once, or not at all.

A urine sample was to be analysed for a drug with basic properties. It was necessary to obtain an extract of the ... 1 ... in a somewhat sim-

$\longrightarrow$

**SAQ 8.4b
(cont.)**

plified matrix in an organic solvent for analysis. An aliquot of the sample was treated with ... 2 ... until ... 3 ... was observed. The sample was then shaken with a suitable volume of ... 4 ... and the ... 5 ... was discarded. The ... 6 ... was then treated with ... 7 ... until ... 8 ... was observed. It was then shaken with ... 9 ... and the ... 10 ... was retained for analysis.

The Phrases

'aqueous phase',
'an alkaline reaction',
'dilute aqueous sodium hydroxide',
'propanone',
'a neutral reaction',
'drug',
'urine sample',
'an acidic reaction',
'diethyl ether',
'organic phase',
'dilute aqueous hydrochloric acid'.

SAQ 8.4c   Devise a scheme in block diagram form to carry
out a broad, rough separation of a complex mixture of organic compounds into the following
fractions (quantitative concentration of individual compounds into a single fraction is not required – rather the composition of each fraction should roughly correspond to the description below).

(*i*)   Neutral (non-acidic or non-basic) extractable compounds.

(*ii*) Acidic extractable compounds.

(*iii*) Basic extractable compounds.

(*iv*) Non-extractable (ie water soluble) compounds.

(NB 'extractable' means extractable from the
aqueous into the organic phase).

Use the following devices to present your block
diagram.

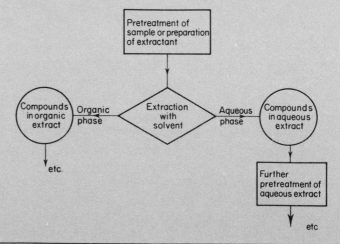

SAQ 8.4c

The second example concerns metal ions in general. Now metal ions, because of their ionic nature, will not under normal circumstances be extracted into organic solvents. The normal procedure adopted is therefore either:

(*a*) to convert the metal ion into a neutral coordinatively saturated chelate complex free of polar groups which is then highly extractable into organic solvents (for example, the extraction into trichloromethane of lead(II) as its complex with dithizone at pH 10); or

(*b*) to convert the metal ion into a complex ion and to form an ion association complex (ie an ion-pair) that may or may not be solvated by the extracting solvent, but that is highly extractable (for example, the extraction into diethyl ether of iron(III) as $FeCl_4^-$ ions from 6 *M*-hydrochloric acid).

With regard to the solvent extraction of metal ions, an important factor is the hydration of the metal ions, making them organophobic in nature. The replacement of some or all of this water of hydration by organophilic ligands is thus an important precursor to solvent extraction. In both these situations, the complex forming reactions and therefore the extractability of the metal ions are governed by general considerations relating to chemical equilibria. So we find extraction depending on the exact chemical conditions (such as pH, choice of complexing agent, and the like) existing particularly in the aqueous phase.

SAQ 8.4d

Copper ions in an aqueous solution of $CuCl_2$ are not extractable into trichloromethane as such. Which *one* of the following might be an effective way of extracting Cu^{2+} into an organic solvent?

(*i*)   Replace the trichloromethane with another organic solvent in which $CuCl_2$ is highly soluble.

(*ii*) Convert the Cu^{2+} ions into a neutral chelate complex with an organic ligand such as 8-hydroxyquinoline.

(*iii*) Render the Cu^{2+} ions insoluble in water by the addition of alkali, and thus by inference soluble in an organic solvent.

(*iv*) Convert the Cu^{2+} ions into a highly stable and bulky organic complex form by using the non-selective organic ligand, EDTA (ethylenediaminetetra-acetic acid).

SAQ 8.4d

8.5. QUANTITATIVE ASPECTS

8.5.1. The Distribution Coefficient and Distribution Ratio

Now let us move on and put our ideas about solvent extraction onto a more quantitative basis. The fundamental law governing the distribution of a solute between two immiscible solvents is known as the *Nernst Distribution Law*.

This can be stated as follows:

> 'At equilibrium the activities in the two solutions of a substance distributed between two immiscible solvents will bear a constant ratio to one another, provided that temperature and pressure remain constant'.

Now, analytical chemists as a rule do not like working with activities, and so we prefer to work with an expression involving concentrations. We therefore define the *Distribution Coefficient*, K_D as follows.

$$K_D = \frac{\text{concentration of substance in solvent A}}{\text{concentration of substance in solvent B}}$$

at equilibrium

Normally for aqueous/organic systems, solvent A is the organic solvent. Now, if we use square brackets to represent molar concentrations and the subscripts aq and org for the aqueous and the organic phase respectively, and if we call the substance being distributed, X, then:

$$K_D \frac{[X]_{org}}{[X]_{aq}} \text{ at equilibrium.}$$

You can see that this has the same form as the expression used for equilibrium processes in general. K_D is thus a ratio and a *dimensionless quantity*, ie it has no units.

It is important to realise that K_D refers to *one specific molecular species only*. If a substance is involved in chemical equilibria in either phase, then each molecular species present will be characterised by its own value of K_D. As such, K_D is a constant provided that temperature, pressure and activity coefficients remain constant (the last is specified because we have defined K_D in terms of concentrations and not activities. This assumption is reasonable at low concentrations provided that ionic strength remains constant).

Also don't forget that K_D applies to concentrations measured under *conditions of thermodynamic equilibrium only*. In solvent extraction, we are normally concerned with systems which reach equilibrium fairly rapidly and so, provided we follow the advice given in Section 8.3 (Practical Procedures in Solvent Extraction) we can be almost certain that the systems we are dealing with are at thermodynamic equilibrium. There are, however, some systems where equilibrium is not reached rapidly, eg systems involving chromium(III) ions. Such systems are said to be kinetically controlled. They might be characterised by high K_D values, but poor extraction efficiency because the system does not reach equilibrium sufficiently rapidly.

SAQ 8.5a	A sample of iodine weighing 0.560 g is dissolved in 100 cm^3 of water. This solution is shaken with 20.0 cm^3 of tetrachloromethane. Analysis of the aqueous phase shows that it contains 0.280 g dm^{-3} of iodine. What is the value of K_D for the distribution of iodine between water and tetrachloromethane?

If we are concerned with the distribution between two phases of a species, X, that exists in solution in a number of chemical forms (which is very likely to be so in practice), then we use the *Distribution Ratio, D*.

$$D = \frac{\text{concentration of X in all chemical forms in Solvent A}}{\text{concentration of X in all chemical forms in solvent B}}$$

at equilibrium

Thermodynamic equilibrium will now refer to the chemical equilibria as well as the distribution process.

The distribution ratio thus gives us information about the ratio of the *total* concentrations of the solute in each phase *irrespective of its precise chemical form*. This is of much more value to anyone using extraction in analysis, and so we shall find that the distribution ratio is of much greater *practical* use to us than the distribution coefficient.

D is not a constant and can vary very widely indeed with experimental conditions, such as pH, that might affect the chemical equilibria. In other words, in this situation a perturbation of any one of the participating equilibria may in turn affect the positions of all the other equilibria and hence the value of D. Nevertheless it is important to remember that the values of the equilbrium constants for each individual equilbrium do remain constant (ie they are constants!). This is the principle behind evaluating D in complex equilibrium systems.

A good example that illustrates the dependence of D on pH, is the extraction of an organic acid from water into an organic solvent. Let us assume that the acid HA is monobasic and that it dissociates into H^+ and A^- in water. Only HA is extracted into the organic solvent. The distribution coefficient for HA is K_D.

$$K_D = \frac{[HA]_{org}}{[HA]_{aq}}$$

The dissociation of the acid in water is governed by the acid dissociation constant, K_A.

$$K_A = \frac{[H^+]_{aq} \cdot [A^-]_{aq}}{[HA]_{aq}} \tag{8.1}$$

Now, according to our definition of D:

$$D = \frac{[HA]_{org}}{[HA]_{aq} + [A^-]_{aq}} \tag{8.2}$$

Remember that D is the ratio of the concentration of the acid in the organic phase to that in the aqueous phase, *regardless of chemical state*.

Rearranging Eq. 8.1,

$$[A^-]_{aq} = \frac{K_A \cdot [HA]_{ag}}{[H^+]_{aq}}$$

Substituting into Eq. 8.2,

$$D = \frac{[HA]_{org}}{[HA]_{aq}(1 + K_A/[H^+])}$$

$$\therefore \qquad \boxed{D = \frac{K_D}{1 + K_A/[H^+]}} \tag{8.3}$$

since $[HA]_{org}/[HA]_{aq} = K_D$

Eq. 8.3 is a very important equation that can be used to predict the variation of D with pH for extractable *monobasic* acids. Similar equations can be derived for polybasic acids.

∏   Now try to derive the following equation for the dibasic acid, H_2A.

$$D = \frac{K_D}{(1 + K_1/[H^+] + K_1K_2/[H^+]^2)} \tag{8.4}$$

where $K_1 = \dfrac{[H^+].[HA^-]}{[H_2A]}$ and $K_2 = \dfrac{[H^+].[A^{2-}]}{[HA^-]}$

According to our definition of D

$$D = \frac{[H_2A]_{org}}{[H_2A]_{aq} + [HA^-]_{aq} + A^{2-}]_{aq}} \qquad (8.5)$$

Let us now derive expressions for $[HA^-]_{aq}$ and $[A^{2-}]_{aq}$ in terms of $[H_2A]_{aq}$, K_1, K_2 and $[H^+]_{aq}$.

$$K_1 = \frac{[H^+]_{aq}.[HA^-]_{aq}}{[H_2A]_{aq}} \qquad (8.6)$$

$\therefore$ by rearrangement

$$[HA^-]_{aq} = \frac{\acute{K}_1.[H_2A]_{aq}}{[H^+]_{aq}} \qquad (8.7)$$

substituting for $[HA^-]_{aq}$

$$K_2 = \frac{[H^+]_{aq}.[A^{2-}]_{aq}}{[HA^-]_{aq}} = \frac{[H^+]^2_{aq}.[A^{2-}]_{aq}}{K_1[H_2A]_{aq}} \qquad (8.8)$$

$\therefore$ by rearrangement

$$[A^{2-}]_{aq} = \frac{K_1K_2[H_2A]_{aq}}{[H^+]^2_{aq}} \qquad (8.9)$$

Substituting for $[HA^-]_{aq}$ and $[A^{2-}]_{aq}$ in our equation for D and factorising the denominator for $[H_2A]_{aq}$, we have

$$D = \frac{[H_2A]_{org}}{[H_2A]_{aq}(1 + K_1/[H^+]_{aq} + K_1K_2/[H^+]^2_{aq})} \qquad (8.10)$$

But since $K_D = [H_2A]_{org}/[H_2A]_{aq}$

$$D = \frac{K_D}{(1 + K_1/[H^+]_{aq} + K_1K_2/[H^+]_{aq}^2} \qquad (8.11)$$

Let us now use Eq. 8.3 to see how D varies with pH for benzoic acid distributed between water and diethyl ether.

Now for benzoic acid,

$K_A = 6.30 \times 10^{-5}$ mol dm^{-3}

$K_D = 720$ for extraction from water into diethyl ether.

Substituting these values into Eq. 8.3 and solving for various values of pH gives the results in Fig. 8.5a and 8.5b. Can you see from these results that at low values of pH, D is very nearly equal to K_D and the extraction of the acid is highly efficient? However, as pH increases, D gradually decreases as more of the acid is ionised. Thus we have now demonstrated quantitatively what we stated qualitatively in Section 8.4, ie that organic acids may be efficiently extracted at low pH but are not extracted at high pH.

pH	D	pH	D
0	720	7	1.14
1	720	8	0.114
2	715	9	0.0114
3	667	10	0.00114
4	442	11	0.000114
5	98.6	12	0.0000114
6	11.3	13	0.00000114

Fig. 8.5a. *Relationship between D and pH for the extraction of benzoic acid from water into diethyl ether. $K_D = 720$, $K_A = 6.30 \times 10^{-5}$ mol dm^{-3})*

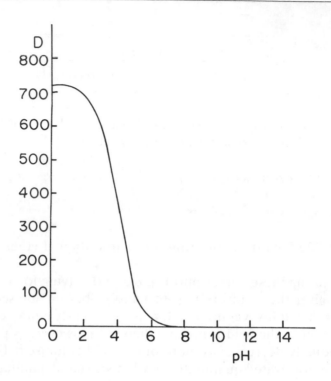

Fig. 8.5b.   *Plot of D vs pH for the extraction of benzoic acid from water into diethyl ether*

Note that this treatment for benzoic acid assumes that there are no other equilibria involved. It is, in fact, known that carboxylic acids can exist in a dimeric form in some solutions, and therefore, for complete accuracy, equilibria should be included to take dimerisation into account. This is likely to increase D but not alter the general form of the relationship between D and pH (this is because dimerisation is likely to be much more significant in the organic phase).

SAQ 8.5b

Only *one* of the following five statements is completely true. Select this statement and identify *all* the errors in the other four statements.

(i) The ratio of the total masses of a substance, X, in a particular chemical form dissolved in two immiscible solvents at equilibrium is given by the distribution coefficient only when the volumes of the two solvents are equal.

(ii) The distribution coefficient is strongly dependent on pH for weak acids and bases but is effectively independent of pH for neutral substances.

(iii) 10.0 cm^3 of water containing 1.00 g of substance X is shaken with 10.0 cm^3 of trichloromethane. The organic layer is found to contain 0.100 g of substance X. Thus it follows that K_D for this system is exactly 0.100.

(iv) The distribution coefficient, K_D, is defined as the ratio of the activities of substance X distributed between two immiscible solvents at equilibrium, and is a constant provided temperature and pressure remain constant and provided X is not involved in any chemical equilibria with other species.

(v) The total range of possible values of K_D goes from infinity for a substance completely retained in the aqueous phase, to 1 for a substance whose concentration in both phases is equal, to 0 for a substance completely retained in the organic phase.

SAQ 8.5b

8.5.2. Percentage Extraction

Although K_D and D are useful quantities that tell us how a particular substance will distribute itself between two solvents, what really interests us is just exactly *how much* material would actually be extracted under a given set of conditions. This is given by E, the *percentage extraction*.

$$E(\%) = \frac{\text{amount of substance extracted}}{\text{total amount present in both phases}} \times 100$$

E will depend on K_D or D and also on the *relative volumes of the two phases*. Let us now see if we can work out an expression relating E to K_D or D. We can do this either for K_D for a simple substance not involved in chemical equilibria or for D for a substance in several chemical forms inter-related by chemical equilibria. The algebra is the same in each case. Let us do it for D. You have a go first.

∏ See if you can derive an expression for E in terms of D and the volumes of the two phases. As a clue, you will need to obtain equations for D and E in terms of the volumes of the phases and weights of solute in, say, the aqueous phase before and after extraction, and then eliminate the weight by combining the two equations.

Assume that you have a solution containing w_0 mg of a particular substance in x cm^3 of water. This is then extracted with y cm^3 of organic solvent. After extraction w_1 mg of the substance are left behind in the aqueous phase.

Concentration (*after* extraction) of substance in organic phase

$$= \frac{(w_0 - w_1)}{y} \text{ mg cm}^{-3}$$

Concentration (*after* extraction) of substance in aqueous phase

$$= \frac{w_1}{x} \text{ mg cm}^{-3}$$

$$\therefore \qquad D = \frac{(w_0 - w_1)}{y} \times \frac{x}{w_1} \qquad (8.13)$$

Since $D = \dfrac{\text{concentration in organic phase}}{\text{concentration in aqueous phase}}$

Now

$$E = \frac{w_0 - w_1}{w_0} \times 100\% \qquad (8.14)$$

ie amount extracted divided by total amount $\times$ 100

These are our equations for D and E which we now have to combine by eliminating w_0 and w_1. If you initially got stuck after this point and are following this response through, have another go on your own from this point on.

We need to rearrange Eq. 8.13 to obtain the expression $100 \times w_0 - w_1)/w_0 \ (= E)$ on the left hand side. Multiplying both sides by $w_1/(w_0 - w_1)$ and then dividing both sides by D gives:

$$\frac{w_1}{w_0 - w_1} = \frac{x}{D.y}$$

Add 1 to both sides:

Then $\dfrac{w_1 + w_0 - w_1}{w_0 - w_1} = \dfrac{w_0}{w_0 - w_1} = 1 + \dfrac{x}{D.y}$

Take reciprocals:

$$\frac{w_0 - w_1}{w_0} = \frac{1}{1 + x/Dy} = \frac{D}{D + x/y}$$

Multiply both sides by 100

$$\frac{w_0 - w_1}{w_0} \times 100 = \frac{100.D}{D + x/y}$$

$= E$ as shown in Eq. 8.14 above.

Thus:

$$\boxed{E = \frac{100.D}{D + x/y}} \qquad (8.15)$$

This equation simplifies to:

$$E = \frac{100.D}{D + 1} \qquad (8.16)$$

when equal volumes of phases are used.

SAQ 8.5c   The neutral, unionised form of an organic acid has a value of K_D of 150 for extraction from water into trichloromethane. At pH 5 the value of D for the acid taking into account ionised (anionic) and unionised (neutral) forms is 12. Thus the percentage extraction of the acid, taking into account both ionised and unionised forms from 100 cm^3 of water at pH 5 into 25 cm^3 of trichloromethane is (to two significant figures):

   (*i*)   97%

   (*ii*)   92%

   (*iii*) 98%

   (*iv*)   75%

8.5.3. Efficiency of Extraction

Sometimes we might find that under a certain set of conditions, the value of E, while perhaps being quite high, is nevertheless noticeably less than 100%.

In other words, the efficiency of extraction is significantly less than 100% under these conditions. If this is so then an important principle to remember is as follows.

> The efficiency of extraction is always improved by carrying out the extraction in several stages with small volumes of organic phase and combining the extracts, rather than by using one large volume of solvent.

This procedure is known as *multiple extraction.*

The extent of the improvement is most noticeable when the single extraction itself is highly efficient. If the extraction is of very low efficiency to start with, then the improvements to be gained by multiple extraction are much less significant.

We shall now work out a simple but very useful equation that will tell us how much of a solute is left in the aqueous phase after a multiple extraction. We assume that the extraction is carried out by dividing the organic phase into n aliquots, shaking the aqueous phase with successive aliquots and finally combing the aliquots. Again, why don't you try it first?

∏   Assuming that:

the volume of the aqueous phase is x cm^3,

the total volume of the organic phase is y cm^3,

the distribution ratio of the substance being extracted is D, the initial weight of substance being extracted is w_0,

the weight of substance left in the aqueous phase after one extraction is w_1,

derive an expression for w_1 in terms of w_0, x, y and D.

Now after extraction:

$$\text{concentration in organic phase} = \frac{w_0 - w_1}{y}$$

$$\text{concentration in aqueous phase} = \frac{w_1}{x}$$

$$D = \frac{\text{concentration in organic phase}}{\text{concentration in aqueous phase}}$$

$$= \left(\frac{w_0 - w_1}{y}\right) \times \frac{x}{w_1}$$

Multiplying through by $w_1 y$ and multiplying out the r.h.s.:

$$w_1 D y = w_0 x - w_1 x$$

Collecting terms involving w_1 on the l.h.s.

$$w_1 D y + w_1 x = w_0 x$$

$$\therefore \qquad \boxed{w_1 = w_0 \times \frac{x}{Dy + x}} \qquad (8.17)$$

This equation can be used to calculate how much substance will remain in the aqueous phase after a single extraction.

Now let us suppose that we carry out the extraction in two stages by using two equal volumes of the organic phase. The volume of each is y cm^3. The amount remaining after the second stage is now w_2. By inference from Eq. 8.17 we can write an equation for w_2 in terms of w_1 exactly analogous to Eq. 8.17:

$$w_2 = w_1 \times \frac{x}{Dy + x} \tag{8.18}$$

We can now combine Eqs. 8.17 and 8.18 to eliminate w_1

$$w_2 = w_0 \times \left(\frac{x}{Dy + x}\right) \times \left(\frac{x}{Dy + x}\right)$$

$$= w_0 \left(\frac{x}{Dy + x}\right)^2 \tag{8.19}$$

I hope that you can see by inference again that were we to shake the aqueous phase with n aliquots of organic phase, each of volume y cm^3 and that if we let the amount remaining be w_n then:

$$\boxed{w_n = w_0 \left(\frac{x}{Dy + x}\right)^n} \tag{8.20}$$

This is an important equation, but we can make just one more modification, and that is to write an equation that tells us what we initially wanted to know. How much of the substance remains in the aqueous phase if we divide our original organic phase of volume y into n equal aliquots and carry out successive extractions with these aliquots?

We can write down this equation directly simply by substituting y/n for y in Eq. 8.20 since y/n is the volume of organic phase used in a single stage of the extraction. If we redefine w_n as the amount remaining under these conditions then:

$$\boxed{w_n = w_0 \left(\frac{x}{Dy/n + x}\right)^n} \tag{8.21}$$

SAQ 8.5d

Test the hypothesis that a more efficient total extraction is obtained by dividing the organic solvent into a number of aliquots, shaking the aqueous phase successively with these aliquots and combining them afterwards. Use the following hypothetical data.

Volume of aqueous phase $= 100$ cm^3
Total volume of organic phase $= 100$ cm^3
Distribution ratio $= 10$
Extraction carried out:

(i) in one stage,

(ii) in 2 to 4 stages, dividing the 100 cm^3 of organic solvent into 2 to 4 equal sized aliquots.

Continuous Extraction

The idea that extraction efficiency can be maximised by carrying out the extraction in many stages with fresh portions of solvent at each stage is the fundamental principle of *continuous extraction*. Here one solvent is made to pass continuously through the other in a countercurrent fashion. Gravity and boiling under reflux are the

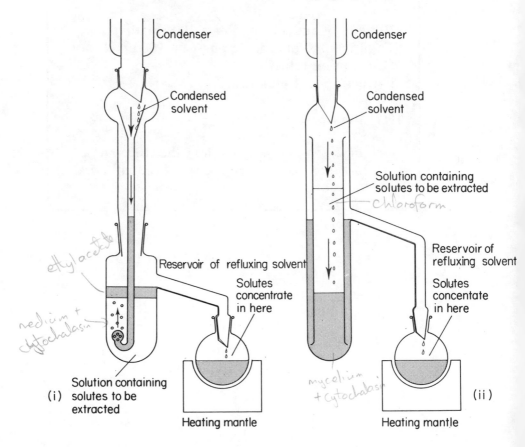

Fig. 8.5c. *Apparatus for continuous extraction*

(*i*) Extracting solvent less dense than the solution from which a solute is being extracted

(ii) Extracting solvent more dense than the solution from which a solute is being extracted

driving forces behind the technique for which special apparatus is available, and shown in Fig. 8.5c and 8.5d. You will note that the apparatus is available in two forms, depending on whether the solvent for extraction is denser or less dense than the solvent containing the sample. In each case, solvent vapour passes from a reservoir of boiling solvent to the condenser where it condenses. Thus pure extracting solvent passes through the sample solution by one means or another. The level of the solution so formed rises and the solution drains back into the reservoir. Thus the solute gradually accumulates in the reservoir. We must assume that the solutes are much less volatile than the solvent so that they do not volatilise with the solvent. We must also assume that the solutes are sufficiently thermally stable not to decompose in the hot, refluxing solvent. Continuous extraction is used to extract solutes from a solvent into another solvent under conditions where the values of D or K_D are unfavourable.

SAQ 8.5e

State whether the following statements, which all relate to *continuous extraction*, are *true* (T) or *false* (F).

(i)   For a solute to be extracted efficiently by continuous extraction, it must be sufficiently volatile to be volatilised along with the extracting solvent.

T / F

(ii)   All other factors being equal, the xylenes boiling at about 140 °C are less useful than diethyl ether boiling at 35 °C, as extraction solvents.

T / F

(iii)   Continuous extraction is invaluable for increasing selectivity in extraction systems.

T / F

SAQ 8.5e (cont.)	(*iv*) Continuous extraction is more versatile if the refluxing liquid is the organic solvent and the solutes are extracted *from* the aqueous phase. T / F

8.5.4. Selectivity of Extraction

So far we have simply been concerned with the overall efficiency with which we can transfer a single substance from one liquid phase to another. This is fine for applications such as in preconcentration, but often we are concerned with the *separation* of two substances. The usual parameter for measuring the separability of two substances in a particular solvent extraction system is the *separation factor*, given the symbol β.

If two substances have distribution ratios D_1 and D_2 (or distribution coefficients K_D' and K_D'') for extraction between the same two solvents under the same conditions, then the separation factor, β_2^1 between these two substances is given by the ratio of the distribution ratios:

$$\beta_2^1 = \frac{D_1}{D_2} \left(\text{or } \frac{K_D'}{K_D''} \right)$$

∏ Under what circumstances might it be appropriate to use K_D values for calculating β, and when should you use D values?

Use K_D values to calculate β when you are concerned with substances *not* involved in any other equilibria affecting the concentrations of the extracted species. Where any other equilibria *are* involved, D values must be used.

Let us now work out a typical value for β in a system involving two substances initially present in equal amounts, but which are separated with about 1% cross contamination. This situation will be achieved if one substance is extracted into the organic phase with a percentage extraction $E_1 = 99\%$ and the other with $E_2 = 1\%$. Let us assume for simplicity that we are working with equal volumes of phases. Then we can rearrange Eq. 8.16 to give Eq. 8.23.

$$D = \frac{E}{100 - E} \qquad (8.23)$$

Then D_1 for substance 1 for which $E_1 = 99$

$$= 99/(100 - 99) = 99$$

And D_2 for substance 2 for which $E_2 = 1$

$$= 1/(100 - 1) = 1/99$$

Thus $\beta_2^1 = 99 \times 99 = 9801 \approx 10000$.

In other words, for solvent extraction to be effective as a technique for quantitative separation, vast differences in extractability between solutes is necessary.

We have just shown that for reasonable conditions, in order to separate two substances present in about equal amounts approximately quantitatively, their distribution ratios should have a ratio of *at least 10000*. This places solvent extraction at a distinct disadvantage to chromatography as a separation technique. Separation factors can be defined in chromatography in much the same way as above (ie as a ratio of D or K_D values) but substances with separation factors as little as 2 between them are well separated in the simplest chromatographic systems, and in high resolution capillary gas chromatography β can be as little as 1.05. Of course when $\beta = 1$, no separation is possible whatever the system. In fact, there are limitations to the usefulness of β as a parameter, as we shall see later.

∏   Why might you suppose that β on its own is not always the best parameter for selecting the optimum conditions for a separation?

Although the maximum value of β might indeed tell us about the *purity* of one or other of the two solutes separated, it does not tell us anything about the *overall efficiency* of the extraction.

For example, suppose we have two substances, A and B with distribution ratios of 10^{-2} and 10^{-7} respectively for extraction into an organic solvent. For this system, $\beta = 10^5$, implying quantitative separation, but *not* efficient extraction. This means in practice that our organic extract will indeed contain only substance A (at least to one part in 10^5) but only 1% of the total amount of substance A present will have been extracted.

In order to assess the overall performance of an extraction with regard to efficiency *and* selectivity, ΔE (*the difference in percentage extractions*) is a more practical parameter. ΔE should be as close to 100% as possible, implying total extraction of one component and no extraction of the other.

There is a very powerful way in which we can improve the separation of two substances in solvent extraction, as follows.

Manipulating values of D via *selective chemical equilibria.*

The values of D are very sensitive to any chemical equilibria that involve the species being extracted. We have already shown this in the extraction of benzoic acid from water into diethyl ether. Now if we have two acids that we wish to separate that have similar values of K_D but have, say, very different values of K_A, then, by selecting an appropriate value of pH, we can maximise the selectivity of the extraction.

Let's try this for a hypothetical situation in which we want to separate benzoic acid from another, weaker acid with a very similar K_D value but with a value of K_A of 3.6×10^{-9} mol dm^{-3}. Let's assume K_D for our second acid is 750.

We shall investigate the problem by trying to find the pH at which the difference in E for the two acids is maximised, rather than the more obvious course of finding the maximum value of β. This is because β doesn't take into account the overall efficiency of the extraction and could well have a maximum value when both D values are either very large or very small (remember the last exercise you tried?) At this stage, we'll assume equal volumes of phases to keep the algebra simple.

Now we need to calculate D for both acids over a range of pH values by using Eq. 8.3,

$$D = \frac{K_D}{1 + K_A/[H^+]} \tag{8.3}$$

and converting D into E by using Eq. 8.16

$$E = \frac{100D}{D + 1} \tag{8.16}$$

We then calculate ΔE (which is equal to $E_2 - E_1$ for the two acids) at various pH values, where E_1 and E_2 are the percentage extractions for benzoic acid and the hypothetical acid respectively. The solution to our problem is the value of the pH at which $E_2 - E_1$ is a maximum. This is the pH at which to carry out the separation.

Now, the mathematics to carry out this calculation is quite straightforward but time consuming and repetitive and is best carried out by using a computer. I've done this for you therefore and summarised the results in Fig. 8.5d.

pH	ΔE (%)	pH	ΔE (as %)
1	0.01	8	89.6
2	0.01	9	98.3
3	0.01	10	95.2
4	0.09	11	67.5
5	0.87	12	17.2
6	8.03	13	2.04
7	46.6	14	0.21

	Acid 1	Acid 2	
K_A	6.30×10^{-5}	3.60×10^{-9}	(mol dm^{-3})
K_D	720	750	

for equal volumes of phases.

Fig. 8.5d.   *Values for ΔE (difference in percentage extraction for two acids) as a function of pH*

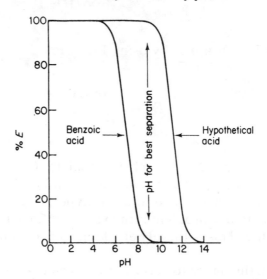

Fig. 8.5e.   *Plot of E vs pH for the separation of benzoic acid from a hypothetical acid by extraction from water into diethyl ether. Constants as in Fig. 8.5d*

Π   From the data in Fig. 8.5d *select* the optimum pH value for the separation of the two acids.

The optimum pH for the separation is *pH 9* (to the nearest 0.5 pH unit) since this pH corresponds to the maximum value of ΔE. We can also see from Fig. 8.5d what a vast effect pH has on the selectivity of the extraction.

Fig. 8.5e shows the data in Fig. 8.5d graphically. The optimum pH is the pH at which the two curves are furthest apart along the y-axis.

SAQ 8.5f

On the basis of the K_D and K_A values Fig. 8.5d, which of the following values (*i*) to (*vi*) is most appropriate for the separation factor, β, between benzoic acid and our hypothetical acid for extraction from water at pH 9 into diethyl ether?

(*i*)   163

(*ii*)   13 100

(*iii*) 88.0

(*iv*)   14 300

(*v*)   1.04

(vi)   18 200

SAQ 8.5f

8.6. THE EXTRACTION OF METAL IONS

Perhaps the most important and useful area of application of solvent extraction in analysis is the extraction of compounds of *metal ions* from water into an organic solvent. Some important applications of the solvent extraction of metal ions are listed below and show the wide scope of the technique.

(*a*) As a general, non-selective *preconcentration* step in, for example, atomic absorption analysis.

(*b*) To improve *selectivity* in spectrophotometric and other forms of analysis by using a selective extraction step.

(*c*) As the *substoichiometric separation step* in a radiochemical isotope dilution analysis.

SAQ 8.6a   For each of the following, indicate whether the statement is *true* (T) or *false* (F).

(*i*)   It is virtually unknown for a metal ion to be extractable from water into a water-immiscible organic solvent as a free hydrated cation.

T / F

(*ii*)   A good method for the extraction of metal ions from water into organic solvents is to convert them, by using suitable organic ligands, into neutral, coordinatively saturated chelate complexes containing the minimum of extraneous polar groups.

T / F

(*iii*)   Metal ions may often be extracted into organic solvents as charged species, provided that this charged species has been somehow incorporated into an ion-association complex. Complex formation and solvation by the extracting solvent are important features of such a system.

T / F

Thus you should by now clearly know and understand that if we wish to extract metal ions from water into an organic solvent, then we must first of all convert them into an extractable complex. These complexes may be broadly classified into two types as follows.

(*a*) Neutral, coordinatively saturated, chelate complexes,

(*b*) Ion-association complexes (ion pairs).

Let us now consider these two extraction systems in a little more detail.

8.6.1. Neutral-chelate Extraction Systems – Qualitative Aspects

In these systems, the metal ion reacts with an organic chelating ligand to form a *neutral chelate complex*. In these complexes the ligand molecules occupy all the metal's coordination sites, so that no water of coordination is present. The neutral complex so formed should be free of polar groups not involved in coordination. One would expect chelate complexes that satisfy these conditions to be extractable into organic solvents with very high distribution coefficients, since one would expect the solubility properties of such a molecule to be similar to those of a relatively non-polar organic compound. This is indeed usually true.

In order to neutralise the charge on the metal ion the ligand should be a weak acid and should complex with the metal as the anion, thus hydrogen ions will be released on complex formation. Normally, the ligands act as weak monobasic acids, and complex with the metal ion in the ratio of one ligand molecule per metal ion per unit charge on the ion. Note also that the uncharged, neutral ligand is often significantly extractable into the organic solvent.

Many such systems are adequately described by the following equilibria in which M^{n+} represents a metal ion in the oxidation state n, HL the neutral form of the complexing ligand, and L^- the anionic form. The top half of the box below represents the aqueous phase and the bottom half the organic phase. Horizontal two-way arrows represent chemical equilibria, vertical two-way arrows represent transfer between phases.

$$
\begin{array}{c}
\text{(Aqueous Phase)} \\
M^{n+} + nHL \rightleftharpoons ML_n + nH^+ \\
\hline
nHL \qquad ML_n \\
\text{(Organic-solvent phase)}
\end{array}
$$

An example of such a system is the extraction of aluminium(III) ions with 8-hydroxyquinoline(HOx) into trichloromethane.

Aqueous phase

Organic phase

(CHCl$_3$)   HOx   ALOx$_3$

∏   List *three* conditions that the aluminium-8-hydroxyquinoline complex (and other such complexes) satisfy, and that account for its extractability into organic solvents.

These are:

(*i*)   the charge on the complex is zero;

(*ii*)   all six coordination sites on the aluminium ion are occupied by ligand atoms;

(*iii*) all the polar groups on the ligands are directed towards the centre of the complex which thus has a non-polar outward aspect and therefore a much lower solubility in water than in organic solvents.

We should also note that because of the release of H^+ on complex formation, these extraction systems are highly pH dependent, complex formation and therefore extraction being suppressed at low pH. However, extraction is also often suppressed at high pH. This, however, is for a different reason, ie that the anionic form of the ligand has to compete with a high concentration of hydroxide ions which will preferentially react with the metal ion to form non-extractable hydroxide complexes if the pH is high enough.

SAQ 8.6b

> If we work at about pH 10, magnesium(II) is only poorly extracted into trichloromethane as a neutral chelate complex with 8-hydroxyquinoline. If butylamine is used in place of 8-hydroxyquinoline, then no extraction of magnesium(II) occurs at all. However, if a mixture of the two reagents is used (8-hydroxyquinoline plus butylamine) then the extraction of magnesium(II) suddenly becomes highly efficient.
>
> Can you offer explanations for these three observations from what you have just learnt about neutral chelate extraction systems? Magnesium(II) has a coordination number of six and so exists in aqueous solution as the $Mg(H_2O)_6^{2+}$ ion.

SAQ 8.6b

Let us now look more generally at the types of metal ions that are well extracted as neutral chelates, some typical chelating reagents and solvents for extraction, and the conditions under which such systems work best in analysis.

(a) Metal Ions Best Extracted as Neutral Chelates

These include low-oxidation-state late transition metal ions, such as Mn^{2+}, Fe^{2+}, Co^{2+}, Ni^{2+}, Cu^{2+}, Ag^+; group IIb and IIIb ions, such as Zn^{2+}, Cd^{2+}, Hg^{2+}, Al^{3+}, Ga^{3+} and other ions such as Sn^{2+}, Pb^{2+}, Bi^{3+} (the total number of metal ions for which neutral chelate extraction systems are known is very large).

Metal ions less successfully extracted by these chelating systems include those of the alkali and alkaline earth metals, because they do not form very stable chelate complexes with the chelating ligands, and high valency transition metal ions, such as Ti^{4+} and V^{5+} because these ions tend to form hydroxide complexes at very low pH values.

(*b*) *Chelating Ligands*

A very large number of chelating ligands have been used and all we can do here is to list some of the more common and important ligands.

(i)

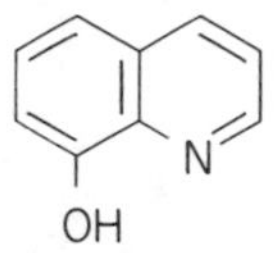

8-Hydroxyquinoline (oxine), a general-purpose, non-selective ligand.

(ii)

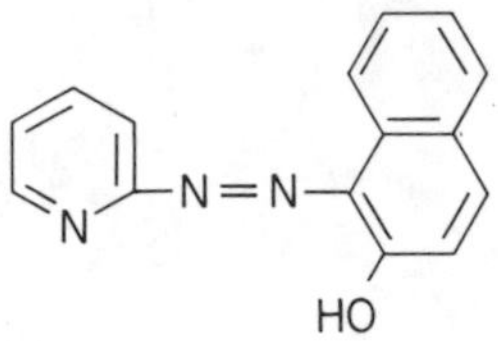

1-(2-Pyridylazo)-2-naphthol, also non-selective, its complexes are highly coloured and useful in extraction spectrophotometry.

(iii)

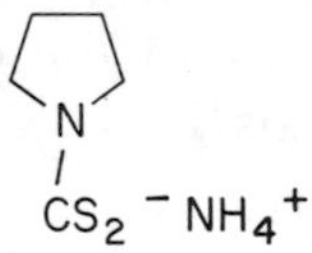

Ammonium pyrrolidine dithiocarbamate (Ammonium *N*-dithiocarbamatopyrrolidine). A non-selective reagent used for preconcentration of metal ions for atomic absorption spectroscopy.

(iv)

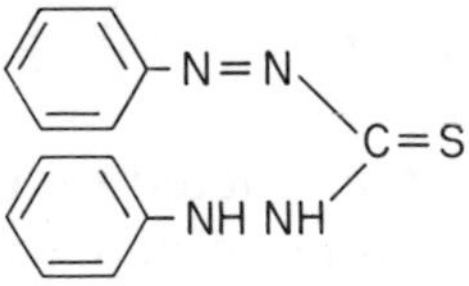

Dithizone. Its complexes are highly coloured and are used in spectrophotometric analysis

$$\text{(thiophene ring)} \quad C\ CH_2\ C\ CF_3$$
$$\qquad\qquad\qquad\ \ \overset{\|}{O}\qquad\overset{\|}{O}$$

Thenoyl trifluoroacetone – one of a number of diketones that exist in tautomeric enolic forms, to produce an acidic hydrogen atom.

(c) *Organic Solvents Suitable for Use in Extraction*

Trichloromethane (chloroform) and other halogenated solvents are popular choices in extraction spectrophotometry although there is a problem concerning toxicity with some of these. Ketones such as 4-methylpentan-2-one (methyl isobutyl ketone, MIBK) are favoured for preconcentration in atomic absorption spectroscopy. Heptan-2-one (n-amyl methyl ketone, MNAK) is perhaps even better than MIBK for preconcentration because it is less soluble in water. Toluene, the xylenes, alcohols and ethers of higher molar mass (obviously only those immiscible with water) are also used.

(d) *Areas of Application*

Neutral chelate extraction systems can be used for any of the applications listed at the beginning of this section. The conditions under which they work best are given below:

(a) At low concentrations of metal ions (therefore for such processes as preconcentration). Their usefulness at high concentrations is limited by the solubility of the complex in the organic phase.

(b) Under conditions where extremes of pH are avoided (because under conditions of extreme pH complex formation is inhibited).

SAQ 8.6c   Below are *three* examples of attempted extraction of a metal ion from an aqueous phase into an organic phase. Only *one* of these truly corresponds to a neutral-chelate extraction that is likely to be successful. The other three might be expected to fail as neutral chelate extractions or else correspond to extraction systems other than neutral chelate. Pick out the one potentially successful neutral-chelate extraction system and explain why the others fail as neutral-chelate extraction systems. Assume that the metal ion has a coordination number of six.

(*i*)   Extraction of high concentrations of iron(III) from a $6M$-hydrochloric acid medium into diethyl ether.

(*ii*) Extraction of aluminium(III) as its 8-hydroxyquinoline-5-sulphonic acid complex from an aqueous medium, pH 9, into trichloro-methane.

(*iii*) Extraction of traces of nickel(II) from an aqueous medium, pH 5, into trichloro-methane (as its 1-(2-pyridylazo)-2-naphthol (PAN) complex.

SAQ 8.6c

SAQ 8.6d   In the following paragraph, for each group of three words or expressions in italics, *delete* the least suitable two in order to produce a description of the *best* procedure for the preconcentration of lead from a 1 dm^3 water sample into 10 cm^3 of organic solvent, before analysis by atomic absorption.

'The sample is first rendered *acidic/neutral/ alkaline* and then evaporated from 1 dm^3 to a volume of *600 cm^3/250 cm^3/30 cm^3*. It is then buffered to *pH 1/pH 5/pH 12* and an excess of *EDTA/dimethylglyoxime/ammonium pyrrolidine dithiocarbamate* is added to complex the lead. Finally the complex is extracted into 10 ml of *MNAK/benzene/MIBK*'.

SAQ 8.6d

8.6.2. Quantitative Treatment of Neutral-chelate Extraction Systems

Most neutral-chelate extraction systems can be described fairly satisfactorily by the relatively simple set of chemical equilibria given in Section 8.6.1. Consequently, we can describe these systems quantitatively in a fairly simple manner.

∏   What important parameter shall we need to evaluate in order to quantify neutral-chelate extraction systems?

The parameter that we need to evaluate is D, *the distribution ratio of the metal ion*, considering all complexed and uncomplexed forms. From this we can obtain E, the percentage extraction as before.

What we shall do is to consider all the equilibria (chemical and distribution) that have a bearing on D, the equilibrium constants associated with them, and then I shall present to you an equation relating D to all these various equilibrium constants. We could have derived this equation, but the algebra involved is rather lengthy and there isn't really time to do this here. The equation depends on a number of assumptions and we must consider these as well. Then we shall use the equation to examine the all-important relationship between D and pH and to see how the efficiency and selectivity of these extraction systems can be optimised.

In this treatment we shall use a number of mathematical terms and symbols. Some of these are defined in the text. The definitions of others are summarised here.

M^{n+} – the free (hydrated) metal ion carrying a formal positive charge of n.

HL – the neutral ligand molecule (a chelating organic ligand).

L^- – the negatively charged anion from the ligand HL, which is the species that actually reacts with M^{n+}.

H^+ – this is actually H_3O^+, but for simplicity we shall simply use H^+ to signify hydronium (hydrogen) ions.

Subscripts

aq, org – when used with chemical formulae or concentrations, these will denote the aqueous or organic phases respectively. When they are absent, the aqueous phase is implied.

Here now, are the *equilibria*, basically four in number, that we must consider in the first instance, together with their respective equilibrium constants. You ought, in fact, to have met *all* these equilibria in earlier parts of this Unit, but I have presented them again for you here for convenience.

(*a*) Distribution of the neutral ligand between the two phases:-

$$HL_{aq} \rightleftharpoons HL_{org}$$

for which the distribution coefficent of HL:

$$K_{DL} = \frac{[HL]_{org}}{[HL]_{aq}}$$

(b) Dissociation of HL in the aqueous phase into H^+ and L^-:

$$HL \rightleftharpoons H^+ + L^- \tag{8.24}$$

for which the acid dissociation constant of HL:

$$K_A = \frac{[H^+][L^-]}{[HL]} \tag{8.25}$$

(c) Complex formation between M^{n+} and L^-. This is a stepwise process and stability constants are defined for each stage. The stability constant for the overall reaction is the product of the stability constants for each stage.

Reaction		*Stability Constant*
$M^{n+} + L^- \rightleftharpoons ML^{(n-1)+}$		$K_1 = \dfrac{[ML^{(n-1)+}]}{[M^{n+}][L^-]}$
$ML^{(n-1)+} + L^- \rightleftharpoons ML_2^{(n-2)+}$		$K_2 = \dfrac{[ML_2^{(n-2)+}]}{[ML^{(n-1)+}][L^-]}$
$ML_{n-1}^+ + L^- \rightleftharpoons ML_n$		$K_n = \dfrac{[ML_n]}{[ML_n^+][L^-]}$

The overall stability constant is β_M, which is what we shall need:

$$\beta_M = K_1.K_2 \ldots K_n$$

$$= \frac{[ML_n]}{[M^{n+}][L^-]^n} \qquad \text{for } M^{n+} + nL^- \rightleftharpoons ML_n \tag{8.26}$$

(d) Distribution of the neutral chelate complex between the two phases:

$$[ML_n]_{aq} \rightleftharpoons [ML_n]_{org}$$

for which the distribution coefficient of ML_n:

$$K_{DM} = \frac{[ML_n]_{org}}{[ML_n]_{aq}}$$

The assumptions that we make are as follows.

(*a*) The concentration of HL is large compared with that of M^{n+}, so that the only complex present at a significant concentration is ML_n. This is a reasonable assumption since in all real analytical situations involving neutral-chelate extraction, M^{n+} concentrations are low and the ligand is always added in excess.

(*b*) The only extractable species are HL and ML_n. This is true because all other species are charged.

(*c*) There are no other equilibria occurring in the organic phase. The only likely equilibrium that we might have to consider is association, as follows.

$$xML_{n\ org} \rightleftarrows (ML_n)_{x\ org}$$

This may occur at high concentrations of ML_n. We, however, are considering exclusively low concentrations.

(*d*) There are no other equilibria occurring in the aqueous phase. Now, this is *NOT* a reasonable assumption and we have to acknowledge here and now that other equilibria can and do occur. We ignore their effect for the time being, but later we must certainly take account of them.

Having established our equilibria and taken into account our assumptions, we are now in a position to consider the equation relating D to the other equilibrium constants. Here it is.

$$D = \frac{\beta_M\ K_{DM}\ K_A^n}{K_{DL}^n} \left\{ \frac{[HL]_{org}}{[H^+]_{aq}} \right\}^n \tag{8.27}$$

This extremely useful equation can be used to predict the effects of changes in a wide variety of parameters on the efficiency of extraction. We can also use it to predict selectivities when two or more metal ions are being extracted. It predicts that D is independent of metal-ion concentration but not of ligand concentration ([HL]$_{org}$, but not [M^{n+}], appears in the equation).

SAQ 8.6e

Explain the following observation which, although at first a little surprising, is in fact true.

'In choosing between two organic solvents for extracting neutral chelate complexes of metal ions where n (the stoichiometric factor) is equal to 2 or more, the solvent giving rise to the smaller value of K_{DM} (distribution coefficient of ML_n), ie the poorer solvent for the extraction of the complex, is actually the solvent producing the higher value of D (the distribution ratio for the metal ion) ie, the better solvent for the overall extraction'.

If you find that you haven't a clue about the explanation to this, don't worry! The response is arranged as a series of 'clues' to the explanation. Look at the response clue by clue and after considering each clue, see if you can come up with the explanation. Perhaps you won't need many of the clues but if you do, at the end you may have a better understanding of some of the factors influencing the overall extraction.

To get you started, I would recommend that you look at Eq. 8.27, which relates D to K_{DM}. Keep it in front of you during your deliberations.

SAQ 8.6e

SAQ 8.6f   Under a particular set of conditions (of pH, etc.), a metal ion in equilibrium with various complexed (chelate) forms is known to have a distribution ratio, D of 60.5 between trichloromethane and water. How many extractions, each involving 25.0 cm^3 of trichloromethane, are required to remove >99.9% of the metal ion from 100.0 cm^3 of water?

The answer to this question can be obtained quite quickly and easily if you can recall from memory an equation that we actually derived earlier in this Unit. However, if you cannot recall this equation, or there is any likelihood of your recalling it incorrectly, *do not* refer back to the text, but instead attempt the problem by working from basic principles, which in this case means essentially working from the definition of D.

SAQ 8.6f

Perhaps the most important parameter relating to the aqueous phase and affecting the extraction is pH. The equation tells us a lot about the relation between D and pH, as follows.

If we hold all parameters except pH constant and put them equal to K', we have:

$$K' = \frac{\beta_M K_{DM} K_A^n [HL]_{org}^n}{K_{DL}^n} \qquad (8.27a)$$

and

$$D = \frac{K'}{[H^+]_{aq}^n} \qquad (8.28)$$

$$\therefore \quad \boxed{\log D = \log K' + n\,pH} \qquad (8.29)$$

Thus the relationship between $\log D$ and pH is a straight line of slope, n. However, we are interested in the relationship between percentage extraction, E, and pH.

∏   Write down an equation for E, the percentage extraction, in terms of D, the distribution ratio and x and y, the volume of the aqueous and the organic phase, respectively.

$$E = \frac{100\, D}{D + x/y} \qquad (8.15)$$

Combining Eqs. 8.15 and 8.28 gives:

$$E = \frac{100\, K'/[H^+]_{aq}^n}{K'/[H^+]_{aq}^n + x/y}$$

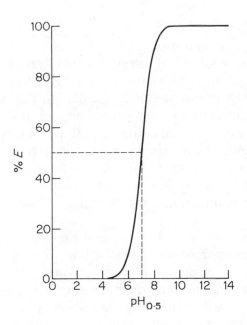

Fig. 8.6a.   *pH-extraction curve for a metal ion with $n = 1$ and $pH_{0.5} = 7.0$, for equal volumes of phases*

A plot of E against pH is known as a *pH-extraction curve* and has the general appearance of a sigmoid curve as can be seen in Fig. 8.6a.

Fig. 8.6a shows that neutral chelate extraction systems are strongly pH dependent, extraction being suppressed at low pH values when the hydrogen-ion concentration is high. This forces the equilibrium in Eq. 8.24 to the left, and thus reduces the concentration of L^- in solution. This in turn forces the equilibrium in equation 8.26 to the left so that complex formation is inhibited. Thus, extraction of the metal ion is inhibited because there is very little neutral chelate complex to be extracted.

The pH at which E is exactly 50% and therefore $D = 1$ (for equal volumes of phases) is known as $pH_{0.5}$. From Eq. 8.29, therefore we can see that:

$$pH_{0.5} = \frac{-\log(K')}{n} \tag{8.30}$$

Under a given set of conditions, $pH_{0.5}$ is a constant and has a characteristic value for each metal ion present. Different metal ions have different values of $pH_{0.5}$ by virtue of having different values of β_M and K_{DM}, although in practice it is usually β_M that varies most significantly from metal to metal. This is because different metal ions can form complexes of vastly differing stabilities, whereas the extractability of these complexes is likely to be governed more by the properties of the ligand and the solvent (we saw a similar effect in SAQ 8.6e where we examined the effect of the solvent on the extraction of the complex and the ligand).

It is because different metal ions have different values of $pH_{0.5}$ under the same extraction conditions that *selectivity* can be introduced into neutral-chelate extraction systems, allowing us under appropriate conditions to separate one ion from another, provided that their $pH_{0.5}$ values differ sufficiently. Fig. 8.6b shows pH-extraction curves for two divalent metal ions with $pH_{0.5}$ values of 4.0 and 10.0 respectively.

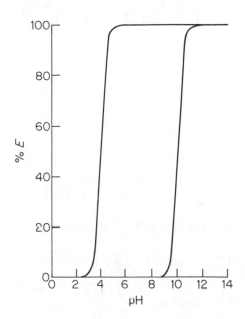

Fig. 8.6b. *pH-extraction curves for two metal ions with* $pH_{0.5} =$ *4.0 and 10.0 respectively (n = 2 for each ion). Equal volumes of phases*

∏   Within what range of pH would the two metal ions in Fig. 8.6b be virtually quantitatively separated?

Between pH 5 and pH 9, the extraction of one metal ion is >99% and the other <1%. The answer is therefore pH 5–9.

Now, the actual *minimum* difference in $pH_{0.5}$ required in order to separate two metal ions with a given, specified value of β, the separation factor (don't confuse β with β_M, the overall formation constant, by the way) can be calculated by considering the *slope* of the pH extraction curve. This is a function of n, becoming increasingly steep as n increases.

Since $\beta = \dfrac{D_1}{D_2}$ where D_1 and D_2 are distribution ratios for two metal ions with *the same value of n,*

$$\log \beta = \log D_1 - \log D_2 \qquad (8.31)$$

Applying Eq. 8.29 to Eq. 8.31 gives:

$$\log \beta = \log_{10} K_1' - \log_{10} K_2'$$

(the terms, n pH, cancel out)

Hence

$$\log \beta = n(\text{pH}_{0.5}'' - \text{pH}_{0.5}')$$

where $\text{pH}_{0.5}'$ and $\text{pH}_{0.5}''$ are the $\text{pH}_{0.5}$ values for the two metal ions.

If we put $\text{pH}_{0.5}'' - \text{pH}_{0.5}' = \Delta\text{pH}_{0.5}$, we have:

$$\Delta\text{pH}_{0.5} = \frac{\log \beta}{n} \qquad (8.32)$$

$\Delta\text{pH}_{0.5}$ then is the minimum difference in $\text{pH}_{0.5}$ values required for the separation of two metal ions with the separation factor, β. For instance, if our required value for β is 10,000 (representing about 1% cross contamination) then for:

$$n = 1 \qquad \Delta\text{pH}_{0.5} = 4.0$$
$$n = 2 \qquad \Delta\text{pH}_{0.5} = 2.0$$
$$n = 3 \qquad \Delta\text{pH}_{0.5} = 1.3$$

In other words the selectivity of a neutral-chelate extraction system increases with increasing charge on the metal ion, because the slope of the pH extraction curve increases with increasing value of n. This formula only applies when n is the same for both metal ions. If we try to separate two metal ions with different values of n, then selectivity becomes a more complex function of pH and $\text{pH}_{0.5}$.

Fig. 8.6c shows pH-extraction curves for three metal ions each with $K' = 10^{-9}$, but with n values of 1, 2 and 3 respectively. You can see how the slope and $\text{pH}_{0.5}$ values depend on the value of n.

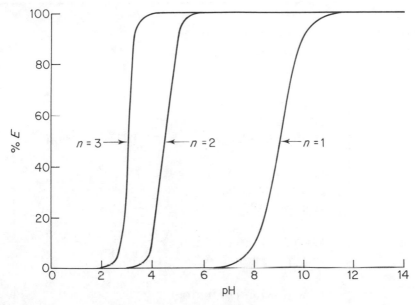

Fig. 8.6c. *pH-extraction curves for three neutral chelate complexes, each with $K' = 10^{-9}$, but with n = 1, 2 and 3 respectively (equal volumes of phases in each case)*

SAQ 8.6g

Two divalent metal ions, A and B, which are present at low and approximately equal concentrations in water are to be separated by solvent extraction by using selective pH control. Neutral, extractable chelates, AL_2 and BL_2 are to be formed with the anion of the ligand, HL and the extraction is to be performed by using equal volumes of phases. The ligand is present initially in the organic phase at a concentration of 0.100 mol dm^{-3}. The following data relating to equilibrium constants are provided

Acid dissociation
constant for the
ligand, HL   = $10^{-5.7}$ mol dm^{-3}

$\longrightarrow$

SAQ 8.6g
(cont.)

Distribution coefficient
for the neutral ligands,
HL $= 10^{3.2}$

Overall stability
constant for the
complex, AL_2 $= 10^{6.9} \; (mol \; dm^{-3})^{-2}$

Overall stability
constant for the
complex BL_2 $= 10^{10.7} \; (mol \; dm^{-3})^{-2}$

Distribution coefficient
for the complex AL_2 $= 10^{4.8}$

Distribution coefficient
for the complex BL_2 $= 10^{5.2}$

Which of the following statements is the only
one that is completely true?

(i)   The $pH_{0.5}$ values for the ions A and B are
        4.05 and 1.95 respectively. These pH val-
        ues represent the limits of the pH range
        within which the separation can reasonably
        be considered as quantitative.

(ii) The difference in $pH_{0.5}$ for the two ions (Δ
        $pH_{0.5}$) is 2.10. Ion A is the more extractable
        ion and is therefore the one found in the
        organic phase after extraction at the opti-
        mum pH.

(iii) The optimum pH for the separation is 3.0.
        This value was obtained by averaging the
        values of $pH_{0.5}$ for each ion. This is a gen-
        eral method for calculating the optimum
                                                            $\longrightarrow$

**SAQ 8.6g
(cont.)**

pH for the separation of two metal ions in neutral chelate extraction systems.

(iv) The separation factor between the two ions at the optimum pH is 16×10^3 (to 2 significant figures). This represents somewhat less than 1% cross contamination between the ions.

As a practical example of the effect of pH on the selectivity of neutral-chelate extraction systems, let us consider the extraction of metal dithizonates. Dithizone is a ligand producing extractable complexes with a wide range of metal ions, but the $pH_{0.5}$ values vary widely from metal to metal. Fig. 8.6d contains some typical values for $pH_{0.5}$ values for the extraction of metal dithizonates into tetrachloromethane. They are not quoted as particularly accurate or as absolute values because they are affected by various experimental conditions such as ligand concentration, but they do serve to illustrate the vital effect of pH on extraction.

Metal Ion	Stoichiometry of Extracted Complex	$pH_{0.5}$ Value
Hg^{2+}	ML_2	0.4
Ag^+	ML	1.1
Cu^{2+}	ML_2	2.0
Bi^{3+}	ML_3	2.6
Sn^{2+}	ML_2	4.7
Pb^{2+}	ML_2	7.5
Zn^{2+}	ML_2	8.5
Tl^+	ML	9.8
Cd^{2+}	ML_2	11.5

Fig. 8.6d. *Approximate $pH_{0.5}$ values for the extraction of metal dithizonates into tetrachloromethane, together with the probable stoichiometry of the extracted complexes*

∏   How might you separate a mixture of Pb^{2+}, Cd^{2+}, and Hg^{2+} ions as their dithizonate complexes?

Looking at the $pH_{0.5}$ values in Fig. 8.6d and bearing in mind (see Fig. 8.6c) that we should be working at a pH of at least two whole pH units away from the $pH_{0.5}$ value for >99% or <1% extraction, then:

(*i*) extraction at a pH between 2.4 and 5.5 would result in quantitative removal of Hg^{2+} into the organic phase, leaving the other ions in the aqueous phase;

(*ii*) then extraction at pH 9.5 would result in removal of the Pb^{2+} leaving the Cd^{2+} in the aqueous phase.

Alternatively, you could extract at pH 9.5, to remove the Hg^{2+} and Pb^{2+}, leaving the Cd^{2+} behind, then back-extract the organic extract at a pH between 2.4 and 5.5 to separate the Hg^{2+} and Pb^{2+}.

Now, there is one final and very important point that we must consider before we leave the subject of neutral-chelate extractions. When we introduced our equation for D (Eq. 8.27) we made various assumptions about the extraction. One of these, we noted at the time, was not valid. We assumed that the only chemical equilibria in the aqueous phase involving the metal ion were those involving complex formation with HL. This, however, is not always true and we must now consider two important situations where other equilibria can affect the extraction.

1. Effect of Hydrolysis at High pH Values

At high pH values, the concentration of hydroxide ions increases and therefore the metal ion may react with the hydroxide ions in preference to reacting with L^- (in other words metal hydroxides are formed). The reactions involved would be of the type:

$$M^{n+} \qquad + \quad OH^- \quad \rightleftharpoons \quad M(OH)^{(n-1)+}$$

$$M(OH)^{(n-1)+} \quad + \quad OH^- \quad \rightleftharpoons \quad M(OH)_2^{(n-2)+}$$

$$M(OH)_{n-1}^+ \quad + \quad OH^- \quad \rightleftharpoons \quad M(OH)_n \downarrow \text{ (in the extreme case)}$$

In addition mixed complexes, such as $ML(OH)^{(n-2)+}$ might be formed. These OH^--containing species, however, are not usually extractable and therefore we might expect extraction to be inhibited at high pH values. We could modify Eq. 8.27 to include terms to account for hydrolysis, but here all we do is to redraw our pH extraction curve to show the decrease in extraction at high pH values due to hydrolysis. I have done this for you in Fig. 8.6e which is a hypothetical case in which hydrolysis has had a rather drastic effect on the pH range in which the extraction can be carried out. Extraction is now efficient only in the narrow pH range of about 5.5 to 7.0. The effect of hydrolysis is often not so severe, but for a number of high-valency transition-metal ions, the pH region in which protonation of the ligand occurs overlaps with that in which hydrolysis occurs, meaning that there is no pH value at which efficient extraction can take place. For these ions neutral-chelate extraction systems are not suitable and an ion-association extraction system must be used.

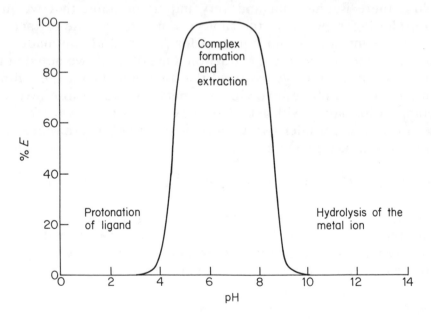

Fig. 8.6e. *Hypothetical pH-extraction curve for a metal ion showing the combined effects of ligand protonation and metal ion hydrolysis*

∏   Can you suggest a simple practical step that you could try
      in order to increase the pH range within which a metal ion
      might be extracted as a neutral chelate, if this range is limited
      by protonation of the ligand and metal hydroxide formation?

Try to lower the $pH_{0.5}$ value, either by choosing a ligand that forms
a more stable and extractable complex, or by increasing the concentration of the ligand, ie, you are trying to increase β_M, K_{DM}, or
$[HL]_{org}$ in Eq. 8.27. But be careful if you change the ligand that you
don't adversely affect K_H or K_{DL} in the process.

2. Presence of Anions and Organic Ligands Forming Water Soluble Complexes with the Metal Ion

Many anions such as fluoride, chloride, cyanide and thiocyanate
react with metal ions to form water-soluble anionic complexes,
such as $FeCl_4^-$, AlF_6^{3-}, $Fe(CN)_6^{4-}$. Likewise we may find it expe-
dient to add to the aqueous phase certain organic ligands such
as EDTA (ethylenediaminetetra-acetic acid) and citrate which also
form water-soluble non-extractable complexes with metal ions.
These ligands compete with the chelating ligand for the metal ion,
and once more, we have to include terms in Eq. 8.27 to account for
their effect. Qualitatively, what happens is that the $pH_{0.5}$ values for
the metal ions move to higher values by an amount related to the
stability of the water-soluble complex and the concentration of the
second ligand.

In practice, we can make good use of this effect to improve *selectivity* in extraction. Let us assume that we have two metal ions that
we wish to separate, but whose $pH_{0.5}$ values are insufficiently different for us to find a pH value where this separation can be carried
out effectively. What we might do under these circumstances is to
see whether we can find a second ligand that reacts strongly with
only one of the metal ions to form a water-soluble complex. The result of this is to move its $pH_{0.5}$ to much higher pH values (perhaps
even preventing extraction at any accessible pH) leaving the other
$pH_{0.5}$ value roughly where it is. If we have succeeded it should now
be possible for us to find a suitable pH range for carrying out the
separation. See Fig. 8.6f for an example.

Ligands used in this way are known as *masking agents* because they mask the effect of the interfering ion, by preventing, or limiting, its extraction. You may well come across masking agents in connection with other analytical techniques. However, their role is always the same, *viz* to form a complex with an interfering ion in such a way as to eliminate the interference from the analysis (in this case by preventing co-extraction in a separation).

A good example of the use of masking agents involves the extraction of lead into trichloromethane as its dithizonate complex. Many transition-metal ions are co-extracted but by using a mixture of cyanide and citrate anions as masking agents it is possible to find a pH where the extraction is adequately selective for lead.

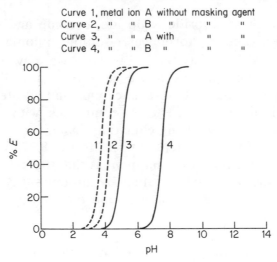

Curve 1, metal ion A without masking agent
Curve 2,   "   "   B   "   "   "
Curve 3,   "   "   A with   "   "
Curve 4,   "   "   B   "   "   "

Fig. 8.6f.   *pH extraction curves for two metal ions in the absence and in the presence of a masking agent. In the absence of the masking agent no pH can be found where the separation is satisfactory. With the masking agent present the ions can be separated at pH 6 to 6.5*

∏   Look back to Fig. 8.6d and find two ions that might be difficult to separate by pH control alone, but for which masking might improve the separation.

The obvious instance is Pb^{2+} and Zn^{2+}, whose $pH_{0.5}$ values are only one pH unit apart. The Zn^{2+} ions, though, can be masked by cyanide ions.

SAQ 8.6h

Given the following information and by using the block diagram approach introduced in SAQ 8.4c, devise a multistage solvent extraction scheme for the complete separation of the following ions which were initially present at low concentrations in an aqueous sample.

The aqueous sample initially contained

$Fe^{2+}, Fe^{3+}, Mn^{2+}, Co^{2+}, Ni^{2+}, Cu^{2+}, Pb^{2+}$.

Here is the information for your work.

(*i*) Neocuproine (2,9-dimethyl-1,10 phenanthroline) is a selective reagent for the chelated ion-association extraction of Cu^+ from a neutral aqueous phase into trichloromethane.

(*ii*) Dithizone reacts with most transition-metal and heavy-metal ions to form extractable complexes. Some $pH_{0.5}$ values for extraction into tetrachloromethane were given in Fig. 8.6d. Here is some more information specifically on the metal ions present in our sample (obtained from reference 16):

Fe^{2+} – extracted quantitatively between pH 7 and 9,

Fe^{3+} – not extracted (oxidises complex),

$\longrightarrow$

**SAQ 8.6h
(cont.)**

Mn^{2+} – extracted at about pH 10 only (the complex is unstable),

Co^{2+} – quantitatively extracted into tetrachloromethane between pH 5.5 and 8.5 or into trichloromethane around pH 8,

Ni^{2+} – extracted into tetrachloromethane at pH 6–9 or into trichloromethane at pH 8–11,

Cu^{2+} – extracted into trichloromethane or tetrachloromethane at pH 1–4,

Pb^{2+} – extracted into trichloromethane at pH 8.5–11.5 or tetrachloromethane at pH 8.0–10.0. $\longrightarrow$

(*iii*) In the presence of 6 *M*-hydrochloric acid the only ion of those in our sample to be extractable into diethyl ether is Fe^{3+}.

(*iv*) Cyanide ion reacts with most transition-metal ions but not with main-group metal ions to form very stable, water-soluble complexes. The metal ions may subsequently be decomplexed by boiling them with acid when volatile HCN is driven off, *provided that* adequate precautions are taken to avoid inhaling the *highly toxic* HCN.

(*v*) In a slightly alkaline medium, and in the presence of masking agents, dimethylglyoxime is a selective chelating agent for the extraction of nickel. Of the ions in our sample likely to interfere, Fe^{3+} is masked

$\longrightarrow$

SAQ 8.6h (cont.)

with tartaric acid, Co^{2+} either with CN^- plus H_2O_2 or formaldehyde, or with NH_3, Cu^{2+} with thiosulphate and Mn^{2+} with hydroxylamine hydrochloride.

(*vi*) In acid solution, ammonium peroxydisulphate is a powerful oxidising agent, oxidising Fe^{2+} to Fe^{3+}, Cr^{3+} to Cr^{6+} and Mn^{2+} to Mn^{7+}. Nitric acid and dichromate ion are milder oxidising agents, oxidising Fe^{2+} to Fe^{3+} only.

(*vii*) The Jones reductor (a zinc/mercury amalgam) reduces Fe^{3+} to Fe^{2+} and Cu^{2+} to Cu metal, whereas the Walden reductor (silver/silver chloride) reduces Fe^{3+} to Fe^{2+} and Cu^{2+} to Cu^+. Both reductors work on a column principle and the reduced solutions so obtained are free from excess of reducing agent. The Walden reductor works best in the presence of hydrochloric acid.

SAQ 8.6h

8.6.3. Ion-association Extraction System

The alternative approach to the extraction of metal ions is to use an ion-association extraction system. Here, we form an *ion-association complex*, normally between a large, bulky organic ion and a suitable counter-ion. The large, bulky ion may simply be added to the system as a component of a reagent, it may be formed by complex formation or it may be formed by solvation. The metal may be incorporated in this large ion or it may be incorporated in the counter-ion. The ionic species containing the metal, as well as the large ion may be cationic or anionic. These complexes are held together by relatively weak electrostatic forces, but are sufficiently stable to be efficiently extracted into organic solvents under appropriate conditions. The stability of these ion-association complexes is generally higher in the organic phase than in the aqueous phase.

These ion-association extraction systems often work efficiently under conditions where the neutral-chelate extraction systems are less than satisfactory.

∏ Under what conditions are neutral-chelate extraction systems generally most effective?

— When extremes of pH (especially low pH) are avoided.

— When the metal ion being extracted is not subject to hydroxide formation at relatively low pH.

— When the concentrations of the metal ions involved are not too high.

— When the metal ions involved are of the type to form stable complexes.

So,

(*a*) Ion association-extractions can be, and often are, made from strongly acidic media.

(*b*) Extractions of high-valency transition-metal ions are often effective, *because* these ions can be extracted from acidic media.

This is possible because the problems associated with hydroxide formation encountered with these ions at the pH values required for neutral-chelate extraction can be avoided.

(*c*) Extractions are available for alkali-metal ions. This is possible because we can have systems where the large bulky ion is the anion and the need to complex the cation is avoided.

(*d*) Extractions are applicable to wide concentration ranges of metal ion. Neutral-chelate extractions are normally effective only at low concentrations.

However, the chemistry of such systems is diverse and often very complex, because of the wide variety of conditions and concentrations met. Thus a systematic quantitative treatment of such systems is beyond our scope. So we content ourselves simply by looking at some representative examples of ion-association extraction systems.

A. *Extractions Involving No Chelation or Solvation*

Here, all we do is to add a bulky organic counter-ion to the system to form ion-association complexes with non-solvated/non-chelated cations or anions. These then can be quite small and relatively simple.

Here is an example of a bulky cation being used in this way.

$$(C_6H_5)_4P^+ + ReO_4^- \rightleftharpoons (C_6H_5)_4P^+ . ReO_4^-$$

Aqueous phase

- -

Trichloromethane

$$(C_6H_5)_4P^+ . ReO_4^-$$

$(C_6H_5)_4 As^+$ is another cation which can be used similarly. Other extractable anions include MnO_4^-, $HgCl_4^{2-}$ and $FeCl_4^-$.

Here is an example of a bulky anion being similarly used:

$$(C_6H_5)_4B^- + Cs^+ \rightleftharpoons (C_6H_5)_4B^-.Cs^+$$

Aqueous phase

- -

Nitrobenzene

$$(C_6H_5)_4B^-.Cs^+$$

B. Liquid Ion Exchanges

Certain acidic phosphate esters such as butyl dihydrogen phosphate ($C_4H_9OPO_3H_2$), dibutyl hydrogen phosphate (($C_4H_9O)_2PO_2H$) and di(2-ethylhexyl) phosphate are liquids, themselves immiscible with water, or they can be dissolved in inert diluents immiscible with water. If these liquids or solutions are equilibrated with aqueous solutions containing certain cations, the phosphate esters can exchange protons with the cations in the aqueous phase, chelating and solvating the cations. Lanthanum(III) and high-valency actinides are examples of such cations to be extracted. These systems are known as liquid cation exchanges.

Similarly, amines of high molar mass that are protonated in the presence of acid, or alternatively tetra-alkylammonium salts of high molar mass, can also exist as liquids immiscible with water, or can be dissolved in diluents immiscible with water. Likewise, if these liquids or solutions are equilibrated with aqueous phases containing certain metal ions present as complex anions, exchange of anions between phases can occur and the metal is extracted as an ion-pair with the tetra-alkylammonium cation. Examples of extractants are trioct-1-ylammonium chloride and Aliquat 336-S (tricaprylmethylammonium chloride). Anions extracted include species such as $FeCl_4^-$. These systems are known as liquid anion exchanges.

C. Chelate Systems

A good example is the extraction of Cu^+ ion with neocuproine into trichloromethane. A bulky organic chelate cation is formed as follows.

$$Cu^+ + 2 \text{ (Neocuproine} \equiv L) \rightleftharpoons [\ldots]^+ \equiv CuL_2^+$$

$$CuL_2^+ + NO_3^- \rightleftharpoons CuL_2^+ . NO_3^-$$

Aqueous phase
— —
Trichloromethane

$$CuL_2^+ . NO_3^-$$

Note that the neocuproine reagent differs from those reagents used in neutral-chelate extraction systems in that it has no acidic hydrogen atoms. The complex still has no polar groups or water of hydration but does have a residual charge, which must be neutralised by ion-association complex formation with the nitrate anion.

∏   Would you expect the extraction of copper(I) as its neocuproine complex to be suppressed at low pH as in neutral-chelate extraction systems?

Yes. Although we do not need to produce an anionic form of the ligand for chelation, the nitrogen atoms of the ligand could be protonated at low pH, thus preventing it from complexing with the metal ion.

D. Oxonium Systems

Here we have a very important group of systems in which metal ions, often in relatively high oxidation states, are extracted from acidic media into oxygen containing organic solvents such as ethers, ketones, or esters. The extraction involves conversion of the metal ion into a complex anionic form by reaction with the anion of the acid. Protons from the acid are then solvated by the oxygen containing extracting solvent. The complex anion and the solvated proton (the 'oxonium' ion) then form an ion-pair (ion-association complex), and this is the species that is extracted. This is a slight over simplification of the reactions involved in practice, since the solvation is sometimes complex. For example, some secondary solvation of the complex anion may be involved.

An excellent example of an oxonium extraction system is the extraction of iron(III) from a $6M$-hydrochloric acid into diethyl ether. The chemical equilibria involved are complex, but can be approximated as follows.

$$Fe^{3+} + 4\,Cl^- \rightleftharpoons FeCl_4^-$$

$$H^+ + 3\,(C_2H_5)_2O + n\,H_2O \rightleftharpoons [(C_2H_5)_2O]_3H^+ . n\,H_2O$$

$$[(C_2H_5)_2O]_3H^+ . n\,H_2O + FeCl_4^- \rightleftharpoons [(C_2H_5)_2O]_3H^+ . n\,H_2O . FeCl_4^-$$

Aqueous phase $\qquad\qquad\qquad\qquad\qquad\qquad\qquad$ (ion-pair)

- -

Diethyl ether

$$[(C_2H_5)_2O]_3H^+ . n\,H_2O . FeCl_4^-$$

Some secondary solvation of the $FeCl_4^-$ anion may also be involved.

Fig. 8.6g gives some indication of the scope of oxonium extraction systems by listing examples of metal ions extracted, acids used for complexing and oxygen-containing extracting solvents. Note the tendency for the metal ions to be in relatively high oxidation states. This fact, the fact that extraction takes place from a strongly acidic medium, and the fact that such extractions work well at high as well as low metal ion concentrations, make the range of applications of oxonium extraction systems complementary to that of chelate extraction systems.

Metals extracted (oxidation state in brackets) Cd(II), Sn(II), Co(II)	Fe(III), U(VI), Th(IV), Au(III), Sn(IV), Nb(V), Ta(V), Sb(III), Ga(III), Tl(III), Ce(IV), Zn(II),
Complexing Acids	HF, HCl. HBr, HI, HSCN, HNO_3
Extracting Solvents	Ethers, esters, alcohols, ketones, (Diethyl ether is the commonest choice)

Fig. 8.6g.   *Examples of Oxonium Extraction Systems*

∏   How might you expect the performance of oxonium extraction systems to change when the acid concentration is
changed?

The acid is present to ensure the formation of the complex anion of
the metal. Thus if the acid concentration is reduced, less complex
formation occurs and so the extraction is suppressed. Thus generally a high acid concentration is required. However, such extraction
systems may be more selective at lower acid concentrations because
then only the most stable complex anions are formed.

E. Extractions Involving Solvation/Coordination by Organic Com-
pounds

Certain ketones (eg 4-methylpentane-2-one), trialkyl phosphates (eg
tributyl phosphate) and phosphine oxides (eg tri-octylphosphine oxide), either used as extracting solvents or as organic reagents diluted
with an inert extracting solvent (eg kerosene) can bring about the
extraction of metal ions by direct coordination and/or solvation of
the metal ion or of the proton. Uranium(VI) can be extracted in
this way from a nitric acid medium by using tributyl phosphate
(TBP). The extracted species vary in composition depending on the
conditions, but include species such as $UO_2^{2+}(TBP)_2 . (NO_3^-)_2$ and

$H^+(TBP)_2 . UO_2(NO_3^-)_3$. Perhaps you can see some similarities between this type of system and the oxonium systems. The difference is the possible direct coordination of the metal ion and the use of inert diluents.

SAQ 8.6i

Match the following extractions with the following general extraction systems on a one-to-one basis.

Extractions

(i) Extraction of uranium(VI) into trichloromethane in the presence of 8-hydroxyquinoline(HOx). The $UO_2(Ox)_3^-$ ion is formed and the tetrabutylammonium cation is present as counter-ion.

(ii) Extraction of the permanganate anion into trichloromethane in the presence of the tetraphenylarsonium cation as counter-ion.

(iii) Extraction of gold(III) from $3M$-hydrochloric acid into diethyl ether.

(iv) Extraction of nickel(II) into trichloromethane at neutral to weakly acidic pH in the presence of 1-(2-pyridylazo)-2-naphthol.

(v) Extraction of lanthanum(III) into dibutyl hydrogen phosphate (a liquid).

(vi) Extraction of uranium(VI) into tri-isooctylamine hydrochloride (a liquid) from a sulphuric acid medium, in which the species $UO_2(SO_4)_2^{2-}$ is formed. $\longrightarrow$

**SAQ 8.6i
(cont.)**

General Extraction Systems

(*a*) Neutral-chelate extraction system.
(*b*) Ion-association extraction system involving
no chelation or solvation.
(*c*) Liquid-cation exchanger extraction system.
(*d*) Liquid-anion exchanger extraction system.
(*e*) Chelate-ion association extraction system.
(*f*) Oxonium extraction system.

Salting-out Techniques

Salting-out is a technique that can be used generally in solvent extraction, but it is especially effective for ion association extraction systems which is why we consider it here. The idea is that if you add a high concentration of an inert, neutral salt (such as NaCl, KCl, NaNO$_3$, or KNO$_3$) to the aqueous phase, then the distribution ratios for the extracted species can in favourable circumstances be significantly increased, thus improving the overall efficiency of extraction.

There are various reasons why salting out may be successful. Here are some of them.

(*a*) In extraction systems involving the formation and extraction of a complex metal anion (eg FeCl$_4^-$), then the addition of a high concentration of KCl will force the equilibrium in the complex formation to the right.

(*b*) The high background electrolyte concentration alters the activities of the extracted species in a manner favouring extraction.

(*c*) The salting-out agent is solvated by water molecules. Because of its high concentration, this means that a significant proportion of 'free' water molecules are effectively removed and the effective concentration of the extractable species in the aqueous phase is increased.

(*d*) The addition of the electrolyte to water reduces the dielectric constant of the aqueous phase. Stability constants for ion association complexes are inversely related to dielectric constant. Thus ion-pair formation is favoured.

∏ Salting-out techniques may be used in the extraction of organic compounds, but not all the above factors will then be operative. Which of the above four factors might you expect to be operative with respect to the extraction of organics?

Only factor (*b*) (concerning activities) and factor (*c*) (concerning available solvent) would be relevant in the extraction of organic compounds. Factors (*a*) and (*d*) are concerned with the formation of complex ions or ion-pairs.

8.7. SOLVENT EXTRACTION – CONCLUSIONS

We have now come to the end of our review of solvent extraction as a separation technique in chemical analysis. You should now have a useful knowledge of a valuable separation technique that is relatively easy to carry out and is applicable both to organic analysis, and to inorganic analysis especially when the analyte is a metal ion.

We have seen the technique used for preconcentration either on its own, when preconcentration factors need not be very high, or in combination with, say, evaporation to get higher preconcentration factors. We have also seen how we can, in favourable circumstances, use solvent extraction to isolate a single analyte or to simplify a matrix.

We have considered the basic practical techniques involved, and the factors that affect extractability and how we can improve the efficiency of extraction. We have also studied selectivity and means for improving this. One factor that was apparent was the importance of *chemical equilibria* in addition to the actual extraction technique in optimising extraction conditions. Because of this, the successful application of solvent extraction involves a good understanding of the chemical as well as the physical principles underlying the process.

SAQ 8.7a

Suggest outline procedures for carrying out the following separations and preconcentrations by solvent extraction. Write a paragraph or two about each, detailing the procedure you would adopt, and explain why you feel that the approach you have selected is a suitable one. Take your time over this question and don't try to do

$\longrightarrow$

SAQ 8.7a (cont.)

it from memory. In fact, I would encourage you to consult the references if possible and to do a little literature 'research' into the problems, although you can produce satisfactory answers from the material I have given you previously. In this respect, you might find reference 16 particularly helpful.

Do please also note that there are no single 'correct' answers to this SAQ and there may well be several quite acceptable approaches to any of these problems. So, if your response does not correspond with mine and you have researched it from a reference, don't worry, your method may well be appropriate. If you can think of more than one adequate approach, then discuss them both (or all of them!) In my responses, I have endeavoured to outline the general approach to the problem and then give one or two practical solutions together with comments about them.

(*i*)   Separation of nickel(II) and cobalt(II) ions present at low concentrations in an aqueous solution.

(*ii*)   Preconcentration by a factor of at least 100 of traces of cadmium(II), mercury(II) and lead(II) ions in a river water sample.

(*iii*)   Separation of iron(III) and chromium(III) ions in an acidic medium.

(*iv*)   Preconcentration of traces of halogenated pesticides (weakly polar organic compounds) in a water sample by a factor of about 1000, before gas chromatographic analysis.

SAQ 8.7a

SAQ 8.7b   We have seen that solvent extraction is a useful
means in chemical analysis for removing certain
substances quantitatively from an aqueous phase
into an organic phase or *vice versa.* We have
seen this applies especially for preconcentration
in trace analysis and for the separation of two
or more interfering substances. However, in order to get the best out of a solvent extraction it
is very important to ensure that all the experimental conditions for the extraction have been
carefully optimised. This applies not only to the
chemical composition of each phase, but also to
the actual physical way in which the extraction
is carried out. $\longrightarrow$

SAQ 8.7b (cont.)

I want you, therefore, to make two lists of chemical and physical factors that must be optimised either generally or in specific situations in order to get the best out of a solvent extraction procedure. The two lists should be under the following headings:

(*i*) *Factors affecting the overall efficiency of extraction* (ie factors affecting the magnitude of E for solutes of interest).

(*ii*) *Factors affecting the selectivity of an extraction* (ie factors affecting the magnitude of ΔE for two particular solutes).

Put your lists side by side and see if you can correlate points in each list. You won't always be able to do this, but I think you will find it instructive to note when this is possible and when it is not.

Try initially to prepare this list from memory without referring to the text, but if this proves impossible then refer to the text and previous SAQ's.

SAQ 8.7b
(cont.)

Objectives

On completion of Part 8 of this Unit, you should be able to:

● understand the basic steps that make up a simple, straightforward solvent extraction, and to describe typical liquid phases used therein;

● recognise the types of separation problem that can be usefully tackled by solvent extraction, and describe practical procedures;

● place a series of unit operations in the correct sequence to produce a viable solvent extraction procedure;

● identify potential sources of error arising practically in solvent extraction;

● understand that the principal factors that affect the extractability of a solute are solubility, solvation by the extraction solvent, and the effects of chemical equilibria;

● apply an understanding of pH control to optimise an extraction (clean-up) procedure for organic/biological samples;

● understand and define the terms 'distribution coefficient', and 'distribution ratio', D, and derive expressions for D for monobasic and dibasic organic acids;

- use these to show the dependence of D on pH for such species;

- understand and calculate 'percentage extraction' and to derive an expression relating it to D;

- demonstrate the increased efficiency of extraction by using multiple batch extraction techniques, and describe techniques of continuous extraction;

- describe selectivity of extraction in terms of the separation factor;

- find the optimum pH for the separation of two acids by solvent extraction;

- understand that metal ions are extractable only when in complexed forms, having appropriate properties, and that metal-ion extraction systems can be classified as neutral chelate and ion-association;

- describe a general model, including the various equilibria involved, for a neutral chelate extraction, and give examples of ions extracted and typical ligands;

- apply a general equation to predict the dependence of neutral-chelate extractions on pH and evaluate the selectivity of neutral-chelate extraction systems *via* pH control;

- describe the effects of hydrolysis and other complexing agents on pH-extraction curves;

- describe qualitatively the various types of ion-association extraction systems for metal ions;

- explain the term 'salting out';

- devise a multistage solvent extraction procedure for separation of the metal ions in a mixture;

- propose methods for carrying out a selection of separations based on solvent extraction, and indicate factors affecting its efficiency and selectivity.

9. Ion Exchange

In this Part, we shall study the important separation technique of *ion exchange*. This, as its name implies is concerned with the separation of ions and one point that we need to get clear right from the start is that in all ion exchange separations, *the species involved in phase transfer must be ionic*. It can be used for simple separations of species with very different properties, for preconcentrations, and, in the form of ion exchange chromatography, for carrying out more complex separations of similar types of ions. It is used extensively in inorganic analysis, but organic species are amenable to ion exchange as well, so long as they are capable of existing as ions. Ion exchange is the most powerful separation technique available to us for separating species as ions.

Our plan of campaign will be to start by considering fairly generally the *basic principles* and *practical aspects* of the technique. We shall then go on to study the *exchange process in more detail* and the various types of *material used as ion exchangers*. We shall follow this with a fairly brief account of *quantitative aspects* and of the *selectivities of ion exchangers*. The *effects of chemical equilibria* on ion exchange processes come next, and we shall finish with some important *applications of ion exchange in analysis*.

As a general phenomenon, ion exchange has been known for a very long time. Many industrial and biological processes involve ion exchange. For instance the exchange of trace metal ions between plants and the soil in which they grow, is believed to involve ion exchange with the soil acting as a cation exchanger. It has even been suggested that the earliest recorded instance of the use of ion exchange is to be found in the Bible – Exodus Chapter 15, verses 23–25, in which one reads that Moses, leading the Israelites through the desert comes to the waters of Marah, which prove to be too bitter to drink. He is told by God to cast a tree into the waters which then become sweet and fit to drink. It has been suggested that the waters were contaminated with magnesium sulphate which was scavenged from the water by the tree, some component of which was acting as an ion exchanger (perhaps carboxylate groups resulting from partial oxidation of the cellulose).

Ion exchange chromatography is an important technique for the resolution of complex mixtures of ions. Ion exchange also contributes to the mechanism of action of *ion-selective electrodes* (eg the glass electrode for pH measurements), a topic dealt with in another *ACOL* Unit.

9.1. BASIC PRINCIPLES

In any ion exchange process we have a two-phase system in which the phases are in contact with each other, and ions can be exchanged between phases, so that an equilibrium is established. This is in accordance with our general model of a separation as described in Part 7. The two phases are as follows.

(*a*) A solution of the ions in a suitable solvent. Most probably this will simply be an aqueous solution, but occasionally we come across aqueous organic solvents. The use of pure organic solvents in ion exchange is rare but not unknown. Unless otherwise stated we assume the solution to be aqueous.

(*b*) A solid material in particle form (although ion exchange papers can be used in paper chromatography). This material is usually a polymer and must have a very low solubility in the solvent (probably water). Its most important feature, though, is that it must have in its structure, ionised groups chemically bonded to the polymer matrix. These groups, which are firmly fixed therefore to the polymer, are known as *ionogenic groups*, and it is at the sites of these groups that the ion exchange is considered largely to occur. We refer to this solid material as the ion exchange material or *ion exchanger*. The ionogenic groups may be positively or negatively charged depending on the particular ion exchanger.

Associated with the ionogenic groups in the ion exchanger will be ions of opposite charge in just sufficient number to secure overall electrical neutrality. These are known as *counter-ions*. These counter-ions are not bound strongly to the ion exchanger, and so when the latter is brought into contact with a solution containing ions, *the ions in the solution of like sign to the counter ions may reversibly exchange with the latter and an equilibrium may be set up*. The basic objective of most ion exchange processes stems directly from this reversible exchange of ions of like sign between the ion exchanger and the solution, to replace certain ions in solution by the counter-ions of the ion exchanger, possibly in order to retain quantitatively a particular ion as counter-ion in the ion exchanger.

Ion exchange processes are classified into two distinct groups as follows:

(*a*) *cation exchange separations* involving cation exchangers, in which cations are exchanged between the ion exchanger and the solution;

(*b*) *anion exchange separations* involving anion exchangers in which anions are exchanged between the ion exchanger and the solution.

∏   Would you expect the ionogenic groups on a cation exchanger to be positively or negatively charged? Likewise for an anion exchanger?

The ionogenic groups should be of *opposite charge* to the ions being exchanged, to preserve electrical neutrality. Therefore, since a cation exchanger is involved in the exchange of cations, which are positively charged, then the ionogenic groups on a cation exchanger must be *negatively charged*, ie anionic.

For exactly the same reasons, the ionogenic groups on an anion exchanger should be *positively charged*, ie cationic.

Let us now see how all this works in practice by considering an actual example of the use of ion exchange in a familiar situation, *viz, domestic or industrial water-softening*, in which calcium and magnesium ions in tap water are replaced by sodium ions by ion exchange. The reasons for doing this are to prevent the deposition of insoluble calcium and magnesium carbonate in, for example, boilers, pipes, heat-exchangers, and to improve the efficiency of anionic detergents (eg soaps) by preventing the precipitation of insoluble calcium and magnesium salts (the scum formed when soap is used with hard water).

∏   Will this be a cation or an anion exchange process?

This will be a cation exchange process, because we are replacing one type of cation (calcium and magnesium ions) in the water by another (sodium ions).

From now on, let us for simplicity assume that only calcium ions, rather than calcium and magnesium ions are involved in the exchange process with sodium ions.

Zeolites are common and suitable ion exchangers for use in this process. These are special types of natural or synthetic clay materials that can be considered as being chemically a kind of polymeric aluminosilicate. Having a macromolecular structure, they are water-insoluble, but at the same time they are anionic in nature and thus accord with our description of a cation exchanger. For use as a water softener we need a zeolite that contains sodium ions as counter-ions (in the sodium form, as we say). If such a zeolite is equilibrated with tap-water containing calcium ions, then the following equilibrium will be set up.

$$\text{Zeolite-2 Na}^+_{(s)} + \text{Ca}^{2+}_{(aq)} \rightleftharpoons \text{zeolite-Ca}^{2+}_{(s)} + 2\,\text{Na}^+_{(aq)}$$

In practice, the zeolites used have a stronger affinity for calcium ions than for sodium ions, and so the above equilibrium lies well to the right. So, a stream of water having passed through a cartridge packed with zeolite particles in the sodium form, will emerge with the calcium ions replaced by sodium ions.

∏   This process of ion exchange will not continue indefinitely. Why not?

The zeolite contains only a limited number of sodium ions and as ion exchange proceeds these are replaced by calcium ions, until eventually all the zeolite will have been converted from the sodium into the calcium form. Then there will be no more sodium ions available for exchange. We then say that the ion exchanger is 'exhausted'.

When the ion exchanger is exhausted we can regenerate it by reversing the above process. This is achieved by passing a highly concentrated solution of sodium ions over the zeolite to force the above equilibrium to the left. When this has been achieved water-softening may be continued, and the cycle repeated as often as needed. Sodium chloride is normally used for this regeneration and this explains why water softeners consume large amounts of common salt.

In the above example we considered an inorganic ion exchanger (a zeolite). In fact, in chemical analysis inorganic ion exchangers are rather rare and organic materials are much more commonly used. These are often in the form of *resins* (cross-linked polymers) onto which ionogenic groups have been specifically chemically bonded. In connection with such materials, we come across the terms, *cation exchange resin* and *anion exchange resin*. From now on, ion-exchange materials can be assumed to be resins unless we say otherwise, and we shall talk freely about resins as ion exchange materials.

Another term that we may need to use in discussion of ion exchange is *co-ion*. This is the ion present in the solution, whose charge is of opposite sign to that of the counter-ion, and therefore of the same sign as that of the ionogenic group. It does not therefore take part in the ion exchange process. In water-softening , co-ions would be anions such as bicarbonate, sulphate, chloride, etc.

Π   Define the following terms.

(*i*)   Ionogenic group,
(*ii*) Counter-ion,
(*iii*) Cation exchange resin and anion exchange resin,
(*iv*) Co-ion.

(*i*)   An ionogenic group is an ionised group chemically bonded to (or part of) an insoluble polymeric material, the two together forming an ion exchanger. It is cationic for an anion exchanger and anionic for a cation exchanger.

(*ii*) A counter-ion is the ion associated with the ionogenic group and whose charge is of opposite sign to that of the ionogenic group. It is not strongly bonded to the resin, and can reversibly exchange with other ions of the same type (ie cations or anions) in solution.

(*iii*) A cation exchange resin is a cross-linked organic polymer insoluble in water, having anionic groups bound to it, and therefore having exchangeable cations associated with it as counterions. An anion exchange resin has chemically bound cationic groups and exchangeable anions associated with it as counterions.

(*iv*) Co-ions are ions present in solution whose charge is of opposite sign to that of the counter-ions and of the same sign as that of the ionogenic groups. They do not take part in the ion exchange process.

SAQ 9.1a   The ion exchange process is briefly described in the paragraph below. Various words and phrases, however, have been omitted. Fill in the missing words or phrases in the paragraph from the list below. Each word or phrase in the list may be used once, more than once, or not at all.

$\longrightarrow$

SAQ 9.1a
(cont.)

'Ion exchange consists of equilibration between a(n) ... 1 ... and a(n) ... 2 ... material which has ... 3 ... chemically bonded to it, which is ... 4 ... in water or other solvents, and which has ... 5 ... associated with it. Ions in the ... 6 ... of the ... 7 ... sign as the ... 8 ... may exchange ... 9 ... with the ... 10 .... The commonest materials used for this purpose analytically are ... 11 ... onto which ... 12 ... have been chemically bonded. They come in two complementary forms: ... 13 ... containing negatively charged ... 14 ... and ... 15 ... containing positively charged ... 16 ... bonded to them. The ... 17 ... is most commonly aqueous. The technique is generally suitable for the separation and preconcentration of ... 18 ... in solution'.

List of words and phrases

inorganic materials   inorganic groups
highly polar molecules   organic polymers
ions   complexing
same   reversibly
solid polymeric   opposite
organic   aqueous
anion exchange resins   insoluble
cation exchange resins   counter-ions
electrolyte solution   soluble
functional groups   chelating
irreversibly

NB There is more than one correct answer.

SAQ 9.1a

Accounts of ion exchange and ion exchange chromatography can be found in most good general textbooks on chemical analysis. For elementary descriptions, see reference 1, p 130–138 and reference 3, p 143–151. For a far more comprehensive and complete treatment of the subject, I have included two specialised texts (references 18 and 19).

9.2. PRACTICAL TECHNIQUES IN ION-EXCHANGE

Ion exchange can be carried out by using various practical techniques, and we must decide which will be most appropriate in a particular analytical situation. These practical techniques correspond essentially to the modes of separation that we discussed earlier in Section 7.5.

We shall always assume that before we attempt to carry out an ion exchange separation, the ion exchange resin, which might typically be supplied in a dry state as beads (particles) of a particular mesh size, has been properly prepared for use. This might involve some or all of the following processes.

(*a*) Allowing the resin to come to equilibrium with the solvent (normally water) with which it is to be used. Dry resin swells considerably on contact with water, because water molecules permeate the resin beads, causing the polymer lattice, and therefore the bead itself to expand.

(*b*) Washing the resin with (*i*) an acid and (*ii*) an alkaline solution. This is to remove trace impurities that are always associated with the resin and may perhaps be troublesome later.

(*c*) Equilibrating the resin with an aqueous electrolyte with which it is subsequently to be used. One consequence of failing to do this is that the degree of swelling which depends on the exact composition of the electrolyte solution, may be wrong, leading to problems if the resin is packed in a column – extremely likely in practice. This is particularly important in ion exchange chromatography.

(*d*) Converting the resin into the correct form. The 'form' of an ion exchange resin refers to the counter-ions present. Thus, for example we can talk about cation exchange resins in the hydrogen-ion or sodium-ion form and anion exchange resins in the chloride-ion or hydroxide-ion form (sometimes the word 'ion' is missed out). If the resin is not supplied in the correct form (commercial resins are commonly supplied in the hydrogen- or chloride-ion form) then it must be packed into a column (in the swollen state) and a concentrated solution of the ion into which form the resin is to be converted must be passed through the column continuously until the effluent from the column is found to be free of the original counter-ions.

∏   How would you convert a cation exchange resin supplied in the sodium form, into the hydrogen form?

After first ensuring that the resin is in the swollen state by equilibrating it with water, pack the resin into a column and pass concentrated hydrochloric acid (selected because it is less reactive in other respects than most other mineral acids) through the column. Test the solution as it emerges from the column for the presence of sodium ions. The concentration of sodium ions will first increase, then pass through a peak and finally diminish as the conversion nears completion. Keep the acid flowing through the column until the effluent is essentially free of sodium ions. Finally remove excess of hydrochloric acid by washing the column with water (de-ionised of course!) or other electrolyte solution, as appropriate.

The order in which you might carry out these preliminary operations is not necessarily the order in which I have discussed them. The important points to remember are that, before a resin is used for ion exchange it should be:

(*a*) properly swollen in the medium in which it is to be used,

(*b*) free of impurities and excess of ions,

(*c*) in the correct form,

(*d*) in equilibrium with an electrolyte solution (in ion exchange chromatography).

Let us now look in more detail at how we could actually carry out some separations by ion exchange.

9.2.1. Batch Separations

We could take a weighed amount of resin in a particular form and equilibrate it in a closed vessel with a measured volume of solution containing the ions that we wish to exchange. If the resin showed a remarkably higher affinity for the ions in the solution than for its own counter-ions, then you might expect the following equilibrium (for a cation-exchange resin),

$$R^-M^+_{(s)} + N^+_{(solution)} \rightleftharpoons R^-N^+_{(s)} + M^+_{(solution)}$$

to lie completely on the right-hand side. However, in practice, the nature of most ion exchange resins is such that this is unlikely to happen. What is much more likely is that the system will reach an equilibrium somewhere between, with both ions present in both phases to different extents. In other words, we have not achieved a clean quantitative exchange of ions. Nevertheless, this simple batch experiment does have its uses, eg in theoretical studies for obtaining information about affinities and selectivities of resins. Selectivity coefficients (the equilibrium constant for the ion exchange process) and distribution coefficients, can be obtained in this way. The experiment is completed quite simply by filtering off the resin and analysing a portion of the solution for both ions.

Separations might possibly be made quantitative by use of a multiple-batch mode, ie equilibrating with fresh portions of resin until all the ion in question has been removed from the solution, but in practice there is a much better and easier way to proceed as we shall now see.

∏ When would you be most likely to use a batch-mode separation in ion exchange?

When you are interested in investigating the selectivity of the resin, rather than in carrying out an analytically useful separation. Equilibrium constants can be determined by measuring the concentration of each ion in the solution phase at equilibrium. However, an analytically useful quantitative exchange is unlikely (though not impossible) under such conditions.

9.2.2. Trapping Techniques (by using a Column)

In this technique, we pack our ion exchange resin into a column and slowly pass the solution containing the ions to be exchanged through the column continuously at a steady rate. Under these conditions we find that we can much more easily get a quantitative exchange to occur, even if the exchange is not quantitative under batch-mode conditions. All that is required is that the equilibrium does not lie *too* far over on the left-hand side.

Consider the equilibrium once again:

$$R^-M^+_{(s)} + N^+_{(solution)} \rightleftharpoons R^-N^+_{(s)} + M^+_{(solution)}$$

Under batch-mode conditions of operation, as we move towards the equilibrium, the concentrations of the species on the left-hand side of the equilibrium decrease and those on the right-hand side increase until their ratio equals the equilibrium constant.

In the trapping-mode experiment on the other hand as the solution moves down the column, it continually meets fresh resin. Thus, in effect the concentration of $R^-M^+_{(s)}$ does not alter. This means that for the ratios to become equal to the equilibrium constant, the equilibrium is continually forced over to the right-hand side and therefore to completion, even if that would not have been true of the batch-mode experiment. You can imagine it as a series of batch equilibria, the solution equilibrating with the resin at the top of the column, moving on to a fresh band of resin, equilibrating again, then moving on again, and so on, although in reality the process is continuous.

The trapping technique, in which you use a column of ion exchange resin and pass the solution continuously through it, would be the technique to use whenever you wished to replace an ion in solution quantitatively by the counter-ion of the resin.

∏   Which practical technique, of those that we met for use in solvent extraction, represents the closest analogy to the trapping technique in ion exchange?

The continuous extraction system described in Section 8.5 for which the apparatus is illustrated in Fig. 8.5c. In this system, solvent is boiled under reflux and a continuous stream of pure solvent passes through the aqueous phase, allowing quantitative extraction of species with unfavourable distribution coefficients. Again, the effect is equivalent to a very large number of batch processes, extractions in this case.

9.2.3. Ion-exchange Chromatography

Ion exchange can be used in classical low-pressure liquid (column) chromatography, modern high-pressure (high performance) liquid chromatography, as well as in paper and thin-layer chromatography. It is the commonest way of separating ions chromatographically (other ways being by ion-pair and partition chromatography).

In an ion-exchange chromatographic separation based on the use of a column, the column is packed with a resin, consisting of very small particles (for good resolution), containing an ion for which the resin does not have too high an affinity. The mobile phase then consists of a solution of these same ions (so you have an excess of a weakly retained ion in both phases). The mixture containing the ions to be separated is placed on top of the column in a narrow band and eluted through the column with the mobile phase. Reversible exchange of ions will occur and each type of ion will move through the column at a rate inversely related to the degree of affinity between the resin and that ion, thus leading to a separation. (Later on, we shall quantify the degree of affinity that a resin shows for one ion over another *via* a selectivity coefficient, derived from the equilibrium constant for the reaction shown in Section 9.2.1).

∏   What practical ion-exchange technique would you use in each of the following procedures?

   (*i*)   Separation of a mixture of chloride, bromide and iodide ions.

   (*ii*) Removal of all metallic cations from a large volume of aqueous solution and their replacement by H^+ in solution.

   (*iii*) Investigation of the selectivity characteristics of a newly prepared type of ion-exchange resin.

The techniques you would use are as follows:

(*i*)   Ion-exchange chromatography,

(*ii*) Trapping technique (by using a column),

(*iii*) Batch-mode experiments.

SAQ 9.2a	*Three* separations based on ion-exchange follow. Describe each separation as batch mode, trapping technique, or ion-exchange chromatography and indicate under what circumstances each might be useful (ie what do you think might have been the objective of the experiment?). Can you suggest how each experiment might have been finished?

> (*i*) A known volume of solution containing a known concentration of a metal cation was shaken in a bottle together with a known amount of cation-exchange resin in the H^+ form. Then the solution was separated from the resin by filtration.

> (*ii*) A column was packed with anion-exchange resin in the nitrate form and a solution of nitrate ions was continually passed through the column. When equilibrium was established a small volume of solution containing a mixture of chloride, bromide and iodide ions was placed on top of the column and the flow of nitrate-ion solution through the column was continued.

> (*iii*) A large volume of water containing a variety of trace metal ions was passed through a small column containing a cation-exchange resin in the H^+ form. This resulted in the complete retention on the column of all the metal ions and the release into the water of an equivalent amount of hydrogen ions.

SAQ 9.2a

9.3. ION-EXCHANGE MATERIALS AND THEIR PROPERTIES

In this section we shall consider the desirable properties of an 'ideal' ion-exchange material. We then consider some of the actual measurable parameters of real ion-exchange materials. We have already noted that in chemical analysis, the most important ion-exchange materials are resins and are based on organic polymers, into which hydrophilic ionogenic groups have been incorporated by one means or another.

We can take this one stage further and note that there is one type of organic resin that has been used much more frequently than any other. This is the polystyrene-divinylbenzene (PSDVB) resin which has almost ideal properties for an ion-exchange resin. It is a cross-linked polymer. Styrene has the chemical formula (I),

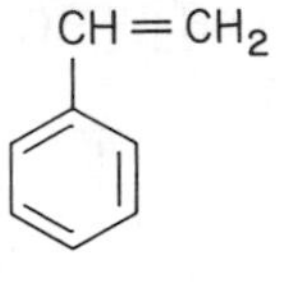

$$CH = CH_2$$

(I)

and polymerises to give a continuous chain polymer (II),

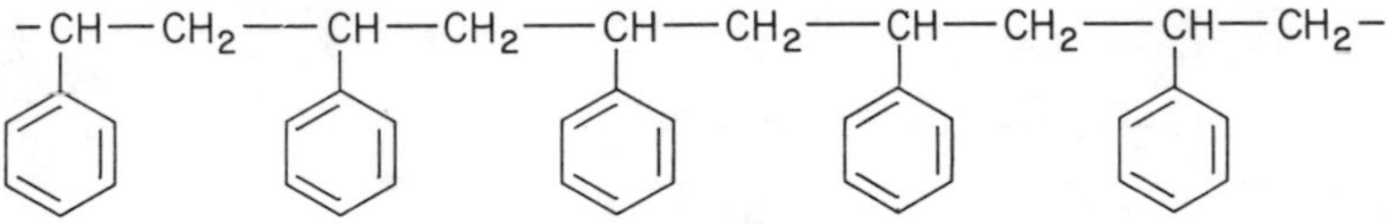

$$-CH-CH_2--CH-CH_2--CH-CH_2--CH-CH_2--CH-CH_2-$$

(II)

divinylbenzene has the formula (III).

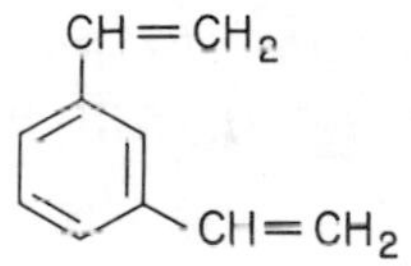

$$CH = CH_2$$
$$CH = CH_2$$

(III)

(In practice commercial divinylbenzene consists of a mixture of isomers, but this does not matter much to us.)

Incorporation of a proportion of *m*-divinylbenzene into the styrene monomer before polymerisation results in the production of a cross-linked polymer in which the continuous chains of styrene units are linked together by divinylbenzene cross-linking groups. The general structure of a PSDVB resin is shown in Fig. 9.3a. The cross-linking is essential in order to produce an insoluble polymer which will retain its resin structure in the presence of an electrolyte solution.

After the PSDVB resin has been prepared, ionogenic groups are incorporated into the benzene rings by substitution reactions. One might expect the final resin to have on average, about one ionogenic group per benzene ring in its structure (normally in the para-position), given the usual preparative method. We shall consider the subjects of cross-linking and types of ionogenic groups that can be

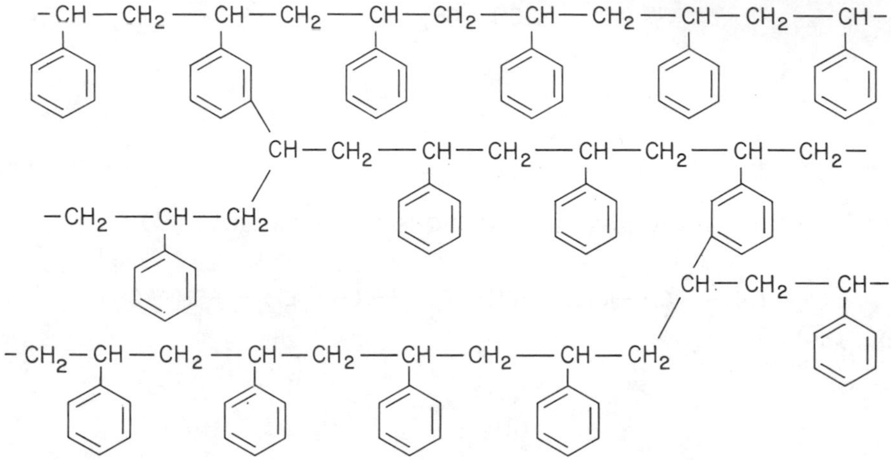

Fig. 9.3a. *General structure of a polystyrene-divinylbenzene (PSDVB) resin before incorporation of ionogenic groups*

incorporated later in this section. Before that, let us consider some of the properties of a resin that are relevant to its performance in ion exchange.

∏   Give as many properties as you can of a cation/anion-exchange resin that might be relevant to its performance in ion exchange.

Here are the factors that I shall consider:

(*i*)   selectivity/affinity towards different types of cation/anion,

(*ii*)   pH-range over which the ionogenic group is ionised,

(*iii*)   exchange capacity of the resin (ie how many ions it can exchange per gram of resin),

(*iv*)   average particle size of the resin beads,

(*v*)   chemical and thermal stability of the resin,

(*vi*)   chemical purity of the resin/monofunctionality of the ionogenic groups,

(*vii*)   physical rigidity of the resin beads,

(*viii*)   solubility of the resin in the solutions used for ion exchange,

(*ix*)   speed of equilibration.

Don't worry if your list was much shorter than mine, because we shall consider each of these in a little more detail in the next part of this section.

9.3.1.   Properties of a Resin Relevant to its Performance in Ion-exchange

Selectivity/Affinity towards different types of Cation/Anion

Some resins may show a high, non-selective affinity to all cations/anions and may thus be suitable in general-purpose applications. Others may show a much lower affinity but greater selectivity, and thus be useful in certain applications only. Yet others may show high affinity for some ions but not for others. Specificity towards a particular ion, although not unknown, is not usual in ion exchange, at least from the action of the resin alone. Choice of ionogenic group and degree of cross-linking (as well as the composition of the electrolyte solution) are important factors here.

pH Range over which the Ionogenic Group is Ionised

The anionic groups incorporated into a cation-exchange resin may be derived from strong acids, in which case they will be strongly ionised at all pH values, or from weak acids when they will be ionised only at high pH. Ion-exchange properties will be exhibited only at those pH values at which the ionogenic groups are ionised. This leads to the idea of a *strongly acidic cation exchanger* and a *weakly acidic cation-exchanger*. The former are used for general-purpose work, whereas the latter show selectivity towards strongly basic cations, and have the advantage in that retained cations can be reliably released by lowering the pH to a range in which the ionogenic groups are no longer ionised. An exactly analogous distinction applies to *strongly basic anion-exchange* and *weakly basic anion-exchangers* except that the latter, which are selective towards strongly acidic anions, display ion-exchange properties only at high pH-values.

Exchange Capacity of the Resins

This is a quantitative measure of the number of exchangeable counter-ions that the resin can accommodate and is normally quoted in units of *milliequivalents* (*Meq*) *of exchangeable counterions per gram of dry resin in the hydrogen or chloride form.* Typical theoretical values are in the range 2 to 10. Meq g^{-1}, with practical values perhaps slightly less than the theoretical because of non-ideal operating conditions. The actual value will depend on the properties of the resin, but high values are generally desired, other things being equal.

Resin-bead particle size

The resin is supplied in bead form of controlled particle size. The smaller the beads the more rapid is equilibration and generally ion-exchange behaviour is nearer to ideal. In column work, however, a high pressure may be needed to maintain an adequate flow-rate when beads of small particle size are used.

Stability of the Resin

This should be high with respect to extremes of heat, pH, oxidising/reducing conditions, etc.

Solubility of the resin

This should be zero in any electrolyte or other solution used in any ion-exchange application.

Chemical Purity of the Resin

As far as PSDVB resins are concerned, residual monomer may

cause problems in certain applications. Also monofunctionality of the ionogenic groups is important (ie all the ionogenic groups should be the same). This is not a problem with PSDVB resins where the groups are introduced by selective substitution reactions.

Physical Rigidity of the Resin Beads

It is desirable that the beads possess adequate physical rigidity, especially if high pressures are to be applied to the system.

Speed of Equilibration

Different types of resin may reach equilibrium with respect to ion exchange in different times. Rapid equilibration is desirable.

SAQ 9.3a	Answer the following questions about the desirable properties of ion-exchange resins.
	(*i*) What sort of selectivity/affinity for ions would you look for in a resin to be used for general preconcentration purposes?
	(*ii*) You are looking for an anion exchange resin suitable for separating anions of strong acids from those of weak acids. What *particular* property would you look for in your chosen resin?
	(*iii*) A new ion-exchange resin has been prepared with an exchange capacity of over 500 micro-equivalents per gram of dry resin. What comment would you make about this figure?
	(*iv*) Solutions resulting from ion-exchange separations by using PSDVB resins appear to $\longrightarrow$

SAQ 9.3a
(cont.)

show a very small absorbance in the ultra-
violet around 250 to 300 nm, which cannot
be explained by any of the ions present.
What is the cause of this absorbance? (NB
This assumes some knowledge of uv spectroscopy. If you have no such knowledge,
start to read the response).

9.3.2. Interaction Between Dry Resin and Electrolyte Solutions

We shall now consider in a little more details what happens when a
dry resin is brought into contact with water or an aqueous solution
containing exchangeable ions.

∏   Two samples of an organic polymer containing ionogenic
       groups and therefore possessing a high affinity for water, are
       brought into contact with water. The first polymer consists
       of continuous chains with no cross-linking. The second poly-

∏ mer also consists of chains, but with a certain degree of cross-linking. What do you think would be the outcome in each case, and what would be the implications in ion-exchange?

The first polymer consisting simply of chains, has no permanent rigid structure and therefore the chains could disperse, with the result that the polymer might dissolve in the water. Such a polymer would be useless for ion-exchange purposes.

However, for the second polymer the chains are to a certain extent 'fixed' to one another *via* the cross-linking groups, and are not free to disperse. The resin has a semi-rigid three-dimensional structure that can be destroyed only by breaking chemical bonds. Such a resin would not dissolve, and would therefore be potentially useful in ion exchange.

The above exercise demonstrates the need for cross-linking in poly mers that are to be used in ion-exchange. However, it tells us nothing about what happens when a cross-linked resin with ionogenic groups comes into contact with water. Remember, this resin has a high affinity for water and would dissolve were it not for the fact that it has this semi-rigid three-dimensional cross-linked structure holding it together. In fact, water molecules penetrate the resin bead causing the latter to *swell*. The chains of polymer are forced apart and the resin acquires a more open porous structure. Water molecules may pass freely from the external solution into the resin beads and indeed throughout them.

What happens if there are ions present in the aqueous solution? These are also able to penetrate the resin beads by entering the pores, and indeed it is necessary for them to do so if ion exchange is to occur throughout the resin bead and not just on its surface, which, in practice, is what happens. Fig. 9.3b shows this diagrammatically.

∏ Not all ions may penetrate the resin bead freely. What type of ion would you expect not to be able to penetrate the resin structure?

Large ions would be excluded from the resin structure because their dimensions would prevent them from entering the pores of the resin.

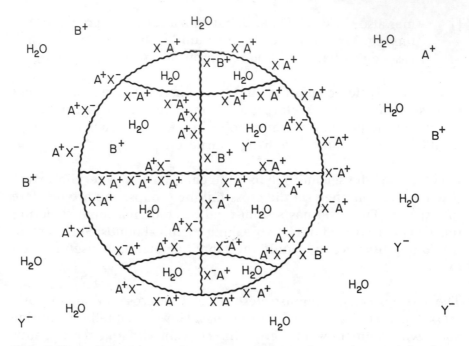

Fig. 9.3b. *Diagrammatic representation of a resin particle showing cross-linking leading to an open porous structure. Note free access of water molecules and other ions into the structure, and some exchange between A^+ and B^+ counterions. (X^- is the ionogenic group, Y^-, the co-ion). This is not to scale, the pores having been drawn large in relation to the bead size*

If you had suggested that only counter-ions and not co-ions would penetrate the resin bead, remember that the bead is electrically neutral, and so ions of both signs should have equal chances of penetrating the resin.

I hope that you can now visualise the resin bead in the ion-exchange process as an extension of the external solution containing water molecules and ionic groups or ions of each sign, whether these are ionogenic groups attached to the resin, counter-ions (perhaps of various types, one counter-ion being exchanged for another) and co-ions to preserve electrical neutrality. Look back at Fig. 9.3b and think about this.

This leads us to describe our two phases in ion-exchange as:

(*a*) *the resin phase*, referring to the resin and all species to be found within the pores of the resin, whether associated with a specific ionogenic group or not;

(*b*) *the external phase* referring to the solution (solvent and ions) found outside the resin beads.

9.3.3. Degree of Cross-linking

The extent to which the resin is cross-linked has several very important effects on its properties. Increased cross-linking produces a more rigid resin structure with smaller pores for the ions to penetrate.

In general practice, increasing the degree of cross-linking has the following effects:

(*a*) decreased swelling of the resin in contact with water,

(*b*) a consequent higher exchange-capacity of the resin on a wet basis,

(*c*) decreased affinity for ions in general, due to decreased penetration by ions into the resin phase,

(*d*) slower equilibration for the ion-exchange process,

(*e*) increased physical stability and rigidity of the resin beads,

(*f*) increased selectivity of the resin.

This last arises because the decreased pore-size means that the affinity of the resin towards larger ions decreases more than that towards smaller ions. Hence we observe an increased selectivity towards ions of smaller size.

Π   How might you control the degree of cross-linking of a PS-DVB resin in practice?

By controlling the ratio of divinylbenzene to styrene used in manufacture.

For PSDVB resins the degree of cross-linking is specified by the proportion of divinylbenzene used in the manufacture of the resin. Typical values fall in the range 2% to 20% of divinylbenzene. For a general-purpose resin 8% of divinylbenzene is the most common choice. Higher proportions of divinylbenzene would lead to more selective resins, whereas lower proportions would produce resins suitable for the exchange of larger ions (eg organic ions).

SAQ 9.3b	State whether each of the following statements is *true* or *false*:

(*i*)   Since ion-exchange resins are invariably solid materials of extremely low solubility with a semi-rigid structure, exchange processes must of necessity be restricted to the surfaces of the resin particles only.

(*ii*)   For a cation-exchange resin, cations are the only species present in the resin phase (besides the resin itself), since the latter contains anions as ionogenic groups.

(*iii*) A PSDVB resin containing 20% of divinylbenzene would show a greater affinity towards small ions than would one containing 8% of divinylbenzene.

(*iv*)   Ion-exchange is possible between two liquid phases without the presence of a solid ion-exchange material.

SAQ 9.3b

9.3.4. Ionogenic Groups in PSDVB and Related Resins

In the preparation of PSDVB resins, after the production of a cross-linked polymer by the copolymerisation of styrene and divinylbenzene, we need to introduce ionogenic groups normally by substitution. One would expect to produce a resin containing on average, about one ionogenic group per benzene ring in the polymer. We can classify the resins that are produced according to the type of ionogenic group introduced. There are essentially five types of ionogenic group, and we have already met, in principle, four of them.

∏   Name the four types of ionogenic group that we have met in principle so far.

(*i*)   Strongly acidic cation-exchangers.

(*ii*)   Weakly acidic cation-exchangers.

(*iii*) Strongly basic anion-exchangers.

(*iv*) Weakly basic anion-exchangers.

as described in Section 9.3.1.

The fifth type is the chelating ion-exchange resin in which the chelate effect is used in conjunction with ion exchange to produce resins with a high affinity and selectivity towards certain types of metal ion.

We now consider each of these five types of ion-exchange material in turn with special reference to PSDVB resins or to similar types of resin as appropriate.

(*a*) *Strongly acidic cation-exchange resin*

These contain anionic groups that are ionised over a wide range of pH. The ionogenic groups almost universally used are *sulphonate ion* $(-SO_3^-)$ *groups*, produced by treating the resin with chloro-sulphonic acid, when on average one $-SO_3^-$ group is introduced (in the para-position) per benzene ring. Because the sulphonic acid groups are *strongly acidic*, the group remains ionised at all normal pH values (cf H_2SO_4).

> *Sulphonated PSDVB resins (ie those containing sulphonate ion groups) are used as general-purpose cation-exchange resins and possess a high affinity for most cations.*

(*b*) *Strongly basic anion-exchange resins*

These contain cationic groups that are ionised over a wide range of pH. The ionogenic groups most commonly used are *quaternary ammonium* $(-CH_2NR_3^+)$ *groups*. The resin is first chloromethylated, and then the $-CH_2Cl$ groups (about one per benzene ring mainly in the para-position) are allowed to react with say, trimethylamine, to produce $-CH_2N(CH_3)_3^+Cl^-$ groups. A quarternary ammonium hydroxide is strongly basic and the chloride also remains ionised at all normal pH values.

> *PSDVB resins containing quaternary ammonium groups are used as general-purpose anion-exchange resins.*

(*c*) *Weakly acidic cation-exchange resins*

These contain ionogenic groups derived from the anions of weak acids and as such are ionised (and thus demonstrate ion-exchange properties) only at relatively high pH values. A common example of such a group is the $-CO_2^-$ *group* (existing as undissociated $-CO_2H$ at low pH values). This resin in fact is not normally a PSDVB resin but is formed directly by copolymerising *acrylic acid* ($CH_2{=}CH$ CO_2H) with divinylbenzene. These resins are more selective than the strongly acidic cation-exchange resins and show preference for the more basic cations. They also have a strong affinity for hydrogen ions, and so washing them with dilute acid will be very effective in removing any retained cations.

(*d*) *Weakly basic anion-exchange resins*

These are prepared in exactly the same way as the strongly basic anion-exchange resins, except that dimethylamine, methylamine or even ammonia, for instance, is used in place of the trimethylamine, so we obtain a resin with *amino-groups* (*primary, secondary, or tertiary*) rather than the quaternary ammonium groups. These amino-groups will be protonated (and thus act as selective anion-exchangers) only at relatively low pH values. They show selectivity towards the more strongly acidic anions.

(*e*) *Chelating ion-exchange resins*

These resins contain familiar chelating groups incorporated into a PSDVB resin and thus combine chelating properties with cation-exchange properties. They show high selectivity towards metal ions, characteristic of the chelating group. An example is the incorporation of aminodiacetate groups ($-CH_2N(CH_2CO_2^-)_2$) (which are related to EDTA) into the resin.

I have summarised all this information for you in Fig. 9.3c.

∏   Here is a short part of a polystyrene chain, containing five benzene rings. Add to each benzene ring an example of an ionogenic group to represent four of the five types of iono-

TYPE	IONOGENIC GROUP	pH RANGE	APPLICATIONS	EXAMPLES OF COMMERCIAL NAMES
Strongly Acidic Cation Exchanger	$-SO_3^-$	Wide (0–14)	General-Purpose Cation Exchange	Amberlite 1R120 Dowex 50 Zerolit 225
Strongly Basic Anion Exchanger	$-CH_2N(CH_3)_3^+$ $-CH_2N(CH_3)_2C_2H_4OH^+$	Wide (0–14)	General-Purpose Anion Exchange	Amberlite 1RA-400 Dowex 1 Amberlite 1RA-410 Dowex 2
Weakly Acidic Cation Exchanger	$-CO_2^-$*	High pH only (5–14)	Selective towards strong bases	Amberlite IRC-150 Zerolit 226
Weakly Basic Anion Exchanger	$-CH_2NH_3^+$ $-CH_2NH_2CH_3^+$ $-CH_2NH(CH_3)_2^+$	Low pH only (0–9)	Selective towards strong acids	Duolite A303
Chelating Ion Exchanger	eg $-CH_2N(CH_2CO_2^-)_2$	Depends on the metal ions involved (pH 6–8 for the group shown opposite)	Selective towards metal ions that react with the chelating group	Dowex A-1 Chelex-100

* Based on acrylic acid rather than styrene.

Fig. 9.3c. *Ionogenic groups incorporated into PSDVB resins (or other resins where indicated)*

Π   genic group discussed above. Then convert the final benzene
       ring into a cross-linking group. Finally draw a small portion
       of polymer structure to represent the ionogenic group *not*
       associated with a PSDVB resin.

$$-CH-CH_2-CH-CH_2-CH-CH_2-CH-CH_2-CH-$$

$$-CH-CH_2-CH-CH_2-CH-CH_2-CH-CH_2-CH-$$

SO_3^-	CH_2	CH_2	CH_2	$-CH_2$
	$^+N(CH_3)_3$	NH_3^+	N	

Cross-linking group

Strongly   Strongly   Example CH_2 CH_2
acidic   basic   of a CO_2^- CO_2^-
cation-   anion-   weakly
exchanger   exchanger   basic   Example of a
                          anion-   chelating ion-
                          exchanger   exchanger

(Rather a strange ion-exchange resin! Don't forget that any real resin
will be monofunctional, ie it will have only *one* of the above types
of group present at any one time).

The resin that we have met, not based on PSDVB was the weakly
acidic cation exchanger based on the copolymerisation of acrylic
acid and divinylbenzene.

∏   Name suitable ion-exchange resins for:

(*i*)   general purpose anion-exchange applications,
(*ii*) general purpose cation-exchange applications,
(*iii*) selective retention of metal ions,
(*iv*) selective retention of strong bases,
(*v*)   selective retention of strong acids.

Suitable ion-exchange resins for the above applications are as follows.

(*i*)   PSDVB resins with quaternary ammonium groups.
(*ii*) Sulphonated PSDVB resins.
(*iii*) PSDVB resins with chelating groups.
(*iv*) Polyacrylic acid-divinylbenzene resins.
(*v*)   PSDVB resins with amino-groups.

9.3.5.   Other Ion-exchange Materials

You have by now, I hope, appreciated the fact that PSDVB resins are the most important types of ion-exchange resin in analysis and indeed, if you are familiar only with these, this will be adequate for most of our purposes. Other materials are used, however, and perhaps we ought to mention a few of them, if only for the sake of completeness.

(*a*) *Inorganic ion-exchange materials*

We have already met the synthetic zeolites as cation exchangers in water softeners. Many other inorganic compounds have been proposed as ion-exchange materials, although practical applications seem rather few. One advantage appears to be when conditions of heat, radiation, etc, might cause organic resins to decompose. Zirconium phosphate is another example of an inorganic cation-exchanger.

(*b*) *Silica-based ion-exchange materials*

These are quite important and have been developed specifically for applications in HPLC (high-performance liquid chromatography). High pressures are used in this technique, and these might cause the swollen resin beads, which we have described as only 'semi-rigid' to become compacted and cause column blockages. Silica is a rigid non-porous substance that can withstand pressure. Very finely divided silica is chemically treated in such a way as to bind chemically onto the surface Si—OH groups, various ionogenic groups as in Fig. 9.3c, *via* organic bridging groups.

Silica particle $\{$ Si — O — [Organic bridging group] — ◇ Ionogenic group ◇

Thus they differ from PSDVB resins in that ion exchange is restricted to the surface of the particle.

Π   Bearing in mind that ion exchange is restricted to the surface of silica-based ion-exchangers, whereas usually it can take place throughout the volume of a PSDVB resin particle, can you think of two ways in which these two materials might be expected to differ with regard to ion-exchange properties?

(*i*) You might expect the exchange capacity of the silica-based material to be less than that of the PSDVB resin, because only the surface of the particles is available, even though the particle size of the former may be much less.

(*ii*) You might expect the silica-based material to be less selective towards large ions, because there are no pores for the ions to permeate. In fact, certain large organic ions can also be retained with good affinity by PSDVB resins by a surface mechanism, but of course, with low exchange capacity.

(*c*) *Cellulose-based materials*

Cellulose is a natural water-insoluble organic polymer, with, nevertheless, a high affinity for water. It contains a high proportion of —OH groups and some of these can be converted by chemical means into a variety of ionogenic groups, thus producing cellulose-based ion-exchange materials. In this way, ion-exchange papers can be produced for use in paper chromatography; also materials for use in tlc. (Paper, of course, is a form of cellulose).

SAQ 9.3c

Here are *five* specifications for an ion-exchange material, together with *five* ion-exchange materials. Match the materials to the specifications.

Specifications

An ion-exchange material suitable for:

(*i*) strong retention of transition-metal cations,

(*ii*) removal of all metallic and organic cations from solution,

(*iii*) high-performance liquid chromatography,

(*iv*) general work in the separation and preconcentration of anionic species,

(*v*) selective retention of more highly basic organic cations.

Ion-exchange materials

(*a*) Silica containing chemically bound ionogenic groups on the particle surfaces.

(*b*) Sulphonated polystyrene divinylbenzene (PSDVB) resin in the hydrogen-ion form.

(*c*) PSDVB resin containing quaternary ammonium groups. ⟶

**SAQ 9.3c
(cont.)**

(*d*) Polyacrylic acid-divinylbenzene resin.

(*e*) PSDVB resin containing iminodiacetic acid groups.

SAQ 9.3d

Imagine that you are trying to select the most appropriate ion-exchange material (resin or otherwise) for a particular purpose and you are looking at specifications of commercially available materials in chemical manufacturers' catalogues. What properties of the material would you look for in the specification? Make a list containing at least *six* important properties, and more if you can.

9.4. SELECTIVITY COEFFICIENTS AND SELECTIVITIES OF RESINS

In this section we consider ion exchange from a quantitative point of view. It is difficult to produce a highly quantitative approach to ion exchange for the simple reason that activities are rather more significant in ion exchange than they are in certain other techniques. This limits the usefulness of concentrations in any thorough quantitative treatment. Nevertheless, we shall be able to develop some useful concepts.

Let us start by once more thinking about the ion-exchange system which, in the last section, we came part-way to visualising for a PSDVB resin system. We shall then elaborate just a little further. Try the next exercise for straightforward revision of what you have learnt (hopefully!) up to now.

∏   Write a short paragraph describing as succinctly as you can a two-phase ion-exchange system based on an organic resin (perhaps a PSDVB resin) and an aqueous electrolyte.

Many answers, of course, are possible, but I hope yours will contain many of the points made in the following paragraph.

'The organic resin will have a *semi-rigid cross-linked structure and ionogenic groups* that make the resin hydrophilic but insoluble in water. Therefore, when it comes into contact with water, instead of dissolving, it swells because water molecules permeate the resin beads and force the polymer strands apart. The resulting swollen polymer has a porous structure, and water molecules can enter and leave the resin bead freely. The ionogenic groups are firmly bound to the polymer, but have associated with them *counter-ions*, of opposite sign, which are not so bound. The aqueous electrolyte also contains ions of either sign, and these may also permeate the resin, provided the ions are not so large that they cannot enter the pores. Ions of either sign may likewise enter or leave the resin beads and ions of the same sign as the counter-ions may be exchanged with the latter within the resin bead. We may talk about a *resin phase*, referring to everything within the interior of the resin bead and an *external phase*, referring to everything outside the resin bead.

Counter-ions are exchanged between the two phases according to the following equilibrium (for an anion exchange)'.

$$R^+X^- + A^+ + Y^- \rightleftharpoons R^+Y^- + A^+ + X^-$$

resin	external	resin	external
phase	phase	phase	phase

If you are now quite clear and happy about this description of an ion-exchange system, I want you to imagine both the phases as electrolyte solutions. You should not have any trouble in doing this with the external phase but for the *resin phase* you need to remember the water that is imbibed into the resin bead and the ionogenic groups plus counter-ions, as well as the free ions that have permeated the resin structure. These all basically constitute a *very concentrated electrolyte solution*. Look back to Fig. 9.3b, and now visualise the water and ions within the resin beads, together with the ionogenic groups as constituting a concentrated electrolyte solution. The *external phase*, on the other hand, is probably (unless we have a very high salt concentration) a more dilute electrolyte solution. This difference in electrolyte concentration between the two phases has two major implications.

(*a*) There will be an osmotic pressure difference between the two phases. The system responds to this pressure difference in a matter that reduces it (Le Chatelier's principle). This occurs by water molecules penetrating the resin, causing it to swell and thus causing the electrolyte concentration in the resin phase to be reduced (also incidentally increasing the concentration in the external phase). Here is the driving force for water penetration of the resin beads.

(*b*) With regard to penetration of the resin by ions, this will tend to be suppressed by the difference in electrolyte concentration between the two phases. In other words, we are saying that, although ions from the external phase can penetrate the resin phase, we would not expect them to do so in large numbers if this big difference in electrolyte concentration exists. Obviously, they do so in sufficient numbers for us to observe ion exchange, but the important point is that if we alter the electrolyte concentration of the external phase, then we expect to alter the amount of electrolyte penetration into the resin phase.

∏   If we increase the electrolyte concentration in the external phase, will this increase or decrease the amount of electrolyte entering the resin phase?

The amount of electrolyte entering the resin phase will be increased. This is a simple application of Le Chatelier's principle. The system is reacting in such a way as to remove the applied perturbation (increased electrolyte concentration in the external phase).

One other important point that we need to appreciate at this stage is that, since concentrations are always high in the resin phase and in practice can vary between very wide limits in the external phase, equilibria cannot be satisfactorily described in terms of concentrations, but *activities* are needed for a proper quantitative description. This is difficult to do in practice and so, as before, we shall in fact *define our equilibrium constants in terms of concentrations, but we have to note that the activity effect means that these will be strongly concentration dependent.*

9.4.1. The Selectivity Coefficient

The affinity that a particular resin shows for one ion over another can be measured quantitatively by the *selectivity coefficient*. Suppose that we have cation A^{a+} exchanging with cation B^{b+}. At *equilibrium*, the concentration of the ions in the two phases, represented as $[X]_r$ for the resin phase and $[X]_e$ for the external phase are related to each other by the selectivity coefficient, K_A^B, which is simply the equilibrium constant for the ion-exchange process. (X is either A^{a+} or B^{b+}).

$$K_A^B = \frac{[B^{b+}]_r^a [A^{a+}]_e^b}{[A^{a+}]_r^b {}_r[B^{b+}]_e^a}$$

Note that the selectivity coefficient is *not* equivalent to the distribution coefficient, since it is defined in terms of the exchange of *two* ions. A distribution coefficient may in fact be derived from the selectivity coefficient (we will do this shortly). We can then derive in turn a separation factor. But let me now warn you *not* to confuse selectivity coefficients with separation factors or distribution coefficients.

∏   In no way can we assume that the selectivity coefficient, K_A^B is independent of concentration. Why is this so?

Because the true thermodynamic constant, which is independent of concentration, must be defined in terms of *activities*. Because of the high concentrations involved, especially in the resin phase, and the large variations in concentration in the external phase large differences between activities and corresponding concentrations will occur. Thus K_A^B may vary quite fundamentally with changes in concentration of either or both ions.

∏   Define a selectivity coefficient for the anion exchange process in which ions Y^{y-} and X^{x-} are being exchanged.

We can do this in a way exactly analogous to the definition of K_A^B above.

$$K_X^Y = \frac{[Y^{y+}]_r^x [X^{x+}]_e^y}{[X^{x+}]_r^y [Y^{y+}]_e^x}$$

The selectivity coefficient must thus always be quoted for the exchange of one ion for another. Values >1 mean that the resin has a greater affinity for cation B or anion Y and *vice versa*. In quoting selectivity coefficients it is convenient for one of the ions to remain constant so that comparisons can be made with regard to the other ion. For cation exchange ion A^{a+} is H^+ and for anion exchange, ion X^- is mainly Cl^-.

9.4.2. The Distribution Coefficient

A distribution coefficient can be defined in ion exchange as follows.

$$K_D = \frac{\text{concentration of a particular ion in the resin phase}}{\text{concentration of the same ion in the external phase}}$$

$$= \frac{[B^{b+}]_r}{[B^{b+}]_e}, \text{ for instance.}$$

This definition is analogous to that for K_D in solvent extraction. You will note that K_D refers only to one of the ions involved in ion-exchange and thus is not as fundamental a parameter as it is in solvent extraction. In fact, you can derive an expression for K_D in terms of K_A^B and the concentrations of the ions in each phase.

∏   Derive an expression for the distribution coefficient, K_D, for ion B^{b+} in terms of the selectivity coefficient, K_A^B, and the concentrations of the various ions involved for the exchange of cations A^{a+} and B^{b+}.

We start with the expression:

$$K_A^B = \frac{[B^{b+}]_r^a [A^{a+}]_e^b}{[A^{a+}]_r^b [B^{b+}]_e^a}$$

and rearrange it to get an expression involving $[B^{b+}]_r/[B^{b+}]_e$ on the l.h.s. This expression gives K_D, by definition.

$$\frac{[B^{b+}]_r^a}{[B^{b+}]_e^a} = K_A^B \frac{[A^{a+}]_r^b}{[A^{a+}]_e^b}$$

$$K_D = \frac{[B^{b+}]_r}{[B^{b+}]_e}$$

$$= \left(K_A^B \frac{[A^{a+}]_r^b}{[A^{a+}]_e^b} \right)^{\frac{1}{a}}$$

It is noteworthy from from this expression that K_D for ion B^{b+} is thus independent of the concentration of ion B^{b+}, but is dependent on the concentration of the ion with which it is exchanging (A^{a+}). Now, if we could arrange conditions such that $[A^{a+}]_r$ and $[A^{a+}]_e$ remained constant during the ion exchange, then K_D would become a useful constant independent of the concentration of the ion to which it referred.

In fact this is easily achieved under analytically useful conditions in *ion-exchange chromatography* in which mixtures of ions are separated (basically various B^{b+} ions) by eluting them through a column containing an ion-exchange resin in the A^{a+} form by using as eluent a solution of A^{a+} ions. Thus:

$$[A^{a+}]_r \gg [B^{b+}]_r \quad and \quad [A^{a+}]_e \gg [B^{b+}]_e$$

under all conditions and so are effectively constant. Therefore, K_D would remain a constant throughout the ion exchange. Since retention in chromatography is a simple function of K_D, different ions will be separated because of different values of K_D. Retention times overall can be controlled by controlling the value of $[A^{a+}]_e$, because of the fixed number of ionogenic groups on the resin.

∏   Our establishment of the fact that K_D for B^{b+} is independent of the value of $[B^{b+}]$ needs one further qualification. What is it?

K_D may still be a function of $[B^{b+}]$ because K_A^B is not independent of $[B^{b+}]$ because of the effect of activities. If conditions can be arranged so that activity coefficients remain constant, then K_D will become a constant. This is likely if $[B^{b+}]$ remains at all times small. Under the conditions specified ($[A^{a+}]_r \gg [B^{b+}]$) this is most likely to be true, so the condition can easily be met, and we can say K_D is independent of concentration of the exchanging ion provided the latter remains small.

SAQ 9.4a   Select the *single true statement* from the following:

(*i*)   Selectivity coefficients in ion exchange are true thermodynamic constants, independent of concentration.

(*ii*)   The selectivity coefficient in ion exchange is the equivalent of the separation factor in solvent extraction.

(*iii*) The value of the distribution coefficient for an ion in ion exchange is independent of the concentration of that ion, but depends on the concentration of the ion with which it is exchanging, under all normal conditions.

(*iv*)   In ion-exchange chromatography, variations of the concentration of the eluting ion in the external phase will not affect its concentration in the resin phase.

(*v*)   Increasing the concentration of electrolyte in the external phase will lead to an increase in concentration of electrolyte in the resin phase.

ION	RESIN			
	Sulphonated PSDVB with 8% cross-linking	Sulphonated PSDVB with 4% cross-linking	PSDVB with $-CH_2N^+(CH_3)_3$ groups with 6–10% cross-linking	PSDVB with $-CH_2N^+(CH_3)_2C_2H_4OH$ groups with 8% cross-linking
Li^+	0.79	0.76		
H^+	1.00	1.00		
Na^+	1.56	1.20		
NH_4^+	2.01	1.44		
K^+	2.28	1.72		
Rb^+	2.49			
Cs^+	2.56			
Ag^+	6.70	3.58		
Tl^+	9.76			
Mg^{2+}	1.15	0.99		
Ca^{2+}	1.80	1.39		
Sr^{2+}	2.27	1.57		
Ba^{2+}	4.02	2.50		
Zn^{2+}	1.21	1.05		
Cu^{2+}	1.35	1.10		
Ni^{2+}	1.37	1.16		
Pb^{2+}	3.46	2.20		
OH^-			0.09	0.5
F^-			0.09	0.08
Cl^-			1.00	1.00
Br^-			2.8	3.5
I^-			8.7	18.0
$CH_3CO_2^-$			0.17	
CN^-			1.6	
NO_3^-			3.8	3.0
SCN^-				4.3
ClO_4^-				10.0

Fig. 9.4a *Selectivity coefficients for various ions on cross-linked PSDVB resins. Cations are exchanging with H^+, anions with Cl^-*

9.4.3. Observed Selectivities of Some Resins

I have collected together for you some experimental values of the selectivity coefficient for the exchange of ions on cross-linked PS-DVB resins. These are summarised in Fig. 9.4a. Study the table for a few moments and then have a go at the exercises which follow. Use the data in the table to help you.

∏　　Study Fig. 9.4a and see whether you can formulate two generalisations concerning selectivities. One of these we have already discussed, the other is new to you.

The first generalisation, which we have already discussed, becomes apparent when we compare the 4% cross-linked cation-exchange resin with the 8% cross-linked cation-exchange resin. The range of values for the selectivity coefficient is greater for the latter. These data support the general principle that increased cross-linking produces a more selective resin.

The second generalisation can be derived by taking any series of ions corresponding to elements from a particular group in the periodic table. We have series for the alkali metals, the alkaline earths, and the halogens. The selectivity coefficient always increases with increasing relative atomic mass (there are actually six examples in the table).

∏　　We have a new resin in the table, the anion exchange resin based on $-CH_2N^+(CH_3)_2C_2H_4OH$ groups. What special advantage do you think it may possess over a resin with $-CH_2N^+(CH_3)_3$ groups?

Note the increased selectivity towards OH^- ions. The low selectivity of conventional anion-exchange resins to OH^- against Cl^- can be a nuisance at times, making it difficult to convert a resin into the OH^- form. This new anion-exchange resin should be easier to convert into the potentially useful OH^- form.

So from this table we have been able to draw some general conclusions about selectivities. However, we could formulate more generalisations if we had more data. I will now summarise for you, *all* the important factors that influence the value of the selectivity coefficient. Study these and then you should be able to attempt the SAQ.

(*a*)　At low concentrations (< 0.1 mol dm^{-3}) selectivity coefficients increase markedly with increasing charge on the ion. So you would expect, for instance, orders of selectivity such as $Al^{3+} >$ $Mg^{2+} > Na^{+}$. This is *not* apparent from Fig 9.4a, since these values were *not* obtained at low concentration. The same principle applies to anions.

(b) Selectivity coefficients increase as the radius of the *hydrated* ion decreases. This can be somewhat confusing at first when you look at the table, since it seems that as the ion gets larger (relative atomic mass increases) its selectivity coefficient increases. The point to remember is that it is not the size of the free ion that is important, but the size of the hydrated ion. A small free ion has a high charge density and has a strongly bound hydration sphere around it. A large free ion with the same charge has a lower charge density, a much more weakly bound hydration sphere and therefore effectively a smaller ionic radius when hydrated than the smaller free ion.

(c) Actual concentrations and relative concentrations of the ions, and exchange capacity of the resin (effectively controlling concentration in the resin phase) can all affect selectivity coefficients *via* activity effects. Perhaps the most important generalisation is that as concentrations increase, the resin increasingly favours ions of lower charge over those of higher charge, so that the factor discussed in (a) is reduced in importance, or an actual reversal of that trend is observed (as in Fig. 9.4a).

(d) The ionogenic groups will have a direct bearing on selectivity coefficients. For instance, greater selectivity is observed when we use weakly acidic or basic exchangers, weakly ionised species thus being discriminated *against*. Also chelating ion-exchangers exhibit selectivities characteristic of the chelating group. With weak exchangers and chelating exchangers pH is also an important factor (as it is if the ions being exchanged are in equilibrium with non-ionised forms).

SAQ 9.4b

Answer the following questions, which all refer to exchange of cations on a sulphonated PSDVB resin in the H^+ form. Give concise explanations for each of your answers.

(i) Which of the following orders will the selectivity coefficients for the alkali metals follow:

$$Li^+ < Na^+ < K^+ < Rb^+ < Cs^+$$

or $Cs^+ < Rb^+ < K^+ < Na^+ < Li^+$?

(ii) Which will show the greatest selectivity towards the separation of Na^+ and K^+:

a 4% cross-linked resin,

an 8% cross-linked resin,

or a 12% cross-linked resin?

(iii) For concentrations < 0.1 mol dm^{-3}, will the selectivity coefficients for the following ions follow the order

$$K^+ < Ca^{2+} < Al^{3+}$$

or $Al^{3+} < Ca^{2+} < K^+$?

(iv) Will *increasing* the concentration of the ions in question (iii) increase or decrease the selectivity of the separation?

SAQ 9.4b

9.5. ION EXCHANGE AND CHEMICAL EQUILIBRIA

In Section 9.4 we considered certain properties of resins that dictated their selectivity. We further considered certain properties of the ions that governed their selectivities on different resins. Fairly central to these discussions was a table of selectivity coefficients (Fig. 9.4a). Have another look now at that table and the range of values for the selectivity coefficients that it contains. Bearing in mind the fact that these are basically equilibrium constants, I think that you will agree that they cover a rather small range and that you do not as a rule observe extremely large or extremely small values (ie all values are relatively close to unity). What this means in practice is that it is very difficult to obtain a quantitative separation of a mixture of cations or of anions by ion exchange unless the high resolving power of a chromatographic technique is invoked. It is quite easy to separate cations from anions, or cations or anions from neutral molecules by ion exchange, but separation of cations from cations or anions from anions generally needs a chromatographic method.

However, there is an extremely important way in which selectivity can potentially be vastly increased in quite simple ion-exchange systems. This we now consider. The basic idea is that we involve the ions that are exchanging, in chemical equilibria in the external phase so that cations for example, are in equilibrium with anions, or alternatively cations or anions are in equilibrium with neutral molecules. If you recall that only cations will be retained on a cation-exchange resin and only anions on an anion-exchange resin, then you should be able to see that, if we involve, say, a cation in an equilibrium with an anionic form, the extent to which such a species will be retained on a cation-exchange resin will depend on the position of that equilibrium. Furthermore this 'cation' could even be retained on an anion-exchange resin in its anionic form. If you have two cationic species in solution and only one can be converted into an anionic form, then their separation is potentially possible by using either type of resin, if you can adjust conditions in the external phase so that one cation is completely in its anionic form. In practice there are many possibilities for doing this sort of thing. This will all become clearer if we look at one or two actual examples.

∏   Can you think of some actual examples of chemical equilibria involving:

(*i*)   a cation in equilibrium with an anion,

(*ii*)   a cation in equilibrium with a neutral molecule,

(*iii*)   an anion in equilibrium with a neutral molecule,

(*iv*)   a cation in equilibrium with a neutral molecule, which in its turn is in equilibrium with an anion?

Here are some actual examples:

(*i*)   Perhaps the best example here is of a metal cation in equilibrium with a complex anion. For example, in the presence of hydrochloric acid, transition-metal cations form complex chloro-anions with various degrees of stability. For example:

$$Zn^{2+} + 3\,Cl^- \rightleftharpoons ZnCl_3^-.$$

(*ii*) Any weakly basic organic (or inorganic) compound can be protonated if the pH is sufficiently low. The protonated form is the cation. Thus, for aniline, for example:

$$C_6H_5NH_2 + H^+ \rightleftharpoons C_6H_5NH_3^+.$$

(*iii*) Any weakly acidic organic (or inorganic) compound can be ionised if the pH is sufficiently high. The ionised form is the anion. Thus, for benzoic acid, for example:

$$C_6H_5CO_2H \rightleftharpoons C_6H_5CO_2^- + H^+.$$

(*iv*) Amino-acids are molecules that can exist in a protonated form at low pH, a neutral (zwitterionic) form at an intermediate pH, and an anionic form at high pH. Here are the equilibria in a generalised form.

$$\underset{\text{Low pH}}{HO_2C-\overset{\overset{\displaystyle R}{|}}{C}H-NH_3^+} \rightleftharpoons \underset{\text{Intermediate pH}}{{}^-O_2C-\overset{\overset{\displaystyle R}{|}}{C}H-NH_3^+} \rightleftharpoons \underset{\text{High pH}}{{}^-O_2C-\overset{\overset{\displaystyle R}{|}}{C}H-NH_2}$$

Note the zwitterionic form at intermediate pH. Although the molecule has positively and negatively charged groups in it, the overall charge on the molecule is zero.

The response to the above exercise provides us with some excellent examples to show how selectivity can be achieved in ion exchange by means of manipulating chemical equilibria. Let us start with a fairly simple example which you can attempt yourself.

∏ How could you separate benzoic acid from benzyl alcohol by ion exchange?

Assuming that we have an aqueous (or perhaps aqueous ethanolic) solution of the species, we first adjust the pH to a high value. The benzoic acid will be converted quantitatively into a salt, whereas the benzyl alcohol will remain as a neutral molecule. We now pass the solution through a column containing a strongly basic anion-exchange resin which will quantitatively retain the benzoate anions, but allow the benzyl alcohol to pass through. If we now pass an acidic solution through the column, the benzoate anions will be reconverted into benzoic acid and can be recovered from the column.

You could devise other simple separations along the above lines. For instance, aniline and benzyl alcohol or aniline and phenol could be separated in a similar way. If you had two acids whose acid dissociation constants were sufficiently different so that a pH could be found at which one exists as an anion and the other as a free acid, then these could be similarly separated. Likewise for two bases of different basicity.

SAQ 9.5a

> Which of the following three ternary mixtures (as solutions in a suitable aqueous organic solvent) is the only one that can be easily resolved into its individual components by using simple ion-exchange methods (ie not ion-exchange chromatography)? Briefly list, in their correct order, the operations which would be necessary to achieve the separation.
>
> (*i*)   Aniline ($C_6H_5NH_2$), phenylmethylamine (benzylamine) ($C_6H_5CH_2NH_2$), and phenylmethanol (benzyl alcohol) ($C_6H_5CH_2OH$).
>
> (*ii*) Benzoic acid ($C_6H_5CO_2H$), phenylethanoic acid (phenylacetic acid) ($C_6H_5CH_2CO_2H$), and aniline.
>
> (*iii*) Phenylmethanol (benzyl alcohol), phenylmethylamine (benzylamine) and phenylethanoic acid (phenylacetic acid).

SAQ 9.5a

Let us now study some more sophisticated ion-exchange separations based on chemical equilibria in solution. Two examples will serve our purpose, the separation of transition-metal ions and the separation of amino-acids.

9.5.1. The Separation of Transition-metal Ions by Anion Exchange

In the presence of hydrochloric acid, many, but by no means all, metal cations react reversibly with the chloride ions present to form complex chloro-anions, such as $FeCl_4^-$.

$$Fe_{aq}^{3+} + 4\,Cl^- \rightleftharpoons FeCl_4^- + aq$$

These complex chloro-anions can be retained strongly on a conventional anion exchange resin. If the resin is subsequently treated with water, the above reaction is reversed, the free metal cation is formed and released from the anion exchange resin. If you have a mixture of two metal cations, one of which forms such a complex chloro-anion and the other does not, then ion exchange can be used as the basis of a simple yet elegant means of separating the ions.

∏   In the presence of $9M$-hydrochloric acid, cobalt(II) is reversibly and quantitatively converted into the chloro-anion, $CoCl_4^{2-}$, but nickel(II) is not. Explain how you could separate cobalt(II) and nickel(II) in solution by using the above information.

Prepare a column containing a strong anion-exchange resin in the chloride form and equilibrated with $9M$-hydrochloric acid. Prepare a solution of the ions to be separated also in $9M$-hydrochloric acid. Note that this solution will acquire a dark-blue colour due to the $CoCl_4^{2-}$ ions, if there is originally a high concentration of Co^{2+} ions present. Pass this solution through the column. The cobalt will be quantitatively retained by the anion exchange resin as $CoCl_4^{2-}$ ions, whereas the nickel will pass straight through as uncomplexed Ni^{2+} and can be collected. When you have collected all the nickel, by passing more $9M$-hydrochloric acid through the column, pass pure water through the column. This causes the $CoCl_4^{2-}$ ions to dissociate back to Co^{2+} and $4\,Cl^-$. The cobalt, being cationic once more, is no longer retained on the anion-exchange resin, and can be collected and so quantitatively separated from the nickel.

Although this is a particularly effective separation as it stands, it by no means represents the limit of the power of this technique for separating transition-metal and other cations. Different metal cations form chloro-anions with widely differing stabilities. For instance, of the first-row late transition-metal ions Fe^{2+} to Zn^{2+}, Ni^{2+} does not form an anion, whereas the most stable anion is that derived from Zn^{2+}. We can define a distribution ratio, D, for these metals in the presence of hydrochloric acid, to describe their distribution between the resin and the hydrochloric acid external phase.

$$D = \frac{\text{total concentration of metal* in resin phase}}{\text{total concentration of metal* in external phase}}$$

*(as cation or anion)

D will depend strongly on the concentration of the hydrochloric acid and, within limits, will increase with increasing acid concentration. Fig. 9.5a shows in graphical form how D varies with acid concentration for a wide range of elements.

∏ You can see from Fig. 9.5a, that for the first-row transition-metal ions, by and large D increases with increasing HCl concentration. However, for some of the ions, the relationship between D and HCl concentration is an inverse one, and for others, D passes through a maximum. Can you offer two explanations: (i) why D should increase with HCl concentration (ii) why D should decrease with HCl concentration?

(i) D increases with HCl concentration because the increasing concentration of Cl^- ions forces the complex-anion formation below

$$M^{n+} + xCl^- \rightleftarrows MCl_x^{(x-n)-} \qquad (x > n)$$

to the right, by a mass-action effect, thus increasing the proportion of M^{n+} present as $MCl_x^{(x-n)-}$.

(ii) D decreases with HCl concentration because Cl^- ions and $MCl_x^{(x-n)-}$ ions are competing with each other for exchange sites on the resin. If the Cl^- concentration is increased in the external phase, then it will tend to increase in the resin phase forcing $MCl_x^{(x-n)-}$ back into the external phase.

Explanation (i) predominates when the extent of anion formation is small, explanation (ii) when it is large. When you see D passing through a maximum, then you can see each explanation predominating at different acid concentrations.

By retaining the ions in a high HCl concentration and then systematically reducing the HCl concentration, we can resolve quite complex mixtures of ions since as the acid concentration is reduced, the chloro-complexes of the metal ions dissociate. The least stable complex dissociates first when the acid concentration is still quite high, and the most stable dissociates only when the acid concentration has decreased to a low value.

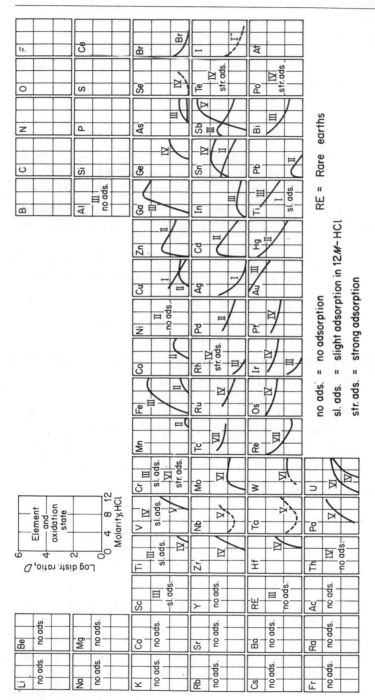

Fig. 9.5a. *Distribution ratios of the elements on Dowex-1 anion exchange resin as a function of the hydrochloric acid concentration*

RE = Rare earths

no ads. = no adsorption

sl. ads. = slight adsorption in 12*M*-HCl

str. ads. = strong adsorption

Fig. 9.5b illustrates graphically how you can separate Mn^{2+}, Fe^{3+}, Co^{2+}, Ni^{2+}, Cu^{2+} and Zn^{2+} by anion exchange by first retaining all but the Ni^{2+} ions in the resin by using $12M$-HCl, then reducing the HCl concentration in stages, thus releasing an ion at a time, until eventually the Zn^{2+} comes off, rather reluctantly, in $0.005M$-HCl.

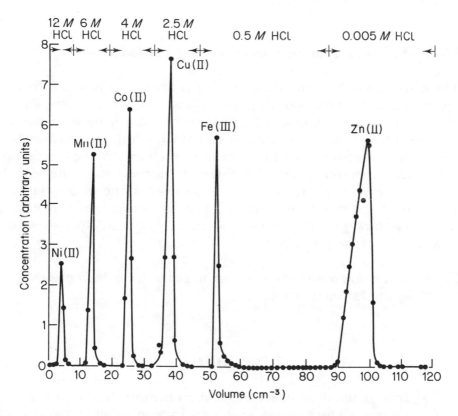

Fig. 9.5b. *Separation of transition elements Mn to Zn (Dowex-1 column; 26 cm × 0.29 cm; flow-rate = 0.5 cm min^{-1})*

∏   Place the ions Mn^{2+}, Fe^{3+}, Co^{2+}, Ni^{2+}, Cu^{2+} and Zn^{2+} in order of increasing stability of their complex chloro-anions.

The correct order is:

$$Ni^{2+} < Mn^{2+} < Co^{2+} < Cu^{2+} < Fe^{3+} < Zn^{2+}$$

In other words, this is the order in which the ions are released from the anion exchange resin as the HCl concentration is reduced. The ions forming the least stable complexes are released from the resin at the highest HCl concentration, those forming the most stable complexes are not released until the HCl concentration has been reduced to quite low values.

9.5.2. The Separation of Amino-acids by Ion Exchange

There are about 25 naturally occurring α-amino acids which form the basic building blocks of all proteins, and a fully hydrolysed sample of a protein will consist of a mixture of most of these in fairly characteristic proportions. Analysis of such a mixture to identify the amino-acids present and to determine their relative proportions is thus an important practical problem in biochemistry. The separation of these amino-acids from one another is one of the classic examples of the successful application of ion exchange in analysis and shows how *pH control* can be used to separate species with different acid/base properties.

∏   Given that α-amino-acids can, in general, can be represented by the formula below:

$$\overset{\displaystyle R}{\underset{\displaystyle HO_2CCHNH_2}{\vert}}$$

draw generalised structures to show in more detail, the different ionised forms an α-amino-acid would adopt in a medium of:

(*i*)   low pH,
(*ii*) intermediate pH
(*iii*) high pH,

and indicate what types of ion-exchange resin, if any, would retain each form.

(*i*) *at low pH*

The molecule has been protonated and is thus cationic in nature. It would thus be retained on a cation-exchange resin.

(*ii*) *At intermediate pH*

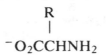

Note the zwitterionic form that the amino acid adopts, rather than the simple covalent form shown in the question. The molecule has both cationic and anionic groups in it, but from the point of view of ion exchange, the important fact is that its net charge is zero. As such it is not retained on either a cation- or an anion-exchange resin.

(*iii*) *At high pH*

$$
\begin{array}{c}
\text{R} \\
| \\
{}^-\text{O}_2\text{CCHNH}_2
\end{array}
$$

The molecule is now anionic. It would thus be retained on an anion-exchange resin.

So we could either:

(*a*) retain the amino-acid on a cation-exchange resin at low pH, and subsequently release it by raising the pH until the zwitterion is formed,

or (*b*) retain the amino-acid on an anion-exchange resin at high pH and subsequently release it by lowering the pH until the zwitterion is formed. In practice, we tend to use the former approach (cation-exchange resin) for separating amino-acids. What we must now consider is the actual value of the 'intermediate pH' at which the amino-acid is converted into the zwitterion and released from the column.

In general, each amino-acid forms its zwitterion at a *different* value

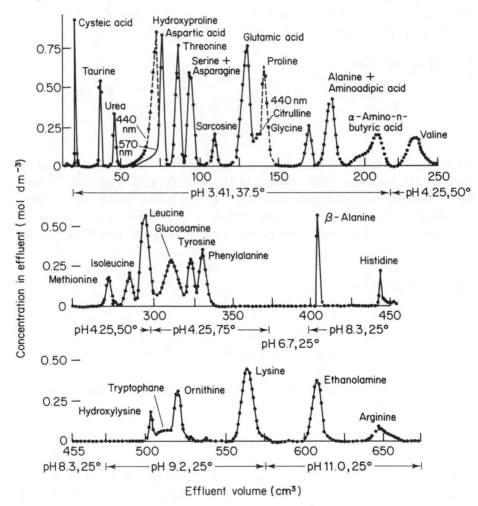

Fig. 9.5c.   *Analytical separation of amino-acids by elution development. Temperature and pH of the eluent are changed stepwise. The effluent is recorded spectrophotometrically at 440 and 570 nm after reaction with ninhydrin. Eluent: sodium citrate buffer for pH 3.41 to 6.7, sodium carbonate–bicarbonate buffer for pH 6.7 to 11.0; resin: Dowex 50-X8 in Na^+ form. Temperatures in $°C$*

of pH, characteristic of that amino-acid, and governed by the nature of the R group present. This value of pH is known as the *iso-electric point* of the amino-acid. Amino-acids containing acidic R groups have low iso-electric points, whereas those with basic R groups have high iso-electric points.

∏   What is the iso-electric point of an amino-acid?

The iso-electric point of an amino-acid is the value of pH at which the amino-acid exists as a zwitterion, rather than as a cation (at lower pH values) or as an anion (at higher pH values). The value varies from amino-acid to amino-acid.

So, if we retain our mixture of amino-acids on a cation-exchange column at a pH *below* the iso-electric point of the most acidic amino-acid present, and then subsequently pass a series of buffers of different compositions through the column in such a way as to raise the pH slowly and in a controlled manner (the increase can either be continuous or in steps) then, the amino-acids will be released sequentially in order of increasing iso-electric points. By this means, it is possible quantitatively to separate virtually all amino-acids from one another. I have illustrated this for you with an example of such a separation in practice in Fig. 9.5c, which shows an almost complete separation of 31 amino-acids and related compounds.

SAQ 9.5b
Six statements are given below about the extent of retention of ions on ion-exchange resins under specific conditions. Only *some* of these statements are true. Identify the true statements and then correct the false statements (correction when required will simply involve converting the statement into its *opposite*).

When you have done this, or as you do it, look at the list of explanations and select one that correctly explains each true and each corrected statement (each explanation may be used once, more than once, or not at all).   ⟶

SAQ 9.5b
(cont.)

The statements

(*i*)   Sulphate ions are retained more strongly than carbonate ions on an anion-exchange resin at neutral pH.

(*ii*) The order of retention of halide ions on an anion-exchange resin at neutral pH is:

$$Cl^- > Br^- > I^-.$$

(*iii*) Nickel ions are retained more strongly than cobalt ions in the presence of 9M-HCl on a cation-exchange resin.

(*iv*) The order of retention of the following anions on an anion-exchange column at high pH and low concentration is:

$$ClO_4^- > SO_4^{2-} > PO_4^{3-}.$$

(*v*)   At a mildly alkaline pH, leucine, an amino-acid, $RCH(NH_2)CO_2H$, with R = $-CH_2CH(CH_3)_2$, is less strongly retained than glutamic acid, an amino-acid with R = $-CH_2CH_2CO_2H$, on an anion-exchange resin.

(*vi*) Magnesium ions are more strongly retained than zinc ions on an anion-exchange resin in the presence of 9M-HCl.

The explanations

A   *Because* differences in the acid dissociation constants lead to differences in the relative proportions of the species present in

$\longrightarrow$

**SAQ 9.5b
(cont.)**

the anionic form, as opposed to the neutral molecular form.

B   *Because* differences in the size of the complex anions formed in the presence of the HCl lead to differences in the degree of penetration of the complex anions into the resin.

C   *Because* differences in the relative stabilities of the complex anions lead to differences in the ratio of cation to anion present.

D   *Because* the retention of the anion decreases in line with increasing radius of the hydrated anion because of decreased penetration of the ion into the resin.

E   *Because* retention of the anion increases with increasing charge on the anion under these conditions.

F   *Because* the radius of the free isolated anion increases along the series of increasing retention. The larger the free isolated anion then the smaller the hydrated anion, because of the reduced extent of hydration.

SAQ 9.5b

9.6. REVIEW OF SOME APPLICATIONS OF ION-EXCHANGE IN ANALYSIS

In this section we shall see how ion exchange can be used practically in chemical analysis. I think that you will see from the examples that we study, that the applications are very widely distributed, and even the fact that ions must be involved in the separation is not quite as limiting as at first it may seem.

Here is a list of the applications that we shall study. Don't consider it an exhaustive list, but just representative of the ways ion-exchange can be of use to the analyst:

(*a*) preparation of de-ionised water,
(*b*) determination of total cation or anion content,
(*c*) preconcentration (mainly of trace-metal cations),
(*d*) in a wide range of analytical separations, both inorganic and

organic, involving chromatographic and non-chromatographic techniques.

Let us now consider each of these areas in more detail.

9.6.1. Preparation of De-ionised Water

Although the preparation of de-ionised water in itself is not a strictly analytical operation, de-ionised water is so widely used in analytical laboratories, that it seems appropriate to include it here. De-ionised water is water which has been essentially freed from all ionic contaminants and is prepared by passing water through a cartridge containing a mixed bed of ion-exchange material, that is *a mixture of a cation-exchanger in the H^+ form and an anion-exchanger in the OH^- form.* All cations are thus exchanged for H^+ and all anions for OH^-, and then we get the reaction:

$$H^+ + OH^- \rightleftharpoons H_2O$$

giving us water essentially completely free from ions (disregarding the self-dissociation of water). The ionic impurities accumulate in the ion-exchanger and eventually 'break through' occurs, when the cartridge is exhausted. At this stage, the cartridge is usually discarded and is replaced by a new one since it is difficult and uneconomic to regenerate mixed-bed ion-exchangers. The quality of the de-ionised water being produced is monitored by continuously measuring its electrical conductivity, as it emerges from the cartridge.

∏   What is the other method commonly used for producing 'pure' water in the laboratory? How do the two methods compare?

The other method is by *distillation*. The main difference between the two methods is that whereas de-ionising removes only ionic impurities, distillation in principle separates the water from all non-volatile impurities. In fact, the process of de-ionisation may actually introduce non-ionic impurities leached out of the ion-exchanger. When these are liable to be troublesome, the de-ionised water should be further treated by passing it through a charcoal filter, say, or distilling it. Sometimes water may be de-ionised and then distilled, although unless special materials are used in the still, distillation may introduce small levels of ionic impurities! It's never a perfect world!

9.6.2. Determination of Total Cation or Anion Content

This is an example of how ion exchange can be used to replace all the cations or anions in a sample by a cation or anion of one type. When the determination of the total concentration of cation or anion is of interest, the ions are exchanged for H^+ or OH^- respectively by methods with which you should be quite familiar by now. When the exchange is complete, the analysis is finished by a simple acid/base titration.

∏   In what units would the result of such an analysis be obtained?

In units of equivalents of H^+ or OH^-, assuming that there was a mixture of cations or anions of different charge present. Under these conditions, it would not, of course, be possible to obtain a result in g dm^{-3} or even in molarity. This may be a limitation to this fairly crude and simple analysis.

9.6.3. Preconcentration

Ion exchange can be used very effectively for preconcentrating ionic species in water, and in this respect has found most application to the preconcentration of *trace-metal ions* in various types of water sample. It is not the only technique that can be used for this purpose. We have already seen how solvent extraction can be used for this, and we have some further techniques to consider, which can be used similarly. When we have studied all the techniques we shall find it interesting to compare them all with regard to their overall performance in trace-metal ion preconcentration.

The basic principles are to take a *large* volume of water containing the metal ions to be preconcentrated, make suitable chemical adjustments to the sample (such as pH adjustment or complexing of the ions) and then to pass it through a small column packed with a suitable ion-exchange material. The chemical adjustments and choice of ion exchanger will have been such as to ensure a high affinity for the metal ions to be preconcentrated. This combined with the trapping-technique effect assures a quantitative retention of the ions on the column. Subsequently the metal ions are released into solution by passing through the column a *small* volume of solution of appropriate chemical composition to release the metal ions from the column (ie reversing the exchange process). The preconcentration factor is equal to the ratio of the original volume of water sample to the volume of solution required to recover the metal ions from the column. In principle, we can obtain very large preconcentration factors this way, because there is no theoretical maximum to the volume of water sample that can be passed through the column, and the volume of solution required to recover the ions can often be kept quite small. This is one of the main advantages of this technique.

One problem that may be encountered is that it may be difficult to get a quantitative recovery of the ions from the column. Sometimes for small amounts of material, an irreversible retention of the ions may occur. Would you like to think about overcoming this problem?

Π   Suggest two ways in which the problem of irreversible retention of metal ions on an organic ion-exchange resin could be overcome.

The two ways that have most commonly been used are:

(*i*)   to *ash* the resin and take up the residual ash in a small volume of acid;

(*ii*)   to analyse the resin directly in the solid state without trying to remove the metal ions. X-ray fluorescence and neutron-activation analysis would be appropriate techniques.

The choice of ion-exchange resin and chemical conditions for retention and recovery are important. Three types of ion-exchange resin can be used to preconcentrate metal ions. Can you suggest what these might be?

∏   What three types of ion-exchange resin, classified according
       to ionogenic group, do you think could be used to precon-
       centrate metal ions?

(*i*)   A strong cation-exchange resin, eg a sulphonated PSDVB
        resin.

(*ii*) A strong anion-exchange resin, when the metal ions must first
        be converted into complex anions.

(*iii*) A chelating ion-exchange resin.

One question that we must consider concerning the sample is
whether it is an aqueous solution of low overall electrolyte content (eg rain water) or relatively high overall electrolyte content
(eg sea water). Bearing in mind that in most sea water samples we
are likely to want to preconcentrate traces of transition- and heavy
metals, whereas the most abundant ions will probably be alkali and
alkaline earth metal ions, then the ion-exchanger should show a selectivity towards the ions we wish to preconcentrate. This may not
be a critical factor with samples of low overall electrolyte content,
but it definitely will be with, for example, sea water with its high salt
content.

Let us now look briefly at the practical implications of the three
types of resin that we could use.

(*a*) Strong cation-exchanger

This might be a sulphonated PSDVB resin in the H^+ form. Retention of metal ions will occur on a fairly non-selective basis, and so
such a resin would be useless for preconcentrating trace-metal ions
in the presence of high concentrations of sodium, calcium etc. Furthermore, recovery of the ions will rely solely on their displacement
by H^+ by passing acid through the column. So for these reasons,
this type of resin is not a satisfactory choice.

(*b*) Strong anion-exchanger

Selective retention of ions is much more possible here. For instance, by adding a suitable concentration of hydrochloric acid to the sample, we could retain selectively many ions that form stable chloro complexes, while sodium and calcium and similar ions would not be retained. Recovery of the preconcentrated ions could more easily be achieved by washing the resin with water (or dilute nitric acid in practice) to dissociate the complexes. Large volumes of reagent are required for these resins, which is a disadvantage.

(c) Chelating ion-exchangers

These are generally the most useful resins for metal-ion preconcentration, because they show a high affinity for metal ions, but only within a certain pH range. Furthermore, they show the selectivity characteristic of the chelating group. For general work, PSDVB resins containing iminodiacetic acid groups are used and the sample is adjusted to pH 6–8. However, calcium is fairly strongly retained and if this is a problem, then often, more selective chelating resins can be used. Chelating resins also score with regard to recovery of the ions, since in the presence of acid they become protonated and lose their chelating properties. $2M$-nitric acid is suitable for recovery of most metal ions from the above resin. These resins tend to be rather expensive; preconcentration represents their most useful application.

| SAQ 9.6a | A sample of sea water is to be analysed for traces of a number of transition- and heavy-metal ions. The method of analysis is to involve a solution technique, and a 1000-fold preconcentration of the ions is required. Below are brief descriptions of *four* preconcentration procedures based on ion-exchange. Only *one* of these is likely to be successful. Which is it and what are the potential snags with the other three methods? For simplicity, assume that the sea-water matrix consists of high concentrations of sodium and magnesium ions and very low concentrations of the analyte ions. $\longrightarrow$ |

SAQ 9.6a
(cont.)

(*i*) The sample (10 dm^3) is passed through a small column packed with sulphonated PSDVB resin in the H$^+$ form. Subsequently the ions are recovered by passing $2M$-HNO$_3$ (10 cm^3) through the column.

(*ii*) The sample (10 dm^3) is passed through a small column packed with sulphonated PSDVB resin in the H$^+$ form. The resin is then suitably ashed and the ash dissolved in 10 cm^3 of $2M$-HNO$_3$.

(*iii*) The sample (10 dm^3) after adjustment to pH 6–8 is passed through a small column packed with PSDVB resin containing iminodiacetic acid groups in the cationic form. Subsequently the ions are recovered by passing 10 cm^3 of $2M$-HNO$_3$ through the column.

(*iv*) The sample (10 dm^3) after adjustment to render it molar with respect to hydrochloric acid is passed through a small column packed with PSDVB resin containing $-CH_2N(CH_3)_3^+$ groups in the Cl$^-$ form. Subsequently the ions are recovered by passing 10 cm^3 of $2M$-HNO$_3$ through the column.

SAQ 9.6a

9.6.4. Analytical Separations

Here we consider separating one species from another, say an analyte from an interferent, by using simple trapping techniques, or perhaps separating a number of species by using either complex trapping techniques or ion-exchange chromatography. The scope of ion exchange and ion-exchange chromatography is very wide, the only limiting factor being that to be retained on the resin, the species must exist in an ionic form. You have of course, met some separations already in the process of learning about ion exchange. What I want to do here is simply to review the scope of analytical separations made possible by ion exchange.

Separation of cations from anions

It is often necessary in simple classical analysis as well as in other situations, to remove certain or all of the cations or anions from a solution and replace them with say H^+ or OH^-, before determination of species of the opposite charge. This is a very simple application of ion exchange and the practical method involves only the simplest of trapping techniques by using a column of strong cation-exchange resin in the H^+ form or anion-exchange resin in the OH^- form.

Metal cations

Alkali-metal and alkaline earth-metal ions can be separated by ion-exchange chromatography on a strongly acidic cation-exchange resin. Other metal ions can be similarly separated, although there may be better alternative methods.

Much more selective separations are possible by using anion-exchange resins and selectively and reversibly converting the metal ions into complex anions. In addition to the chloro complexes of transition-metal ions, with which we have dealt at length, there are separations involving other complexing species, and heavy-metal cations as well as many radionucleides.

The lanthanides and actinides form two series of metal ions with very similar properties. It is very difficult to separate the components of mixtures of these metal ions to give pure samples of individual ions. Ion-exchange chromatography, with complexing eluents (eg citrate media for the lanthanides), is the classical approach to solving such problems.

Anions

The separation of anions by ion-exchange chromatography on a strong anion-exchange resin in dilute carbonate bicarbonate buffers has recently become of great importance as an HPLC technique, especially in water analysis. The technique is known as *ion chromatography*. It can also be used for cations, but most actual applications seem to be for anions.

Organic and biochemical species

We have seen how organic molecules can be fractionated according to acid-base properties by ion-exchange. On a more sophisticated basis we have seen how amino-acids in complex mixtures can be separated by using what is in effect ion-exchange chromatography, with gradient-pH elution. There are many more applications of ion exchange to the separation of organic compounds, not all of which are even to compounds that we normally think of as ionic. For instance, aldehydes and ketones can be separated by ion-exchange, as well as simple sugars.

∏   How is it possible to separate, for instance, aldehydes and ketones, or simple sugars by ion exchange, even though these are not ionic species?

By derivatisation. Aldehydes and ketones react reversibly with the bisulphite ion to give anionic bisulphite complexes. Likewise, sugars react with boric acid to give strongly acidic (therefore anionic) sugar-borate complexes.

Many large biochemical molecules, (ie biological macromolecules) are water soluble and ionic. These can be separated by ion exchange by using special resins designed to cope with large molecules.

9.6.5. Conclusions

I hope that you have come to appreciate ion exchange as a separation technique of great versatility, and wide applicability. We have seen it used for quite simple separations as well as for complex chromatographic separations. It is used in preconcentration, reagent preparation, all types of separations and analysis. Its scope extends to most inorganic cations and anions, organic acids and bases, and combined with derivatisation, to certain non-ionic organic species, as well as to many molecules of biochemical interest. It has been described as the analyst's 'panacea' with regard to separation. I wouldn't put it as strongly as that, but, together with solvent extraction and, embracing its chromatographic applications, it certainly represents the most powerful technique that we have for separating ionic species. We now take our leave of the topic of ion exchange with a look at a table summarising some of the representative applications of ion exchange in analysis (Fig. 9.6a) and a couple of SAQ's on ion-exchange applications.

APPLICATION	RESIN USED	SOLVENTS USED (ELUENTS)
Replacement of all cations by H^+	Strong cation-exchanger in H^+ form	Sample medium
Replacement of all anions by OH^-	Strong anion-exchanger in OH^- form	Sample medium
Removal of all ions (preparation of di-ionised water)	Mixed cation/anion-exchanger in H^+ and OH^- forms	Sample medium
Preconcentration of metal cations	Chelating ion-exchanger	pH 6–8 then $2M$-HNO_3
Separation of cations from one another, eg		
Alkali metals,	Strong cation-exchanger in H^+ form	$0.7M$-HCl
Alkaline earths	Strong cation-exchanger in H^+ form	$1.2M$-ammonium lactate
$Cd^{2+}/Bi^{3+}/Cu^{2+}$	Strong cation-exchanger	$0.2M$-HBr/$0.5M$-HBr/ $2.5M$-HNO_3
Zr^{4+}/Hf^{4+}	*Either* strong anion-exchanger *or* strong cation exchanger	3.5% H_2SO_4 $0.9M$-citric acid/ $0.45M$-HNO_3
Transition-metal ions	Strong anion-exchanger	$12M$-HCl $\rightarrow$ $0.005M$-HCl
Lanthanides	Strong cation-exchanger	Citrate buffers
Actinides	Strong cation-exchanger	Lactate buffers
Separation of anions from one another,		
F^-, Cl^-, Br^-, I^-	Strong anion-exchanger such as $-CH_2N(CH_3)_2C_2H_4OH$	M-$NaNO_3$ + NaOH to give pH 10.4
Common inorganic anions in water	Strong anion-exchanger	$0.003M$-$NaHCO_3$/ $0.0024M$-Na_2CO_3

ortho-, pyro-, tri-, tetra-phosphate	Strong anion-exchanger	$M \rightarrow 0.005M$-NaCl
Organic Separations		
Acidic/Neutral/ Basic compounds	Cation/anion exchanger	Acidic or Alkaline buffers
Aldehydes, ketones, alcohols	Anion exchanger	Ketones and aldehydes held as bisulphate adducts. Eluted with hot water and NaCl respectively.
Sugars	Anion exchanger	As borate association complexes. Eluted by pH-gradient.
Amino-acids	Cation exchanger	Stepwise or pH-gradient elution

Fig. 9.6a. *Some representative applications of ion-exchange in analysis*

SAQ 9.6b

Which of the following separations could be achieved by ion exchange or ion-exchange chromatography?

(*i*) Separation of U^{235} from U^{238} as gaseous UF_6.

(*ii*) Resolution of a mixture of rare-earth metal ions.

(*iii*) Resolution of a mixture of simple sugars.

(*iv*) Resolution of a mixture of halogenoalkanes.

SAQ 9.6b

SAQ 9.6c   Propose schemes based on ion exchange for carrying out the following:

(*i*)   separation of Fe^{2+} and Fe^{3+},

(*ii*)   separation of Co^{2+}, and Pb^{2+} at low concentrations,

(*iii*) preconcentration of all the metal ions
present in a 1 dm^3 rain-water sample, by
a factor of 1000,

(*iv*)   separation of a carboxylic acid, an aldehyde, and an alcohol,

(*v*)   preconcentration of common anions in a
sample of tap-water.

SAQ 9.6c

Objectives

On completion of this Part, you should be able to:

● understand qualitatively the basic phenomenon of ion exchange, in terms of an ion-exchange resin in contact with a solution of an electrolyte;

● distinguish between cation-exchange and anion-exchange resins;

● write a chemical equation to show the ion-exchange process, and to understand that this is a reversible process which may reach equilibrium;

● describe how an ion-exchange process may be carried out in:

(i)   batch mode,
(ii)   multiple-batch mode,
(iii)   trapping mode (by using a column),
(iv)   continuous mode (chromatography);

- describe the various types of ion-exchange material available:

  eg

  (*i*)   PSDVB-SO$_3^-$
  (*ii*)   DVB-Acrylic acid,
  (*iii*) PSDVB-CH$_2$NH(CH$_3$)$_2^+$,
  (*iv*)   PSDVB-CH$_2$N(CH$_3$)$_3^+$,
  (*v*)   PSDVB-CH$_2$N(CH$_2$CO$_2^-$)$_2$,
  (*vi*)   Inorganic exchangers,   ⎫ Non-resinous
  (*vii*) Silica-based materials for hplc; ⎬ materials

- describe their properties in terms of:

  (*i*)   cation-exchange resins *versus* anion-exchange resins,
  (*ii*)   the pH range suitable,
  (*iii*) exchange capacity;

- appreciate that increase in the degree of cross-linking affects the properties of resins in various important ways:

  (*i*)   greater selectivity,
  (*ii*)   greater stability,
  (*iii*) reduced speed of attaining equilibrium;

- understand the ion-exchange process in more detail:

  (*i*)   that resins have a macroreticular porous structure, and that solvent and ions of each sign may penetrate into the resin;
  (*ii*)   that this leads to the definition of a two-phase system (resin phase and external phase) which is not simply related to exchange at inorganic sites;
  (*iii*) that because of high concentrations in the resin phase, activities will be very important;

- define selectivity coefficient as the equilibrium constant in concentration terms, for the ion-exchange process;

- consider values of selectivity coefficients and to predict their dependence on some of the following:

(*i*) the properties of an ion such as its hydrated ionic radius and charge,

(*ii*) the degree of cross-linking of the resin,

(*iii*) the nature of the ionogenic groups,

(*iv*) the concentrations of ions involved;

- appreciate that another very important factor affecting the apparent selectivity of an ion-exchange system is the existence of any chemical equilibria involving the exchanging ions in the external phase;

- define distribution coefficient and distribution ratio in the context of an ion-exchange system;

- describe some applications of ion exchange:

(*i*) removal from solutions of all ionic species – preparation of de-ionised water;

(*ii*) replacement of all ions of one charge type (cations/anions) by a single ion (eg domestic water softeners); replacement of all cations by H^+ followed by titration to give total cation concentration;

(*iii*) preconcentration of cations and anions, and recovery from the resin;

(*iv*) selective separations of ions of like sign by using chemical reactions to convert some into neutral molecules or ions of opposite charge;

(*v*) ion-exchange chromatography.

10. Other Separation Techniques

10.1. INTRODUCTION

In Parts 8 and 9 we have examined in some detail two of the most important separation techniques in analysis, namely solvent extraction and ion exchange. In this Part we conclude our study of separation by looking, rather more briefly, at a limited number of other separation techniques of perhaps somewhat less importance, but nevertheless of use in specific areas of analysis.

Before we proceed any further, may I refer you back to Section 7.6, in which we reviewed very briefly *all* the separation techniques normally encountered in analysis? It would be a good idea to revise that Section and answer the associated questions again before embarking on this Part.

The separation techniques that I have selected for study in this Part are as follows.

(*a*) *Adsorption/Desorption*, because of their importance in the sampling and preconcentration of trace organic compounds (perhaps present as pollutants) in air and water.

(*b*) *Precipitation and Coprecipitation*, because of their application to the separation and preconcentration of metal ions. You will have studied precipitation already perhaps in the *ACOL; Clas-*

sical Methods. Furthermore, the techniques are in some ways comparable to solvent extraction and ion-exchange, at least in their areas of application.

(*c*) *Gas- and Vapour-Phase Separations*. Distillation is a time-honoured separation technique which, although not now as important as it used to be in complex analytical separations since the advent of gas-liquid chromatography, still finds use in preliminary sample pretreatment, simple batch separations, crude fractionations, etc. Furthermore, also worth considering are some separations relying on conversion of a particular species into a volatile derivative and selectively volatilising (or distilling) this from the matrix.

These techniques are thus largely concerned with preliminary separations, associated with the sample pretreatment stage of the analysis, and thus fit in with the theme of sample pretreatment and separation of this Unit. The other extremely important area of separation in analysis is chromatography and there are a number of other ACOL Units devoted to this subject.

10.2. ADSORPTION/DESORPTION

When a fluid, such as a gas, liquid or solution, comes into contact with certain types of solid, molecules in the fluid phase can be held or retained, normally *reversibly*, on the surface of the solid by a variety of physical or chemical means. This is the process of *adsorption*, and the reverse process, in which the adsorbed molecules are released into the solution or vapour phase, is known as *desorption*. You can imagine the solid material as having a layer of adsorbed molecules temporarily 'stuck' onto its surface. Solids that are able to adsorb gas or solution-phase molecules in this way are known as *adsorbents*. In general, they are characterised by having:

(*a*) a large surface-area-to-mass ratio so that a relatively small mass of adsorbent can adsorb a relatively large number of molecules;

(*b*) specific adsorption sites on the surface, where adsorption will occur.

∏   Indicate for each of the following properties of a particulate
        solid, whether it would be consistent with producing a large
        surface-area-to-mass ratio:

        (*i*)   all particles are spherical in shape;
        (*ii*) the particles are small;
        (*iii*) the particles are irregular in shape with a porous sur-
                face structure.

(*i*)   *Not consistent*. A sphere is the shape consistent with producing
        the smallest surface-area-to-mass ratio. Think of the shape of
        a drop of water, in which surface tension works to minimise
        this ratio.

(*ii*) *Consistent*. The smaller the particle size, the larger will be the
        surface-area-to-volume ratio, and hence surface-area-to-mass
        ratio .

(*iii*) *Consistent*. An irregular surface with many pores (indenta-
        tions) will have a larger area than a smooth surface. However,
        if the surface is very irregular with deep pores there could be
        a steric factor. Large molecules, or oddly shaped molecules,
        might find access to adsorption sites in the pores difficult. See
        Fig. 10.2b.

Remember, we are talking about *adsorption* (note very carefully
the spelling), involving adsorption of molecules onto the *surface*
of a solid. Do not confuse this process with *absorption* whereby
molecules are absorbed into the *bulk* of an absorbent phase (a solid
or a liquid). Look at Fig. 10.2a for a light-hearted illustration of
this distinction. Fig. 10.2b shows some examples of adsorbent surfaces with specific adsorption sites and illustrates some of the ideas
presented in the last exercise.

The actual forces involved in adsorption differ in different situations. For simplicity we can consider two types of adsorption.

(*a*)   *Physical Adsorption*. This is a fairly general form of weak ad-
        sorption in which the molecules are held on the surface of the

ABsorption   ADsorption

Fig. 10.2a. *The difference between absorption and adsorption (taken from Reference 13)*

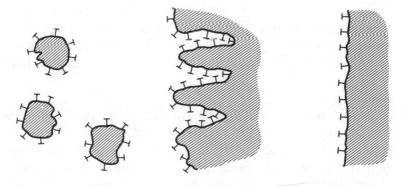

Fig. 10.2b. *A variety of adsorbent surfaces, showing the effect of particle size and the porosity of the surface*

solid merely by the general attractive forces between all molecules, known as van der Waals forces.

∏   What are van der Waals forces?

You should have come across van der Waals forces in physical chemistry. They are a group of general non-specific attractive forces between molecules that can be divided into three types.

(*i*)   Dipole – dipole interactions between a dipolar molecule and dipolar adsorbent.

(*ii*)   Dipole – induced dipole interactions between a polarisable molecule and a polar adsorbent (or *vice versa*).

(*iii*) Induced dipole – induced dipole interactions between polarisable (non-polar) molecules and adsorbents (these interactions exist because momentary dipoles are formed in non-polar molecules by the random movement of electrons in the molecules).

Reference 13 explains this quite clearly. You need not concern yourself with the equation presented in this text.

(*b*)   *Chemical Adsorption* (Chemisorption). Here the molecules are held more firmly and specifically by reversible chemical bonding onto the surface of the adsorbent. *Hydrogen bonding*, for instance, plays a very important role here, as does charge transfer.

Desorption occurs by several mechanisms. If the adsorbent bearing adsorbed molecules is treated with a *good solvent*, the molecules will effectively be 'dissolved' off the particle surfaces. Alternatively, adsorbed molecules may be displaced from adsorption sites by a solvent that itself is strongly adsorbed. Finally, volatile compounds may be removed from an adsorbent by heat (*thermal desorption*).

Adsorption and desorption are reversible processes and in any given system may be operating continuously in opposition to each other. We thus have an equilibrium system which for molecule X can be represented as

$$X_{gas,vapour,soln} \underset{desorption}{\overset{adsorption}{\rightleftharpoons}} X_{adsorbed\ on\ solid\ surface}$$

For this equilibrium, we can define a distribution coefficient:

$$K_D = \frac{\text{amount of adsorbed X per gram of adsorbent}}{\text{concentration of X in solution or vapour phase}}$$

Not all compounds are equally strongly adsorbed on all adsorbents. Not all solvents are equally effective at desorption. The basic property of an adsorbent, solvent, or adsorbed molecule that controls the strength of the adsorption (or *activity*) is *polarity* with polarisability also being of importance. Here are some general principles governing strength of adsorption.

(*a*) Polar adsorbents show the highest activity.
(*b*) Polar adsorbents adsorb polar solutes (or gases) more strongly than non-polar solutes (or gases).
(*c*) Polar solvents are the most efficient for desorption.
(*d*) Polar solvents desorb polar solutes more effectively than non-polar solutes.
(*e*) The trends are reversed for non-polar adsorbents and solvents, but the interactions and range of interactions are more restricted.
(*f*) In thermal desorption, volatility and temperature become the significant factors.
(*g*) Molecular shape and molecular size may have some significance due to steric effects at the irregular surface of an adsorbent.

Adsorption occurs at specific adsorption sites on the solid surface, as we have already seen. The activity of these sites is not constant, because of the chemical nature of the adsorption sites as well as to steric factors. If you look at Fig. 10.2b, you can see most clearly for the porous surface (the *middle* example) that the adsorption sites are far from equal to each other with regard to accessibility. The chemical nature of the adsorption sites may vary from site to site, affecting polarity or polarisability and thus activity.

∏   Consider the range of activity of the adsorption sites as discussed above, and state how would you expect the activity of the adsorbent as a whole to alter as increasing amounts of material are adsorbed onto its surface.

You would expect it to *decrease*, since the most active and accessible sites will be occupied first, less active sites being subsequently filled. A plot of amount of adsorbed material *versus* concentration of material in solution (or in the gas phase) at equilibrium is known as an *adsorption isotherm*. Typically they take the shape shown in Fig. 10.2c.

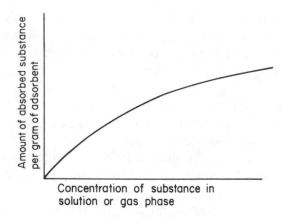

Fig. 10.2c.   *A typical adsorption isotherm*

At high concentrations, of course, eventually all the adsorption sites will be occupied, and the adsorbent becomes saturated.

Adsorption is a fairly commonly encountered phenomenon and the number of solid materials that show adsorbent properties is very large indeed. However, a much more limited range of adsorbents find routine use in analysis. Here are some typical examples of analytically useful adsorbents.

Charcoal – a powerful and non-selective adsorbent for a wide range of organic compounds, which is highly porous in nature.

Silica (finely divided particulate). Silica has a three-dimensional structure based on polymeric SiO_2. However, on the surface there are Si—OH groups. These are the adsorption sites, the mechanism of adsorption involves hydrogen bonding.

Alumina (finely divided particulate). This is rather like silica but based on polymeric Al_2O_3 with Al—OH groups as the adsorption sites. Both alumina and silica are strong adsorbents, with silica showing acidic properties and alumina basic properties.

Porous organic polymers, such as those based on polystyrene and divinylbenzene, as well as polyethers. These are milder adsorbents relying on the polarisability of the benzene ring or the more weakly polar ether linkages.

Adsorbents often have to be 'activated' before use. This involves heating them to a particular temperature for a particular length of time to desorb any molecules that have been adsorbed from the atmosphere. For example, silica and alumina are normally activated before use by heating them to 100–150 °C or 300–350 °C respectively before use, to desorb large amounts of water adsorbed from the atmosphere.

∏   Bearing in mind the fact that silica and alumina are very powerful adsorbents, which would be preferable for the adsorption of acidic substances and which for basic substances?

Use acidic silica for the acidic substances and basic alumina for the basic substances. You might have suggested the opposite, in order to increase the adsorption by introducing acid/base interactions, but in practice these would be so powerful as to cause irreversible adsorption.

You may well have come across adsorption in practice by way of the use of charcoal filters to remove noxious vapours (smells!) from the atmosphere and of decolorising charcoal for purifying organic compounds.

SAQ 10.2a

Which of the following processes can correctly be described as an adsorption?

(*i*)   Collection of trace-metal ions in water by passing the water through a column packed with a sulphonated polystyrene-divinylbenzene resin.

(*ii*) Collection of traces of an organic vapour in the atmosphere by passing air through a glass trap maintained at liquid nitrogen temperatures.

(*iii*) Collection of traces of a toxic gas in the atmosphere by passing air through a solution with which the gas reacts.

(*iv*) Collection of traces of organic compounds in a river water sample by passing the water through a column packed with charcoal.

SAQ 10.2b

Offer explanations for the following observations.

(i) The more polar an organic compound is, the less efficiently is it adsorbed onto a non-polar absorbent from an aqueous medium.

(ii) Adsorption of a volatile compound from the vapour phase is more pronounced at lower temperatures.

(iii) Silica, which in a finely divided form adsorbs water very readily normally does not strongly adsorb compounds in general until after it has been heated at 100–150 °C for at least 30 min.

(iv) The adsorption capacity (the maximum possible amount of material that can be adsorbed per gram of adsorbent) of an adsorbent increases as the particle size of the adsorbent decreases.

SAQ 10.2b

Applications of Adsorption in Analysis.

Adsorption can be used in analysis in a number of ways, largely, but not entirely, connected with the collection, preconcentration, and separation of analytes. Here are some of its more important applications.

(*a*) The collection and preconcentration of organic vapours from the atmosphere.

(*b*) The collection and preconcentration of organic compounds (volatile or non-volatile, but not too polar) present at trace levels in water.

(*c*) The permanent removal of certain fractions of material from organic liquid or solution samples, as a preliminary clean-up step.

(*d*) As an important branch of chromatography. Adsorption chromatography covers tlc and hplc, with, for example, silica or alumina as stationary phase and various organic solvents as mobile phase. A wide range of not too polar organic compounds can be separated. It also embraces gas–solid chromatography for the separation of gases and very volatile organic vapours.

These are its main uses, although you may also recall the use of adsorption indicators in titrimetry. Furthermore, coprecipitation, which we shall study later in this Part, involves adsorption. However, we shall now be selective and just consider applications (*a*) and (*b*), ie the use of adsorption for collection and preconcentration. Adsorption chromatography will be fully covered in the appropriate chromatography Units.

10.2.1. Adsorption for the Preconcentration of Organic Compounds in the Atmosphere and in Water Samples

The basic idea is to pump air or water at a constant, known flow-rate for a known length of time through a tube packed with a suitable adsorbent, which quantitatively removes any organic compounds present by adsorption. The equipment involved, which normally includes a small battery-operated pump, is easily portable. Consequently the technique can really be considered as a *sampling* technique, since it is possible to set up the pump and adsorption tube to sample air or water at any chosen location. The volumes of air or water that can be sampled are theoretically unlimited and in practice are very large. This leads to very large preconcentration factors.

The most common applications for these air or water samplers is for monitoring environmental pollutants. One especially important example is the personal air monitor. This consists of a small, battery operated pump that fits into the pocket of a worker's laboratory coat and is connected to the adsorption tube, which is attached to the lapel of the coat, in order to monitor the air that he is breathing. Sampling rates might be of the order of 1 dm^3 of air per minute and sampling times might be 8 hours (ie a complete working shift).

∏   If a personal air-monitor samples at the rate of 1.50 dm^3 per minute, what would be the total volume of air sampled over an 8-hour working shift?

The total volume of air sampled would be

$$1.50 \times 60 \times 8 = 720 \text{ dm}^3.$$

∏   How would you classify the above separation process?

    (*i*)   Batch,
    (*ii*) Multiple batch,
    (*iii*) Continuous,
    (*iv*) Trapping.

(*iv*) Trapping. The technique is very comparable with say, the use of ion exchange for preconcentrating metal ions in water.

The result of this combined sampling and preconcentration is to obtain a sample of an organic compound from a known and possibly very large volume of air or water, conveniently preconcentrated onto a few grams or less of adsorbent. However, before we can carry out the final analysis, we need to *desorb* these compounds efficiently and quantitatively by one means or another. We can either use solvent desorption or thermal desorption.

In *solvent desorption* we could shake the adsorbent with a good solvent, usually organic, for the compound in question (batch technique), or we could pass the organic solvent through the adsorption tube (reverse of the trapping technique). Lastly we could use a Soxhlet extraction technique, in which the adsorbent is placed in a Soxhlet thimble and solvent is continuously passed through the adsorbent by reflux distillation. (See Section 2.5.2 and Fig. 2.5a, for a fuller description and diagram of the Soxhlet extraction technique.)

In *thermal desorption*, which is suitable only for volatile molecules, the adsorbent is heated in a stream of inert gas sufficiently strongly to release the adsorbed molecules into the vapour phase. The vapour may then be condensed in a cold trap or otherwise treated as required.

∏   One practical problem in using an adsorption tube is that of breakthrough. This occurs when the adsorbent has become saturated with adsorbed molecules. Further molecules passing through the adsorption tube are no longer retained. The cure for breakthrough is simply to shorten the sampling time. However, can you suggest a simple practical means for *monitoring* whether breakthrough is occurring or not?

The usual way of monitoring for breakthrough is to use an adsorption tube in which the adsorbent is divided into two sections as below.

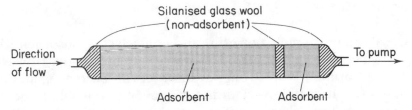

Because of the direction of flow, saturation will occur progressively starting at the left-hand end of the tube, the last adsorbent to be saturated being at the right-hand end of the tube. So we analyse both charges of adsorbent. If any adsorbed molecules are found in the small charge of adsorbent at the right-hand end of the tube, breakthrough has occurred from the main charge (left and centre of tube). Such analyses are rejected, the only acceptable analyses being those where *no* adsorbed molecules are found in the right-hand side (second) charge.

∏ A common method of desorption is simply to shake the adsorbent with a measured volume of organic solvent and then separate the adsorbent and solvent by filtration or decantation. How can you check that the desorption has resulted in 100% recovery?

Treat the adsorbent with a further volume of solvent and analyse both extracts separately. No desorbed compounds should be found in the second volume of solvent if the recovery was 100% initially. However, if there is analyte present in the second extract, then the two extracts can be combined and the adsorbent tested with a third volume of solvent (a multiple batch procedure). Repeat the procedure until no further material is desorbed.

Choice of Adsorbent

The most traditional adsorbent used for preconcentrating organic compounds in this way is *charcoal*. Charcoal is a highly porous material with a very complex and somewhat variable molecular structure, consisting primarily of carbon in the form of fused aromatic rings (the graphite structure), but with many residual organic groups attached. These tend to be polar in nature and provide the necessary adsorption sites.

Charcoal is an extremely powerful and non-selective adsorbent that traps a wide range of organic compounds very efficiently. However, because adsorption is so strong, desorption can become something of a problem. It is often very difficult or even impossible to desorb certain molecules, whereas others can be desorbed only with the possibility of chemical change occurring. Whenever this is so, and it often is, then charcoal is clearly not a very appropriate material to use. The alternative is to use a milder adsorbent, and a number of synthetic *porous organic polymers* have been produced for this purpose. They may not be quite such efficient adsorbents as charcoal, but their performance is likely to be more efficient overall because of the likelihood of obtaining 100% desorption. Two such materials are *XAD-2*, a porous polystyrene-divinylbenzene copolymer, used mainly for adsorbing moderately polar molecules from water samples, and the *Tenax* series of polymers, based on the following structure.

These are particularly useful for adsorbing molecules from atmospheric samples. Recently, the trend does seem to be towards the use of these milder adsorbents in preference to charcoal.

Choice of Organic Solvent for Desorption

Basically, this should be a good solvent for the adsorbed compounds, and at the same time an appropriate solvent for the subsequent stages of the analysis, which may well involve further preconcentration by evaporation, or, say, analysis by chromatography or spectroscopy. Common practical choices include:

(*a*) diethyl ether (ethoxyethane) (good desorbent and volatile),

(*b*) trichloromethane,

(*c*) methanol (polar, transparent in the uv),

(*d*) carbon disulphide (doesn't interfere in gas chromatography),

(*e*) pyridine (good desorbent).

∏ What is the alternative to the use of a solvent for desorption, and for what types of compound would it be suitable?

Thermal Desorption, which is suitable for desorbing *volatile* organic compounds.

Examples of Sample Collection/Preconcentration by Adsorption

(*a*) *From River Water*

Here is a possible procedure for obtaining a preconcentrated sample, suitable, say, for chromatographic analysis of a mixture of dissolved non-volatile organic compounds (eg pesticides) present at trace levels in river water. The adsorption step can be carried out as part of the sampling procedure at the actual site.

Water is pumped at a known flow-rate and for a known time, first through a filter to remove suspended solids and then through a small tube containing 2 g of activated charcoal. The total volume of water sampled (measured by multiplying flow-rate by time) is very flexible and could be, for example, anything between 1 dm^3 and 100 dm^3. The organic compounds are trapped on the charcoal by adsorption and the tube is taken back to the laboratory to complete the analysis. The organic compounds are desorbed by slowly passing 25 cm^3 of diethyl ether through the tube. The ethereal solution, which will be wet, is dried with, say, magnesium sulphate and is evaporated to dryness. The residue of non-volatile organic compounds is then dissolved in 0.1 to 1 cm^3 of diethyl ether or other solvent ready for further pretreatment or analysis by chromatography.

∏   What is the maximum overall preconcentration factor implied by the above procedure (as described).

1 000 000, ie 10^6, assuming that analytes from 100 dm^3 of water are preconcentrated into a final volume of 0.1 cm^3 of organic solvent.

∏   Suggest *two* possible reasons for obtaining less than 100% recovery from the above procedure.

Here are the two possible reasons:

(*i*)   'breakthrough' in the adsorption tube, due to the charcoal adsorbent becoming saturated with adsorbed molecules,

(*ii*) non-quantitative desorption and recovery of the adsorbed molecules when the charcoal is treated with the diethyl ether.

In practice the second cause of loss (incomplete desorption) is likely to be the more serious cause of error.

(*b*) *From the Atmosphere in a Factory Working Environment*

A tube containing a porous polymer of the Tenax type is connected to a small battery-operated pump and set up at a suitable sampling site in the factory. Air is sucked through the tube at the rate of 1.25 dm^3 min^{-1} for 8 hours. The tube is then removed and placed in an automated thermal desorption (ATD) unit connected directly to a gas chromatograph (GC). In the ATD unit the tube is strongly heated to release the adsorbed compounds which are swept in a stream of nitrogen into a small cold trap where they are quantitatively retained by freezing them. Subsequently the cold trap is rapidly heated to vaporise the compounds which are swept as a narrow band of vapour into the gas chromatograph for separation and analysis.

∏ For what types of organic compound would this method be suitable?

Volatile and thermally stable organic compounds. They are vaporised twice and analysed by the GC.

∏ Why have a cold trap between the ATD and GC? Why not simply allow the desorbed compounds to pass straight into the GC?

Release of the molecules by thermal desorption is not necessarily a very fast process. Thus the molecules would not be swept into the GC in a narrow band, which is necessary for good separation. This is much more easily achieved by using a cold trap.

∏ The chromatogram from such an analysis showed a peak which corresponded to 60 μg of diphenyl ether. What was the average concentration of diphenyl ether in the atmosphere at the sampling point over the 8-hour period?

The volume of air sampled

$$= 1.25 \times 60 \times 8 \text{ dm}^3$$

$$= 600 \text{ dm}^3.$$

This contained 60 μg of diphenyl ether.

∴ the average concentration of diphenyl ether in the atmosphere at the sampling point over the 8-hour period

$$= 60/600 \ \mu\text{g dm}^{-3}$$

$$= 0.1 \ \mu\text{g dm}^{-3}.$$

SAQ 10.2c

Here is a description of an attempted sampling/preconcentration procedure, but unfortunately the analyst who carried this out made a number of mistakes. Can you spot four mistakes and suggest in each case what he should have done?

'Water, which was to be analysed chromatographically for a wide range of organic compounds of various properties, was passed slowly through a short column in two sections, each packed with granulated charcoal. The flow-rate and sampling time were not recorded, although the total sample volume was measured as 10 dm^3. The charcoal from the two sections of the tube was combined for desorption. The wet charcoal was dried by heating it at 150 °C for 30 min and the compounds were then desorbed by shaking the charcoal with 10 cm^3 of hexane and filtering off the charcoal.'

The standard textbooks give this topic scant coverage but in reference 20, pp 565–575 the use of charcoal for the recovery of pollutants from water is discussed.

10.3. PRECIPITATION

In this section, I am assuming that you are reasonably familiar with the basic principles of precipitation reactions and their application to gravimetry. If this is not so, or if you feel that your knowledge in this area has become a little 'rusty', then the *ACOL; Classical Methods,* covers the relevant material, as do many excellent standard textbooks on analysis such as references 1 and 2.

In particular, you should be familiar with the following concepts.

(*a*) Solubility products and their application to predicting solubilities.

(*b*) The effect of pH and complex formation on solubilities. Multiple equilibria involving precipitation, acid-base equilibria, and complex formation. (A qualitative appreciation will be adequate).

(*c*) The mechanism of precipitate formation. Formation of nuclei and crystal growth.

(*d*) Factors affecting the purity of the precipitate such as adsorption, post-precipitation, occlusion, and coprecipitation.

That said, in this section we shall consider two applications of precipitation specifically to separation, *viz* the use of coprecipitation as a preconcentration technique for trace-metal ions in water, and precipitation as a means of carrying out relatively simple batch separations before an analysis.

10.3.1. Coprecipitation for the Preconcentration of Trace-Metal ions in Water

We have already met a number of techniques for trace-metal pre-concentration, to which we now add coprecipitation. However, I think it fair for me to point out straight away that coprecipitation, · although having its uses in certain situations, is by no means as important an analytical preconcentration technique as the other techniques we have studied.

∏   Name *two* techniques other than coprecipitation that can be used for the preconcentration of trace-metal ions in water.

We have in fact met *three* such techniques:

— Solvent extraction,
— Ion exchange,
— Evaporation (of the water or solvent from an extract). We have met this in conjunction with the other techniques, but it can be used on its own.

Can I ask you now to recall a common source of error in gravimetric analysis, namely the contamination of precipitates by other species present in the solution that may or may not be precipitated under the prevailing conditions? This contamination may occur through one or other of several mechanisms, collectively known as *coprecipitation*.

∏   Briefly describe *three* ways in which precipitates can be contaminated by other species present in solution, ie in which coprecipitation can occur.

There are actually *four* distinct mechanisms for precipitate contamination, ie coprecipitation. Note that I have not included amongst these the simple mechanism of the simultaneous formation of two precipitates. None of these mechanisms requires the coprecipitated species to be quantitatively insoluble in its own right.

The four mechanisms are:

(*a*) surface adsorption,
(*b*) isomorphic inclusion (mixed-crystal formation),
(*c*) non-isomorphic inclusion (solid-solution formation),
(*d*) occlusion (ion entrapment).

Now, although in gravimetry coprecipitation is nothing other than a nuisance and a source of error, under certain conditions *trace species* can be coprecipitated *quantitatively* and thus we have a means of collecting or scavenging such material at concentrations at which it would not otherwise be possible by precipitation. In other words, we have the basic means for a potentially viable preconcentration procedure. Now let us look at the mechanisms of coprecipitation in more detail.

(*a*) *Surface Adsorption.* If a precipitate is formed from two ions of opposite charge and one of these ions is present in excess, the precipitated particles will tend to carry an excess of these ions on their surfaces, and thus will be charged. Other ions of the opposite charge will then be attracted to the surface of the particles and will be held there by electrostatic forces. For example, if you precipitate oxalate ion with calcium chloride, when calcium ions are in excess, the precipitate will be contaminated with calcium chloride. Colloidal hydrous oxides of iron, aluminium, and certain other cations are precipitates that strongly adsorb heavy metals by this mechanism, the excess ions on the particle surface being hydroxide ions.

(*b*) *Isomorphic Inclusion* (mixed-crystal formation). Here the coprecipitated ion has dimensions and a chemical composition such that it can fit into the crystal structure of the precipitate without causing appreciable strain or distortion. For example, traces of chloride ion are coprecipitated with silver bromide. Sometimes both ions can be coprecipitated. For example, potassium permanganate forms a mixed crystal with barium sulphate.

(*c*) *Non-isomorphic Inclusion* (solid-solution formation). Here the bulk material being precipitated acts as a 'solid solvent' for the trace components. The rule for liquid solvents that 'like dissolves like' applies here too. Here is a typical example. The trace-metal ions are first complexed with an organic reagent, such as 8-hydroxyquinoline. An indifferent organic compound that is insoluble in water but dissolved in a solvent that is miscible with water (such as β-naphthol in ethanol or acetone) is added. The β-naphthol is then precipitated, carrying down the 8-hydroxyquinoline complexes with it.

(*d*) *Occlusion* (ion entrapment). The precipitate is formed so fast, that ions are physically engulfed in the solid before they can diffuse away. Unlike the other mechanisms, this is somewhat unspecific.

If you think about the above four processes you should be able to see that, even if in some of the examples insoluble species are coprecipitated, it is not actually *necessary* for the coprecipitated trace component to be insoluble on its own.

Adsorption and both inclusion processes [processes (*a*), (*b*) and (*c*)] are all potentially suitable for preconcentration purposes. Occlusion [process (*d*)] on the other hand, is too unpredictable and unspecific to be of any practical use.

Now, let us put all this to good use to devise procedures for preconcentrating trace-metal ions in water by coprecipitation. We always have to specify a *carrier* or *collector*. This is the material which, added to the solution at fairly high concentration, is subsequently precipitated, and with which the trace components are coprecipitated. This leads us to think about certain problems that are always associated with coprecipitation.

∏   Preconcentration by coprecipitation involves addition to the sample, of a carrier or collector at fairly high concentration, and precipitation of this under such conditions that the trace components are coprecipitated with the carrier. Separation is then by filtration. What problems are associated with this procedure? See if you can think of two such problems.

The two problems that are associated with this procedure that first come to mind are as follows.

(*i*) The preconcentrated analytes are obtained in the solid state rather than in solution. Since most subsequent analytical techniques require the sample to be in solution, a dissolution step must usually follow the coprecipitation.

(*ii*) The preconcentrated analytes are obtained in the presence of a large excess of carrier. This may well interfere in the subsequent analysis. If this is so, a subsequent separation of trace analytes from the carrier may be required. For metal ions this is often achieved by ion exchange.

∏   Can you name any analytical techniques that are suitable for the identification or determination of metals in the solid state, ie techniques that could be applied directly to the precipitate without dissolution?

(*i*)   Arc/spark emission spectroscopy,
(*ii*) X-ray fluorescence spectroscopy,
(*iii*) Spark-source mass spectrometry,
(*iv*) Neutron-activation analysis.

The above are just four such techniques. Don't worry if you are not familiar with these. Let me assure you that, perhaps with the exception of the first, they are all complex specialised techniques, involving high-cost instrumentation. Hence our interest in getting the coprecipitated analytes back into solution.

Here are two examples of preconcentration of trace-metal ions by coprecipitation.

(*a*) Heavy-metal ions and transition-metal ions from sea-water

We are particularly fortunate here in that we do not need to add a carrier, since there are sufficient magnesium ions naturally present in sea-water to serve this purpose. So, all we have to do is to make the sample of sea-water alkaline and the magnesium ions are precipitated as $Mg(OH)_2$. The trace metal ions are adsorbed onto the $Mg(OH)_2$ precipitate so long as there is an excess of OH^- ions in the solution, which is likely in practice. The precipitate is filtered off and dissolved in hydrochloric acid.

Π   Now suggest how the trace metal ions might be separated
        from the Mg^{2+} ions in hydrochloric acid solution.

A distinct possibility is anion exchange, since if the HCl concentration is high enough, most transition- and heavy-metal ions form
stable chloro-complexes that will be retained on the anion-exchange
resin, whereas the Mg^{2+} will not. Some ions, such as Ni^{2+}, are not
retained in this way. To separate Ni^{2+}, neutralise the acid and carry
out a neutral-chelate solvent extraction.

(*b*) *Heavy-metal ions and transition-metal ions from rain-water*

Adjust the pH of the rain-water sample to a neutral to slightly alkaline value (to ensure that subsequent complex formation is quantitative) and then add to it a solution of 8-hydroxyquinoline. Then
prepare a solution of β-naphthol in ethanol or acetone. Add this solution to the sample. The β-naphthol will be precipitated and will
carry down the complexed trace-metal ions in the form of a solid
solution. Filter off the precipitated β-naphthol.

Π   How might you obtain an aqueous solution of the precon-
        centrated metal ions free of organic material?

You could wet-ash the precipitate with strong mineral acids in a
Kjeldahl flask. Alternatively, you could redissolve the precipitate in
an organic solvent immiscible with water and back-extract the metal
ions into an acid medium.

SAQ 10.3a   Propose *two* outline schemes for preconcentrat-
                 ing the metal ions Cu^{2+}, Zn^{2+}, Cd^{2+} and Pb^{2+}
                 present at trace levels in 1 dm^3 of tap-water. The
                 main preconcentration step should be by copre-
                 cipitation. The final sample should consist of 5
                 cm^3 of aqueous solution, free of carrier.

                 Make a few critical comments about the relative
                 merits of your two schemes.

SAQ 10.3a

For further reading on this topic, which gets scant coverage in most standard analytical textbooks, I recommend reference 20, pp 486–491.

10.3.2. Precipitation as a Separation Technique

The basic idea here is to look for conditions under which one species or group of species forms an insoluble derivative, whereas other species do not. However, before we do that, here are a couple of revision questions for you.

∏ What equilibrium constant tells us whether, under given conditions, a particular ionic compound will be precipitated or not? How does it tell us and what other information does it give us?

The solubility product. For an ionic compound $bA^{a+} . aB^{b-}$, the solubility product, K_{sp} is given by:

$$K_{sp} = [A^{a+}]^b . [B^{b-}]^a$$

where $[A^{a+}]$ and $[B^{b-}]$ represent the concentrations of the ions in solution *in equilibrium* with precipitated solid.

You can predict whether material will be precipitated or will redissolve, since at equilibrium the product, K_{sp}, for the solution cannot be exceeded. If it is, then material will be precipitated until the appropriate product of the molar concentrations of the ions in solution is equal to K_{sp}. On the other hand, if a solution in which the product of the concentrations is less than K_{sp} is in contact with solid, then dissolution of the solid will proceed until K_{sp} is achieved. The solubility product also gives us information about the completeness of a precipitation since it tells us about the concentration of the ions remaining in solution above a precipitated solid. A small value of K_{sp} implies a highly insoluble substance whereas a higher value implies greater solubility. The range of values of K_{sp} for inorganic substances normally considered as insoluble is from *ca* 10^{-4} to very nearly 10^{-100}!

Note also that K_{sp} refers to systems in thermodynamic equilibrium. It tells us nothing about the possibility of the formation of supersaturated solutions or about the rate at which solids will dissolve.

∏   If we consider only solutions in thermodynamic equilibrium with solids, K_{sp} is not the whole story when we come to calculate actual solubilities of many ionic compounds under specific chemical conditions. What other factor needs to be taken into consideration?

The possibility that one or other of the ions is involved in other chemical equilibria. For example, if the anion of the salt is also the anion of a weak acid, then it may be in equilibrium with the free acid. The relative concentrations of anion and free acid will then be a function of pH. The solubility of such a salt will then become pH dependent, solubility decreasing with increasing pH.

Other examples involve complex formation, such as the dissolution of $Cu(OH)_2$ or AgCl in aqueous ammonia, the increased solubility of AgCl in solutions of high chloride content, or the redissolution of certain metal hydroxides [such as $Al(OH)_3$] at high pH. In each of these examples the metal cation forms a complex ion with ammonia or the anion present in excess (Cl^- or OH^- in these examples).

Thus precipitation can be used to separate two ions in solution, provided the individual solubility products of the two ions with a common ion of the opposite charge are sufficiently different, notwithstanding any relevant complex-formation reactions. There are many examples of this, and we can now consider how some of them can be used in practice.

(*a*) *Separations of Cations by Hydroxide/Oxide Precipitation*

The values of K_{sp} for metal hydroxides differ tremendously. Moreover [OH^-] can itself be varied quite simply by a factor of 10^{15} or more and monitored by pH control. Thus certain groups of cations can be separated from one another simply by adjustment of pH and precipitation of the hydroxides or oxides. I have given you some examples of how this works in practice in Fig. 10.3a.

∏   What practical problem that we have already met, might limit the efficiency of some of these separations?

Coprecipitation. We have seen already that some metal hydroxides act as efficient carriers for the coprecipitation of other ions. This means that the precipitate could be contaminated with other ions under these conditions unless we take care to avoid coprecipitation (avoid an excess of OH^- ions in solution).

(*b*) *Separations Based on Metal-sulphide Formation*

Most metal ions, apart from those of the alkali and alkaline earth metals, form sparingly soluble sulphides, although their solubility products differ greatly. If we could vary the sulphide-ion concentration over a very wide range, then it should be possible to separate metal ions *via* their sulphides.

∏   How can we vary the concentration of free sulphide ion (ie S^{2-}) over a very wide range?

Reagent	Species forming precipitates	Species not precipitated
Hot concentrated HNO_3	Oxides of W^{6+}, Ta^{5+}, Nb^{5+}, Si^{4+}, Sn^{4+}, Sb^{5+}	Most other metal ions
NH_3/NH_4Cl buffer	Fe^{3+}, Cr^{3+}, Al^{3+}	Alkali metals and alkaline earths, Mn^{2+}, Cu^{2+}, Zn^{2+}, Ni^{2+}, Co^{2+}
$CH_3CO_2H/CH_3CO_2NH_4$ buffer	Fe^{3+}, Cr^{3+}	Common dipositive ions
$NaOH/Na_2O_2$	Fe^{3+}, most dispositive ions, rare earths	Zn^{2+}, Al^{3+} oxyanions of Cr^{6+}, V^{5+}, UO_2^{2+}

Fig. 10.3a.   *Separations based on control of pH*

Hydrogen sulphide (H_2S) is a weak acid, ionising to give HS^- and S^{2-} ions in solution. The concentrations of HS^- and S^{2-} relative to the concentration of H_2S are a function of hydrogen-ion concentration. At low pH, there will be a predominance of H_2S, whereas at high pH, the S^{2-} ion will predominate. So, the answer to the question is not to concern yourself with the concentration of sulphide reagent added (either H_2S or $(NH_4)_2S$) but rather to vary the pH of the solution.

Thus separations are possible by treating the solutions of a mixture of metal cations with H_2S or $(NH_4)_2S$ at different controlled values of pH. Fig. 10.3b shows some potential separations.

(c) Separations by Using Organic Reagents

The neutral chelate complexes that we used for the solvent extrac-

Reagent and Conditions	Metal Ions Precipitated	Metal Ions Not Precipitated
H_2S + $3M$-HCl	Hg^{2+},Cu^{2+}, Ag^+ oxyanions of As^{5+},As^{4+}, Sb^{5+},Sb^{3+}	Bi^{3+},Cd^{2+}, Pb^{2+},Sn^{2+}, Sn^{4+},Zn^{2+}, Cd^{2+},Ni^{2+}, Fe^{2+},Mn^{2+}
H_2S + $0.3M$-HCl	Hg^{2+},Cu^{2+}, Ag^+,As^{5+}, As^{3+},Sb^{5+}, Bi^{3+},Cd^{2+}, Pb^{2+},Sn^{2+}, Sn^{4+}	Zn^{2+},Co^{2+}, Ni^{2+},Fe^{2+}, Mn^{2+}
H_2S + acetate buffer (pH 6)	Hg^{2+},Cu^{2+}, Ag^+,As^{5+}, As^{3+},Sb^{5+}, Sb^{3+},Bi^{3+}, Cd^{2+},Pb^{2+} Sn^{2+},Sn^{4+}, Zn^{2+},Co^{2+}, Ni^{2+}	Fe^{2+},Mn^{2+}
$NH_3/(NH_4)_2S$ buffered to pH 9	Hg^{2+},Cu^{2+}, Ag^+,Bi^{3+}, Cd^{2+},Pb^{2+} Sn^{2+},Zn^{2+}, Co^{2+},Ni^{2+}	As^{5+},As^{3+}, Sb^{5+},Sb^{3+} Sn^{4+}

Fig. 10.3b. *Separations based on sulphide precipitation*

tion of metal ions were effective for that purpose because of their low solubility in water. In the absence of an extraction solvent, therefore, these reagents can in many cases be used as precipitants for the metal ions. Selectivity can be introduced into these precipitations in much the same way as it was in solvent extraction.

∏   Suggest *three* ways in which selectivity can be introduced
       into the precipitation of metal ions as neutral complexes.

(*i*)   By selecting a specific reagent,
(*ii*)   by the use of pH control,
(*iii*) by the use of masking agents.

All three methods were also found useful for increasing selectivity
in the solvent extraction of metal ions.

Fig. 10.3c gives some illustrative examples of separations by precipitation using organic reagents.

(*d*) *Electrolytic Precipitation (Deposition) of Cations*

If a solution containing a mixture of cations is electrolysed under
conditions of controlled potential, then those metal ions that can be
reduced to the metal at the chosen potential will be deposited as free
metal at the cathode (this is a type of precipitation). The removal of
the metals can be controlled by adjusting the potential of the cathode. A common choice of cathode is the mercury electrode, since
the metals that are removed will usually form an amalgam with the
mercury. The metal ions most conveniently removed from solution
include transition- and heavy-metal ions, those ions in other words
that are more easily reduced than zinc(II), leaving ions such as those
of the alkali metals, alkaline earths, and aluminium(III) in solution.

∏   In what kind of situation might an electrolytic separation
       prove useful?

In the situation where you are analysing for metals such as the alkali metals, alkaline earths, or aluminium(III) but where traces of
transition- or heavy-metals might interfere. Electrolytic precipitation is then used to remove the interferents. Titrimetry, polarography, spectrophotometry, and spectrofluorimetry are all techniques
where such interferences may arise.

One final point is that there is an electrochemical technique known
as *anodic stripping voltammetry* in which traces of heavy-metal ions

Reagent and Conditions	Metal Ions Precipitated	Metal Ions Left in Solution
1-Nitroso-2-naphthol in slightly acid solution	$Co^{2+}, Fe^{3+}, Pd^{2+}, Zr^{4+}$	$Pb^{2+}, Cd^{2+}, Be^{2+}, Al^{3+}, Ni^{2+}, Mn^{2+}, Zn^{2+}, Ca^{2+}, Mg^{2+},$ oxyanions of As(V) and Sb
Dimethylglyoxime in slightly alkaline ammoniacal solution containing citric or tartaric acid	Ni^{2+}	$Fe^{3+}, Al^{3+}, Cr^{3+}, Mg^{2+}, Bi^{3+}, Cd^{2+}, Zn^{2+},$ oxyanions of As(V) and Sb
Dimethyglyoxime in sodium acetate buffer	Ni^{2+}	$Co^{2+}, Zn^{2+}, Mn^{2+}, Pb^{2+}$
8-Hydroxyquinoline in ammonium acetate/acetic acid buffer	Al^{3+}	Be^{2+}
8-Hydroxyquinoline in ammoniacal buffer	Mg^{2+}	Other alkaline earth metals
2-Methyl-8-hydroxyquinoline	Many metal ions	Al^{3+}

Fig. 10.3c. *Illustrative examples of separations of cations using organic reagents as precipitants*

are actually electrolysed into a single drop of mercury for the purposes of *preconcentration*. The actual electrochemical determination that follows is of the heavy-metal ions themselves.

(e) Organic and Biochemical Separations Based on Precipitation

Precipitation may also be used to remove certain types of organic or biochemical material from a sample. A most important example of this comes from the clinical analysis of blood, serum, or plasma where it is normally necessary to remove the protein from the sample before analysing for other components. Here are two ways by which protein can be precipitated from such a sample.

Folin-Wu Method

One volume of blood, serum, or plasma is mixed with 7 volumes of water and 1 volume of $0.33M$-H_2SO_4 and allowed to turn brown. Then 1 volume of sodium tungstate solution (10% w/v $Na_2WO_4 . 2H_2O$), is added and after 2 minutes the precipitated protein is removed by filtration or centrifugation.

TCA Method

One volume of blood, serum, or plasma is mixed with 9 volumes of 5% aqueous trichloroacetic acid (TCA). After the proteins are precipitated, the mixture is filtered or centrifuged.

SAQ 10.3b	Suggest how the following separations could be achieved by means of precipitation: (*i*) copper(II) and silver(I) from cobalt(II) and nickel(II), (*ii*) iron(II), chromium(III), and aluminium(III) from calcium(II), copper(II) and zinc(II), (*iii*) iron(II), cobalt(II), and nickel(II) from calcium(II), magnesium(II) and aluminium(III), $\longrightarrow$

SAQ 10.3b
(cont.)

(*iv*) nickel(II) from iron(III), chromium(III) and zinc(II).

SAQ 10.3c

In our discussions of separation of metal ions by precipitation we have proposed a number of separations, largely based on considerations of relative magnitudes of solubility products and on the possible existence of other chemical equilibria affecting the concentrations of one or other of the precipitating ions. However, if these separations are attempted in practice you may come face to face with two or more problems which mean that these separations must be evaluated pratically before being finally accepted as feasible. Suggest *two* of these practical problems.

SAQ 10.3c

10.4.   GAS- AND VAPOUR-PHASE SEPARATIONS

We finish our studies in this Unit on separation techniques with a very brief look at how equilibria between the *gas or vapour phase* and liquids, solutions, or even perhaps solids can be used for separation purposes in analysis. We shall also consider briefly *distillation* and *fractional distillation* as separation techniques. Following that, *volatilisation*, whereby certain components of a mixture are chemically converted into volatile derivatives and then distilled out of the sample, will be our concern. Then we shall consider the use of *cold trapping* for the collection of analytes present in vapour form. Finally, we shall note how *evaporation* of solvents has been used for preconcentration purposes (we have, of course, already used this in earlier sections of this Unit).

The most important analytical technique for separating species by way of gas- or vapour-phase equilibria, however, will not be considered by us in this section. This is, of course, *Gas Chromatography*. This technique merits a whole *ACOL* Unit.

I shall assume that you have some knowledge of the basic principles of distillation, fractional distillation, and associated techniques. The following SAQ is designed to test your knowledge of such techniques. Work through it. If at the end, some points still confuse you, then perhaps you need to revise the topic of distillation from standard textbooks of physical or analytical chemistry (eg reference 13). But from the point of view of this Unit, do not make a big issue of this. Our evaluation of distillation in analysis will show it to have limited application as a separation technique, because of the power of gas chromatography when used for the separation of volatile species.

SAQ 10.4a   Indicate which of the following statements are *true* and which are *false*. Where the statements are false, explain clearly what the true statement should be.

(*i*) Distillation is a technique for separating species on the basis of differences in boiling-point. All species that can be volatilised are theoretically separable by distillation as long as their boiling-points differ, the ease of separability depending on, and solely on, differences in boiling-point.

(*ii*) Fractional distillation is a variant on normal distillation in which the condensed vapours are collected in separate fractions of different boiling-range.

(*iii*) Fractional distillation offers the only feasible means of separating a very complex mixture of volatile liquids on an analytical scale. $\longrightarrow$

**SAQ 10.4a
(cont.)**

(*iv*) If you attempt to purify a high-boiling liquid by distillation in a conventional still but find that it starts to decompose below its boiling-point, then you will need to seek some technique other than distillation for purifying it.

(*v*) Steam distillation is a special technique whereby the liquid being distilled is held in a vessel over a steam bath. This technique is designed to ensure that distillation takes place exclusively at 100 °C, thus permitting the separation of those species that are more volatile than water from those that are less.

(*vi*) Azeotropic distillation means distillation under an atmosphere of nitrogen. This technique is used for the distillation of species sensitive to oxidation at high temperatures (from the French 'l'azote', meaning nitrogen).

10.4.1. Applications of Distillation Techniques in Analysis

Distillation has many varied applications to the separation of liquids on an industrial scale. One only has to think of the production of spirits (of the intoxicating variety!) and the fractional distillation of mineral oil to produce petrol, paraffin, lubricating oil, vaseline, and paraffin wax for two very familiar examples.

We, however, need to turn our attention to the application of distillation in the analytical laboratory. In years gone by, it was used very extensively for separating liquids in complex mixtures and, in favourable cases, identifying them by their boiling-points. However, more recently, distillation has lost favour to gas chromatography in this field, as we have already noted. This is because, as long as you are working on an analytical or semi-preparative scale, gas chromatography can be used to resolve quite complex mixtures into pure compounds, whereas even the most powerful fractionating columns may fail to do this. Distillation does, however, have its uses for resolving complex mixtures into fractions of different boiling range, which are then simpler to analyse by gas chromatography or other techniques. This is particularly useful in the analysis of hydrocarbon mixtures and mixtures of organic solvents. Indeed, if you were presented with a complex sample matrix for analysis of the organic solvents present, you would do well by starting the analysis with a separation of the solvents by distillation.

Another very obvious use is in the purification of solvents for use in analysis. Every laboratory has its still for the preparation of distilled water, but distillation is also used for purifying many organic solvents, a middle fraction being taken to eliminate impurities that are both more or less volatile than the solvent itself. This middle fraction should have a constant boiling-point.

Distillation may be used to remove water quantitatively from samples, say, of biological origin. An azeotropic distillation, in which the water forms a constant-boiling mixture with toluene (added to the sample), is one such example. The liquids separate from one another in the distillate and the volume of water distilled can be measured.

Finally, distillation is used in conjunction with volatilisation, whereby a chemical reaction is used to convert a particular component of a mixture into a volatile derivative which can then be quantitatively removed by distillation (if it is not actually gaseous). This use of chemical reactions will be the subject of the next section.

10.4.2. Volatilisation

Here, we look at ways of selectively removing a particular species from a sample matrix by employing a selective chemical reaction to convert it into a volatile derivative. This can them be distilled quantitatively out of the sample. This is largely used to eliminate non-metallic elements from a matrix, although some metals can be treated in the same way. We might be carrying out this process simply to remove an interfering element, in which case collection of the vapours is of no analytical importance, or alternatively we could collect the vapours and subsequently analyse for the eliminated element. Really, all that needs to be added to this is a table of examples to show how this is done in practice. Fig. 10.4a gives some of them. Read it through carefully and study the examples given, but do not try to commit it to memory. You will, however, find it useful for reference, eg for the next SAQ.

Elements Volatilised	Conditions for Volatisation	Other Elements
Carbon (as inorganic carbonate)	From acid solution as CO_2	
Nitrogen	Reduce to NH_3 with Devarda's alloy, or *via* the Kjeldahl Method	
Sulphur (as sulphides or sulphites)	Acid treatment to yield H_2S or SO_2	
Fluorine (inorganic)	Steam distil as H_2SiF_6 from acid solution containing SiO_2	

Boron (as Borate)	Heat the acidified sample with H_2SO_4 and CH_3OH to give volatile $B(OCH_3)_3$	
Silicon (from silicates)	Distil as SiF_4 from HF solution	
Arsenic	Distil as $AsCl_3$ at 110 °C from a solution containing H_2SO_4 and HCl	Antimony and tin are left behind
Antimony	Distil as $SbCl_3$ at 155–165 °C from a solution containing H_2SO_4 and HCl	Tin is left behind if H_3PO_4 is also present
Tin	Distil as $SnBr_4$ at 140 °C after addition of HBr.	
Bromine	Oxidise to Br_2 with KIO_3 or telluric acid and volatilise	Chlorine is left behind
Iodine	Oxidise to I_2 with HNO_2 and volatilise	Bromine and chlorine are left behind
Chromium	As CrO_2Cl_2 on heating the sample with $HClO_4$ and HCl	
Osmium and Ruthenium	As volatile tetroxides by distilling them from HNO_3 solution	
Mercury (as inorganic Hg^{2+})	As mercury vapour on reduction with $NaBH_4$	
Se,Te,As,Sb,Bi, Ge,Sn,Pb	As volatile hydrides on reduction with $NaBH_4$ in aqueous solution	

Fig. 10.4a. *Volatilisation of selected elements from solution*

SAQ 10.4b	If you study Fig. 10.4a, you should be able to spot possibilities for the sequential separation of two groups of elements by volatilisation. What are these two groups? There are also two possibilities for volatilising elements in a suitable form for atomic absorption spectroscopy, a technique applicable to metals and metalloidal elements and requiring the element either in the atomic state or in a form readily convertible into the atomic state. What are these two possibilities?

10.4.3. Cold Trapping

An alternative to the use of an adsorption tube for collecting vapours from the atmosphere, would be to pass air through a cold trap, ie a small vessel cooled with, say, liquid nitrogen or solid carbon dioxide. Provided the substances in the air that are to be trapped have boiling-points well above that of nitrogen or carbon dioxide, which is likely in practice, they should be quantitatively retained in the cold trap.

One advantage of a cold trap over an adsorption tube is the certain, quantitative, and rapid recovery of the trapped substances, merely by subsequent heating of the cold trap. Disadvantages though, are that for any monitoring, large amounts of water vapour will be collected in the trap along with trace organic vapours, as well as the apparatus involved being somewhat more complex and less portable than an adsorption tube apparatus. It would hardly be suitable for a personal air-monitor, for instance.

∏   We have, in fact, already used a cold trap as part of another preconcentration system. Do you remember what this was?

Thermal Desorption. Thermal desorption is not necessarily a rapid process. The thermally desorbed vapours can be reconcentrated in a cold trap, which can subsequently be rapidly heated to release the vapours more or less instantaneously, for gas chromatography for example (see Section 10.2.1).

10.4.4. Solvent Evaporation

In principle, this is the simplest way to *preconcentrate* a species in solution. All that is implied here is that a large volume of liquid sample that contains an *involatile* analyte is reduced to a small volume by evaporating, or boiling off, the solvent. Note carefully the need for the species being preconcentrated to be *involatile*. We have, in fact, already used this technique as an adjunct to other preconcentration procedures given in previous Sections of this Unit, for example in the solvent extraction of metal ions to increase the overall preconcentration factor.

It finds particular use for the preconcentration of metal ions in water samples. A large volume of the water sample is gently heated on a water bath and allowed to evaporate slowly. The water samples should be slightly acidified, since glass surfaces act as cation exchangers and can therefore retain trace-metal ions. So the acid maintains a high concentration of H^+ ions, which are preferentially retained. One disadvantage of this technique is that it is lengthy and time-consuming.

Solvent evaporation may also be used to preconcentrate trace organic material or organic complexes in organic solvents. These often evaporate more easily than water, and so the operation is not so time-consuming. However, you must take great care to ensure that you do not lose any of the analytes themselves through volatilisation. There are also some safety implications in that you will generate large volumes of potentially toxic organic solvent vapours that could also present fire hazards. These vapours must be conducted away safely.

Here is an example of solvent evaporation being used to good effect in water analysis. The analytes in this example are trace organic acids. The initial sample which has a volume of 3–4 dm^3 is made alkaline with NaOH so that the acids are converted into involatile salts. The water sample is then *evaporated* down to a volume of 30 cm^3, made acid and shaken with an approximately equal volume of diethyl ether. The organic extract is then *evaporated* down to a volume of 1 cm^3 to give an overall preconcentration factor of 3000–4000. The analysis is completed by using chromatography on the 1 cm^3 of ether extract.

SAQ 10.4c	In the course of this Unit, we have studied *four* separate techniques suitable for preconcentrating trace-metal ions in water. Name these four techniques and then briefly compare them under the following headings, highlighting those techniques that you think are best, or worst, under each heading. $\longrightarrow$

**SAQ 10.4c
(cont.)**

(*i*) Preconcentration factors obtainable.

(*ii*) Metal ions that can be preconcentrated.

(*iii*) Suitability of the form of the preconcentrated sample for subsequent analysis.

(*iv*) Speed, ease, and simplicity of the preconcentration procedure.

(*v*) The effect of other ions present at high concentrations, which are not to be preconcentrated.

(*vi*) Actual number of applications in practice.

10.5. CONCLUSIONS

In this Part, we have discussed some further separation techniques, including adsorption, for the preconcentration of organic compounds from the atmosphere and from water samples. In practice, this is probably the most widely used of the separation techniques discussed here, and therefore the most important. We also considered the use of coprecipitation as a rather specialised technique for trace-metal ion preconcentration, as well as looking at some separation possibilities by conventional precipitation techniques. Finally, we considered the potential of gas- and vapour-phase separations, and noted the possibilities for selective separations by converting certain elements into volatile forms. We also noted how, despite all our sophisticated preconcentration techniques, it is sometimes possible to carry out an effective preconcentration simply by evaporating off the solvent!

Objectives

On completion of this Part of the Unit, you should now be able to:

- understand the basic principles underlying adsorption of organic molecules from solution or the vapour phase onto a solid adsorbent, and their desorption;

- describe the properties of selected adsorbents;

- explain how adsorption/desorption can be used for the collection/preconcentration of organic compounds from water samples or from the atmosphere;

- understand the mechanisms of coprecipitation of trace components onto a carrier precipitate;

- explain how coprecipitation can be used to preconcentrate trace-metal ions in water;

- explain how in certain circumstances precipitation can be used for simple separations and to describe some examples of these;

- discuss generally the role of distillation techniques in analysis;

- give examples of separations involving the selective volatilisation of an element from a matrix *via* a chemical conversion;

- appreciate the use of solvent evaporation as a preconcentration technique.

Conclusions to the Unit

By the time you have reached this point, you should have completed the Unit and should have attempted all the exercises and SAQs, and, I hope have answered them all satisfactorily, even if only after several attempts. If this is so, our congratulations are in order in at least having the perseverence to complete the Unit! However, perhaps more to the point is to ask yourself the question – 'How much have I learned from the Unit?'

You will no doubt recall that right at the beginning of the Unit, there was a multiple-choice test that you should have attempted, that covered the syllabus of the Unit. You should have recorded your mark for that test, and it was quite probable that that mark was a low one. I said at the time that when you had completed the Unit, I would ask you to return to this test and do it once more. I hope your mark will be higher this time, the amount by which your mark increases indicating, only in a very rough qualitative way, the extent of what you have learned from the Unit.

So, now go back to the beginning of the Unit and repeat the Preliminary Multiple-Choice Pre- and Post-Test. After completing the Test, record your answers and marks in the grid provided and compare them with your first attempt.

Finally, may I draw your attention to the list of references and books for further reading in the Bibliography. This includes all the references mentioned in the text of the Unit, and basically contains a selection of texts for further reading on the various topics covered in the Unit. At the time of writing, some of these books are no longer in print. I have retained these, however, since copies are still likely to be found on Library bookshelves. I have, however, indicated which these are. Some of the references are to fairly general and not too advanced analytical textbooks that you might well find useful for a number of the Units (references 1–3, 9, 13, 14) whereas others are of a more advanced, specialist nature (references 5, 7, 8, 12, 15–20). Nevertheless, you should be able to find useful sections in them. These are for borrowing from a Library. Items 10 and 11 are highly specialised works for reference only whereas references 4 and 6 are sources of physical data.

Self Assessment
Questions and Responses

SAQ 1.1a | List, from memory, the stages of the analytical sequence as we have described them for a qualitative and for a quantitative analysis.

Response

This question simply involves straight recall from Section 1.1. If you were unable to answer the question satisfactorily, then you need to reread and learn the contents of the section more carefully. Were you able to describe an analysis in terms of the stages of the analytical sequence in the spaces provided? This should give you more practice at grasping and retaining a knowledge of these stages.

Your lists should have looked like this.

	Quantitative		**Qualitative**
(*i*)	Definition of Problem.	(*i*)	Definition of Problem.
(*ii*)	Choice of Method.	(*ii*)	Choice of Method.
(*iii*)	Sampling.	(*iii*)	Sampling.
(*iv*)	Sample Pre-treatment	(*iv*)	Sample Pre-treatment.
(*v*)	Measurement.	(*v*)	Qualitative Tests.

(*vi*)	Calibration.	(*vi*)	Tests on Reference Materials
(*vii*)	Evaluation of Results.	(*vii*)	Interpretation of Tests.
(*viii*)	Action.	(*viii*)	Action.

****

SAQ 1.2a

You are required to choose between an atomic absorption and a solution spectrophotometric method for the routine determination of lead ions in tap water samples. The concentrations of the lead ions in the samples are likely to be in the region of 0.01 to 0.1 mg dm^{-3}, and we need to know the concentration to the nearest 0.005 mg dm^{-3}. Here are some facts about the two methods.

	Atomic Absorption	Solution Spectrophotometry
Accuracy at 0.1 mg dm^{-3}	± 5%	± 0.2%
Limit of detection	0.005 mg dm^{-3}	0.002 mg dm^{-3}
Sample Pretreatment	Addition of Reagents	Buffering, Addition of Reagents, Solvent Extraction, Back Extraction,

$\longrightarrow$

**SAQ 1.2a
(cont.)**

Re-extraction
(different
conditions
each time).

Which method would you recommend? Explain
your choice.

Response

I would expect you to recommend the *atomic absorption* method
because of its much simpler sample pretreatment stage. Admittedly
the solution spectrophotometric method offers a better limit of detection and better accuracy, but these are better than required. The
performance of the atomic absorption method is quite adequate,
bearing in mind the specification for the analysis. The involved solvent extraction procedure (required to overcome interferences in
spectrophotometry) would mean that the analysis would be slow and
labour intensive, compared with the simple sample pretreatment in
atomic absorption (which, incidentally would involve electrother-
mal atomisation for that particular limit of detection).

****

SAQ 2.2a   Which of the following statements are true (T)
and which are false (F)? Explain your choice of
answer in each case.

(*i*)   It is advantageous to grind solid samples
to a small particle size, not only to reduce
subsampling errors, but also to aid attack

$\longrightarrow$

SAQ 2.2a
(cont.)

by solvents or reagents used to decompose the samples.

(T / F)

(*ii*) A diamond mortar is a device for pulverising especially hard samples. It is so-named because the tool used to shatter the sample is diamond-tipped.

(T / F)

(*iii*) Samples which are too hard to grind with a pestle and mortar can be more effectively ground with a ball-mill.

(T / F)

(*iv*) After a sample has been ground, its maximum particle size can be ascertained by passing the sample through sieves of known mesh number. The higher the mesh number of the sieve through which the sample completely passes the smaller is its maximum particle size.

(T / F)

Response

(*i*) (T). Subsampling errors are reduced by grinding provided the sample is well mixed after the grinding. Grinding increases the surface area: volume ratio, which means that the total surface area available to attack by solvents or reagents is increased.

(*ii*) (F). The diamond mortar is all of hard-steel construction. A diamond tool is not used in the diamond mortar.

(*iii*) (F). A diamond mortar would be more appropriate here. A ball-mill is suitable only for fairly soft samples. It saves 'elbow-grease' needed with the pestle and mortar and is somewhat more convenient for grinding samples in quantitative analysis.

(*iv*) (T). Remember, high mesh-number means finer particles.

****

SAQ 2.3a

> Name *three* analytical techniques that can be used directly on solid samples, and write one sentence per technique describing the form in which the sample might be presented in the technique.

Response

There are a number of such techniques and I do not intend to list them all. Rather, I have listed a selection of the more common of them and, more specifically, those that I have discussed in the text. The sampling techniques described are also not the only ones possible, but rather illustrate the sample presentation techniques that we have just discussed.

(*i*) *Infra-red Spectroscopy*. The sample is ground to a fine powder with an excess of KBr and compressed into a flat disc.

(*ii*) *Spark Emission Spectrometry*. The sample (a metal or alloy) is chill-cast into the shape of an electrode, with a flat, smooth polished surface for the most accurate results.

(*iii*) *X-ray Fluorescence Spectroscopy*. The sample is presented in the form of a flat disc with a smooth surface, a few cm in diameter.

(*iv*) *Microscopy*. The sample is presented as a thin film of material.

Mass spectrometry, thermal analysis, combustion techniques, activation analysis, and certain qualitative classical tests can all be applied to solid samples, but the form of the sample is less critical here. Microprobe analytical techniques as well as some less-known spectroscopic techniques are also applicable to solids and require smooth flat surfaces. X-ray powder diffraction analysis is applicable to powdered solid samples, provided that they are crystalline and not amorphous.

As I have said, the list is not meant to be exhaustive, but you do have a good selection of examples here.

****

SAQ 2.4a   Select from the following the procedure most likely to be adopted in practice for the preliminary pretreatment of samples of plant material, such as grasses or mosses before trace-metal determination.

(*i*)   Store in a sealed container over water until the water content of the sample has been equilibrated with that of the air above it. Check that this has been achieved by weighing the sample and finding when it has come to constant weight.

(*ii*)   Freeze the samples and then dry them to constant weight under vacuum, and keep them in the frozen state.

(*iii*)   Dry the sample to constant weight in a furnace at 1000 °C. Store in a desiccator and check the weight before analysis.

$\longrightarrow$

SAQ 2.4a
(cont.)

> (*iv*) Dry the sample to constant weight in an oven at 105 °C. Store in a desiccator. Check the weight before analysis.

Response

Procedure (*iv*) is the most likely to be adopted in practice, because it is simple and in it the most generally available apparatus is used. Large numbers of samples can be so handled.

Let us now look at the problems raised by the other approaches.

Procedure (*i*) is, in fact, not a drying operation as such but an attempt to standardise the water content at some high level. It is generally easier to standardise a value at zero (ie by drying the sample) or at least at a low value than at some arbitrary high value. It is very unlikely that standard conditions could be achieved unless temperature and pressure were carefully controlled and all the samples were fairly similar in nature. Even then, equilibration could take a very long time.

Procedure (*ii*) is a possibility, but the apparatus required is not generally available (freezing under vacuum). This is a more specialised technique used where loss of volatiles or thermal decomposition are likely to be significant. This is unlikely in this particular example.

Procedure (*iii*) is inapplicable. Plant material would be well and truly ashed at this temperature and the analytical result would refer to plant ash, not dried material. In fact, a subsequent part of the analysis may well be such an ashing procedure, when loss of metals through volatility will have to be considered. These conditions might, however, be needed for driving off traces of water from minerals such as aluminates or silicates.

****

SAQ 2.5a	Explain fully how you would pretreat a soil sample in order to determine the cobalt content available to plants growing in the soil. The pretreatment should result in a clear, aqueous solution suitable for a final analysis by, say, atomic absorption spectrometry.

Response

My procedure, described below, is based on the assumption that plant-available cobalt can be leached out of the sample by using an appropriate leaching agent that has similar activity to plant roots. We need to dry the sample first to standardise the water content, but we do not grind the sample. After all, plant roots cannot grind soil! We might need to pass the soil through a coarse sieve to remove stones, again to standardise the sample. One large stone in a weighed aliquot of soil could make nonsense of our results. I would expect your answer to include all the points below, except that I would not expect you to recommend an actual leaching agent for cobalt without going to a source-book on soil analysis, which I did in preparing my answer.

Here, then, is my procedure.

(*i*)   First, dry the sample at 105 °C to constant weight.

(*ii*) Pass the sample through a coarse sieve (low mesh-number) to remove large stones. There is no need to grind the sample because we are not looking for complete dissolution. You can record the weights before and after sieving and the number of stones removed.

(*iii*) Now leach this sample with an aqueous solution with similar dissolution powers towards cobalt as plant roots. A recommended leaching agent is 2.5% aqueous acetic acid (ethanoic acid), pH 2.5 (ref. 4).

(*iv*) Separate the solid and solution phases by filtration and wash the residue with more 2.5% aqueous acetic acid to obtain the extractable cobalt in quantitative yield. The filtrate is the solution to be used in the final analysis by atomic absorption spectrophotometry.

****

SAQ 2.5b

You are provided with a sample of rubber which is to be analysed for residual accelerators and antidegradants. The analytical technique to be used is high-performance liquid chromatography which requires a solution of the analytes in a non-viscous, non-uv absorbing solvent. The accelerators and antidegradants that are possibly present appear to be soluble in a range of halogenated solvents as well as esters and ethers, but not in alcohols or water. You have some doubts about the thermal stability of some of these compounds.

How would you set about obtaining a solution for analysis in this situation?

Response

It should be obvious to you that we must carry out some leaching or extraction process on the sample. Should you have considered any other procedure, I suggest that you read no further, but have another go, thinking about how you would carry out the extraction and what sort of solvent you would use.

It will be of great help in speeding up the extraction if care is taken to ensure that the rubber is in the form of fairly small particles. Indeed, with large particles, the extraction may never be quantitative. You could cut the rubber sample up into small cubes or alternatively immerse it in liquid nitrogen when it will become brittle and can be shattered by using a pestle and mortar (work fast, before it warms up again; or you can pour liquid nitrogen into the mortar). The small particles will thus present to the solvent for extraction, a large surface area per volume of sample.

The actual extraction could be carried out either by simply shaking the sample with a solvent in a sealed bottle for a suitable length of time. This is an attractive idea because there will be no problems due to thermal instability. If you are concerned that this will not result in a quantitative extraction, you could try a Soxhlet extraction, but then questions of thermal degradation arise.

In conclusion, let us recommend that we use the Soxhlet extraction apparatus to ensure quantitative extraction, but that we use a solvent with a low boiling-point to minimise thermal degradation. Two excellent solvents are available for this purpose from the classes of solvents in which we know that the solutes are soluble, *viz* dichloromethane (bp 40 °C) or diethyl ether (bp 35 °C).

****

SAQ 2.6a	Explain how you could distinguish between the total iron concentration and the soluble iron concentration in a sample of tap-water.

Response

Before any actual analytical measurements are made, the water should be filtered to isolate any suspended solids and give a particle-free solution. The iron content of this solution will represent soluble Fe.

Now, extract the Fe from the filter paper, which will correspond to the Fe in the suspended solids. Hydrochloric acid would probably be a suitable reagent for this, if we assume that the suspended iron is mainly iron oxide or insoluble iron complexes with organic acids. Determine the Fe in the extract and equate this to Fe in the suspended solids. It will be important to perform a blank analysis on an unused filter paper, because there may be a small amount of iron extracted from the filter paper itself. Now, add the value for Fe in suspended solids (corrected for any blank value) to the soluble-Fe value to get the total Fe concentration.

You may have suggested that one aliquot of the water be filtered and another boiled with HCl to solubilise any suspended particles. Provided you are certain that all the iron is solubilised in this way, this would be a satisfactory alternative approach.

****

SAQ 3.1a   Which of the following statements are *true* and which are *false*? Give reasons for your answers.

(*i*)   With the advent of modern instrumental elemental analytical techniques, the need to obtain the sample in solution form has been very considerably lessened.

(*ii*) Broadly speaking, most solid inorganic samples that one encounters for analysis require rather drastic chemical decomposition techniques to bring them into solution.

(*iii*) A desirable feature of any dissolution procedure used in the elemental analysis of inorganic samples is that the sample should dissolve without undergoing any chemical change.   ⟶

SAQ 3.1a (cont.)	(*iv*) Of the procedures used for obtaining solutions from intractible solid inorganic materials, far and away the majority involve either the use of hot concentrated mineral acids or fused acidic or basic inorganic electrolytes.

Response

(*i*)   *False*. If you thought otherwise, have a look back at the very first exercise in this Part of the Unit, ie in Section 3.1. There you will find a list of all the common techniques available for elemental analysis. Clearly, from this list, you can see that the majority involve solutions. Although the small number that are applicable to solids are admittedly instrumental, you will note the presence of many extremely important instrumental techniques that require samples in solution. Furthermore, the most popular and rapidly developing techniques (flame and plasma techniques, for instance) require solutions. It would therefore perhaps be true that certain developments in instrumental chemical analysis have lessened the need for dissolution techniques, but nevertheless solution techniques are still of extreme importance.

(*ii*) *True*. Unless your sample is a relatively simple metal salt soluble in water, or one of a fairly limited range of the more electropositive metals or their oxides or carbonates, or such compounds, then you will need fairly drastic conditions to bring it into solution, ie conditions more severe than treating it simply with water or dilute acid.

(*iii*) *False*. Chemical change is almost always essential in dissolution techniques for inorganic solids unless they are simple water-soluble metal salts. Since we are concerned with elemental analysis, there is absolutely no reason for retaining the original chemical form of the sample. It is more important that the elements in the sample are converted into soluble forms that respond suitably to the final measuring technique.

(*iv*) *True*. We shall meet one or two other methods, but far and away the majority of the procedures that we shall study for 'difficult' samples will indeed involve acids or fused electrolytes.

****

SAQ 3.2a

Following are five samples that are to be taken into aqueous solution. Only *one* is directly soluble in water, without any chemical change. Which is it?

(*i*)   Sodium metal (for determination of traces of potassium).

(*ii*)   Potassium perchlorate (for determination of traces of caesium).

(*iii*) Limestone ($CaCO_3$) (for determination of Fe and other impurities).

(*iv*)   Sodium dihydrogen phosphate (for determination of the total phosphorus content).

(*v*)   Calcium sulphate ($CaSO_4.2H_2O$ – gypsum) (for determination of trace metals therein).

Response

The correct answer is (*iv*), sodium dihyrogen phosphate. This salt, like most sodium salts, is freely water-soluble.

Item (i) is incorrect, because although sodium metal reacts with water to give a solution of sodium hydroxide, the question asked for samples that dissolved without chemical change. Solutions of sodium metal can be obtained under special conditions, but not in water (eg in liquid NH_3).

Item (ii) is incorrect. Potassium salts are water-soluble, with very few exceptions. Unfortunately one of these is potassium perchlorate, together with rubidium and caesium perchlorate.

Item (iii) is incorrect, because although calcium hydrogen bicarbonate [$Ca(HCO_3)_2$] is soluble in water, being the major contributor to water hardness, $CaCO_3$ is not and is the main component of boiler scale and 'fur' in kettles. Limestone is, after all, $CaCO_3$ deposited by decomposition of $Ca(HCO_3)_2$ in natural waters. Limestone is soluble in dilute acids but only on decomposition with liberation of CO_2.

Item (iv) is incorrect. Calcium sulphate is only sparingly soluble in water. For trace analysis, one would need to dissolve a fairly substantial sample. In fact, calcium sulphate is taken into solution with hot dilute hydrochloric acid.

****

SAQ 3.3a

For each of the following, first of all indicate whether the named substance can be brought into solution (with chemical change if necessary) by simple treatment with dilute aqueous acid. Then, when the answer is 'yes', suggest an appropriate acid, indicate the chemical form of the species in solution and whether any component of the sample is lost in the dissolution process.

(i) Brass. $\longrightarrow$

SAQ 3.3a
(cont.)

> (*ii*) Pearls.
>
> (*iii*) Silver bromide.
>
> (*iv*) Diamonds.
>
> (*v*) A mixture of alkaline earth oxides and car-
> bonates.

Response

Let us take each substance in turn and explain why each might or
might not be expected to dissolve in dilute acid. Where the answer
is 'yes' we can then answer the other questions regarding choice of
acid, form of analyte in solution, and any potential losses.

(*i*) *Brass* – Yes.

Brass is an alloy of copper and zinc. Zinc is an electropositive element that might be expected to dissolve in most acids with the
release of hydrogen. Copper on the other hand is slightly more electronegative than hydrogen. However, as we have seen, dilute nitric
acid is sufficiently oxidising to dissolve copper, but with the release
of oxides of nitrogen rather than hydrogen. Brass may thus be dissolved in dilute nitric acid (in practice a $50:50$ $HNO_3:H_2O$ mixture
is recommended). The brass yields copper(II) and zinc(II) ions and
no component is lost (assuming no losses occur through splashing).

(*ii*) *Pearls* – Yes.

These consist of nothing more than calcium carbonate in a particu-
larly attractive form. Thus, should you wish to dissolve pearls, which
I suspect is unlikely in practice, dilute acid should be quite effective.
Dilute hydrochloric acid will suffice, the reaction being:

$$CaCO_3 + 2\,HCl \rightarrow CaCl_2 + H_2O + CO_2 \uparrow$$

from which you can see that the pearls are converted into calcium chloride, and CO_2 is lost from the solution. Legend tells us that Cleopatra dissolved pearls in vinegar and drank the mixture in order to enhance her beauty. You now know that all she did was to drink an expensive solution of calcium acetate!

(iii) Silver Bromide – No.

This is used extensively in photographic emulsions as the active ingredient in recording the latent image. Photosensitised AgBr is reduced to silver by the developer and the remaining AgBr is dissolved by the fixer. This is aqueous sodium thiosulphate. Silver bromide is quite insoluble in water and is unaffected by normal dilute acids. So the answer here is 'No'.

(iv) Diamonds – No.

Diamonds consist of carbon, nothing more or less. Carbon is a non-metallic element that under appropriate oxidising conditions could be converted into CO_2. As far as diamonds are concerned, this is an expensive and rather non-productive process that, anyway, could not be carried out with dilute acid.

(v) Alkaline earth oxides and carbonates – Yes.

The alkaline earth elements are electropositive, and both oxides and carbonates are readily soluble in dilute acids to yield the appropriate metal salts of the acids. However, not all acids are equally appropriate, since certain alkaline earth salts are of low solubility. Nitric acid, hydrochloric acid or perchloric acid would be suitable as soluble salts are produced, but sulphuric acid, hydrofluoric acid or phosphoric acid would be less suitable because of the insolubility of such salts as $BaSO_4$, CaF_2, and $MgHPO_4$. CO_2 is lost from the solution when carbonates are involved.

****

SAQ 3.3b	Consider only the acids HCl, HNO$_3$, H$_2$SO$_4$, HClO$_4$, HF and H$_3$PO$_4$.

    (*i*)   Which acid has the highest boiling point?

    (*ii*)   Which acid is the most powerful oxidising agent?

    (*iii*) Which acid is likely to be the most effective for dissolving refractory oxides?

    (*iv*) Which acid's anion is the least likely to form stable complexes with the metal cation?

    (*v*)   Which acid forms water-insoluble salts with the heavy alkali-metal cations?

Response

The answers to all these questions are to be found in the text of Section 3.3.2, where the properties of these acids are fully discussed. If you had difficulty answering these questions, go through this section again and see if you can glean the answers.

The answers that you should have obtained are as follows.

(*i*)   *Sulphuric Acid* – boiling point 330 °C.

Boiling-points of the other acids:

100% H$_3$PO$_4$		-213 °C
72% HClO$_4$	(azeotrope)	-200 °C
67% HNO$_3$	(azeotrope)	-121 °C
36% HF	(azeotrope)	-111 °C
6M-HCl	(azeotrope)	-109 °C

(*ii*) *Perchloric acid* – especially when hot. (Other oxidising acids are HNO_3 and H_2SO_4. HCl, HF, and H_3PO_4 are more useful as acids and complexing agents than oxidising agents).

(*iii*) *Hydrofluoric acid* – because of the strong complexing properties of the fluoride ion. Other acids may have some dissolving power with respect to the refractory oxides, but most recipes for dissolving such materials in acid involve the use of HF.

(*iv*) *Perchloric acid*. The ClO_4^- ion shows only the very weakest tendency to form complexes. All the anions of the other acids can, under appropriate circumstances, form complexes which may as with HCl and HF, be quite stable.

(*v*) *Perchloric acid*. The perchlorates of K, Cs and Rb are of low solubility whereas almost all other common salts of these elements are extremely soluble. Incidentally, almost all other perchlorates are also highly soluble in water.

****

SAQ 3.3c

Recommend the most suitable acid from the following list for each of the applications below:

HCl, HNO_3, H_2SO_4, $HClO_4$, HF, H_3PO_4.

Each acid may be recommended once, more than once, or not at all. For one of the applications, no single acid is suitable. Which is this?

(*i*) An acid for dissolving an intractable stainless steel and which oxidises many of the metals present to their highest oxidation states. $\longrightarrow$

SAQ 3.3c
(cont.)

(*ii*) An acid for 'fuming a sample to dryness' without loss of metal ions as volatile compounds.

(*iii*) An acid for bringing refractory metals and their compounds into solution.

(*iv*) An acid for bringing the common transition-metal ions into solution.

(*v*) An acid for bringing the noble metals into solution.

(*vi*) An acid for dissolving an acid-soluble alloy with the elimination of Sn from the matrix.

Response

(*i*) For this application we want the most powerfully oxidising acid from our list, which is $HClO_4$. In fact, only Mn of the common alloying elements in steel is not taken to its highest oxidation state.

(*ii*) In our studies of the various acids, we noted that the metal sulphates were of low volatility. Therefore, H_2SO_4 is recommended here. $HClO_4$ is a possible alternative, but HCl and HF would be very poor choices as many chlorides and fluorides are volatile.

(*iii*) The acid we need here is HF, because of the strong complexing power of F^-, especially with refractory elements. Their oxides are attacked and soluble fluorides and fluoro complexes are formed.

(*iv*) Several acids will serve here including HNO_3, H_2SO_4, and $HClO_4$. HCl is a bit marginal, whereas HF and H_3PO_4 are weak acids and less suitable. Because some of the transition-metal ions are not very electropositive we need an oxidising acid. HNO_3 provides sufficient oxidising power and is somewhat safer than H_2SO_4 or $HClO_4$.

(*v*) No one acid is suitable for this application. The noble metals require greater oxidising power than is provided even by $HClO_4$. We shall see shortly how this can be provided.

(*vi*) HNO_3 is recommended here, because in its presence, Sn forms an insoluble hydrated oxide that can be removed by filtration. The other metals form soluble nitrates (provided Sb is absent and passivation is not a problem).

****

SAQ 3.3d

Here are *four* dissolution treatments involving the use of mixtures of acids or of more than one acid in sequence.

Explain briefly why the mixed-acid treatment or the use of acids in sequence is to be preferred to the use of a single acid.

(*i*) The sample is dissolved with heating in a mixture of 3 parts concentrated HCl and 1 part concentrated HNO_3.

(*ii*) The sample is heated with concentrated HNO_3. As the HNO_3 boils off, $HClO_4$ is added and heating continued until dissolution is complete.

(*iii*) The sample is heated with HF and H_2SO_4 in the ratio of about 1 : 10. It is then taken to fumes of SO_3 and subsequently treated with a further small amount of H_2SO_4.

$\longrightarrow$

SAQ 3.3d
(cont.)

> (*iv*) The sample is heated with concentrated
> HCl and then concentrated HF is added
> dropwise to complete the dissolution.

Response

Here are brief explanations as to why these acid mixtures or sequential acid treatments are more effective than individual acids.

(*i*)   When HNO_3 is mixed with HCl it oxidises it to produce a number of species including nitrosyl chloride and free chlorine. These are far more powerful oxidising agents than either acid on its own and are capable of oxidising, and therefore bringing into solution, many highly electronegative elements that are unattacked by the acids on their own. This mixture is well-known under the name of 'aqua regia'.

(*ii*) Perchloric acid is an extremely powerful oxidising agent when hot, and there is an explosion hazard should the hot acid come into contact with easily oxidised material. Nitric acid is a weaker oxidising agent. Therefore if the sample is first treated with HNO_3 and only the more intractable components of the sample left for $HClO_4$ treatment, the explosion hazard is averted.

(*iii*) Hydrofluoric acid is a powerful acid for bringing refractory elements into solution as fluoro-complexes. However, in order to avoid subsequent interference in the analysis it is sometimes necessary to break down the fluoro-complexes and remove all traces of F^- from the solution. This is done by 'taking to fumes' with H_2SO_4 several times, thus driving off F^- as HF. Perchloric acid can also be used in this context, or an alternative approach is to add H_3BO_3 which 'masks' the fluoride by forming BF_4^- ions which are very stable.

(*iv*) Here perhaps we have a sample consisting of a mixture of substances, the greater part of which is soluble in HCl (an oxide or a carbonate, perhaps) while a smaller part (say a silicate) requires the HF treatment. You might wonder why we do not just use HF for the whole sample. HF is a most toxic and unpleasant material to handle. By this procedure, we keep the amount of HF involved to a minimum.

****

SAQ 3.3e

In each of the following dissolution procedures one of the components of the dissolution medium is not a mineral acid. Identify this component and explain its function.

(*i*) *Zinc–aluminium alloy*. Dissolve in concentrated HCl and then treat with 10-volume H_2O_2.

(*ii*) *Tin–lead solder*. Dissolve in a mixture of 10 parts HCl, 5 parts HNO_3 and 50 parts of 10% aqueous tartaric acid.

(*iii*) *Oxygen-containing compounds of arsenic and antimony*. Heat strongly with concentrated H_2SO_4 to which Na_2SO_4 has been added.

(*iv*) *Aluminium metal*. Dissolve in concentrated HCl to which a small quantity of Hg^{2+} ions has been added.

Response

(*i*) The non-acid component here is H_2O_2 which has been added as an *auxiliary oxidising agent*, aiding the dissolution of the aluminium (the zinc should dissolve in HCl without difficulty).

(*ii*) Here *tartaric acid*, the non-mineral acid component, is added to act as a *complexing agent* to keep the tin and lead ions in solution and prevent their precipitation by hydrolysis.

(*iii*) The addition of Na_2SO_4, the non-mineral acid component, is simply to *raise the boiling-point* of the H_2SO_4, and thus its reactivity, so that it can more effectively bring the sample into solution.

(*iv*) The addition of Hg^{2+} *ions catalyses* the dissolution of the aluminium, increasing its rate of dissolution. In practice, small amounts of Hg are precipitated onto the surface of the aluminium. These set up a galvanic cell which hastens the dissolution of the aluminium.

****

SAQ 3.4a

Explain how each of the following electrolytes functions as a flux in terms of the Lewis acid/base theory. Give an example of the type of material thus brought into solution.

(*i*) K_2CO_3,

(*ii*) $KHSO_4$,

(*iii*) KHF_2,

(*iv*) $CaCO_3 + NH_4Cl$.

Response

(*i*) K_2CO_3.

This behaves in a manner exactly analogous to that of Na_2CO_3, which was described in the text. Carbonate fusions are particularly useful for silica and silicates, although we can illustrate this behaviour with any metal oxide that can act as a Lewis acid. Re-read the description of the use of Na_2CO_3 to solubilise silica and apply the same argument to K_2CO_3.

Let us now go through the argument again with another metal oxide. K_2CO_3 is particularly useful for Nb_2O_5 and Ta_2O_5 because the resulting potassium niobate or tantalate is soluble in dilute KOH solution, unlike the sodium salts.

On heating K_2CO_3 we get the reaction below.

$$CO_3^{2-} \rightarrow CO_2 \uparrow + O^{2-}$$

Then, say, the following reaction.

$$Nb_2O_5 + O^{2-} \rightarrow 2NbO_3^-$$

This is a Lewis acid/base reaction where the Nb_2O_5 is the Lewis acid (electron acceptor) and O^{2-} the Lewis base (electron donor). On dissolution in dilute alkali we get $KNbO_3$.

In fact the stoicheiometry of the above reaction is somewhat more complex than this, because Nb_2O_5 can react with O^{2-} in a variety of different proportions producing a variety of complex oxyanions, probably as mixtures.

However, the principle is valid, ie of an acidic oxide reacting with basic O^{2-} to produce a negatively charged oxyanion whose potassium salt is soluble. As long as this idea is illustrated in your example, then you understand the basic principle.

(*ii*) $KHSO_4$.

On heating $KHSO_4$, the following changes occur

$$2\,KHSO_4 \rightarrow K_2S_2O_7 + H_2O \uparrow$$

$$K_2S_2O_7 \rightarrow K_2SO_4 + SO_3$$

Sulphur trioxide is a Lewis acid and can react with metal oxides to produce the corresponding sulphate.

$$SO_3 + O^{2-} \rightarrow SO_4^{2-}$$

So, for instance, for beryllium oxide we have

$$SO_3 + BeO \rightarrow BeSO_4$$

As long as you illustrate this process with a metal oxide being converted into a soluble metal sulphate, then you have understood this mechanism correctly. You could have used MgO, Al_2O_3, Fe_2O_3, $Cr_2O_{3<}$, etc. But BaO would *not* have been a good example!

(*iii*) KHF_2.

Here, the melt results in the production of F^- ions which can act as a Lewis base to give fluoride and fluoro-complexes with various refractory metals.

In the text some examples of KHF_2 fusions with refractory metal oxides, including ZrO_2, are mentioned.

$$ZrO_2 + 6\,F^- \rightarrow ZrF_6^{2-} + 2\,O^{2-}$$

(Again the stoichiometry given is intended to be illustrative rather than definitive, because the range of complex fluorides may be greater than this.) The assumption then is that K_2ZrF_6 is water soluble. If it is not, treatment of the flux with H_2SO_4 converts the fluoro-complex into a soluble metal sulphate (HF is driven off).

(*iv*) $CaCO_3 + NH_4Cl$.

This is a more specific flux, used to extract alkali metals from silicate matrices. You need to understand that on heating the flux, it reacts to give $CaCl_2$ and CaO. It is in fact the Cl^- ions which are the reactive species, forming water-soluble and extractable alkali-metal chlorides. Here the flux acts as the Lewis base and the analyte as the Lewis acid.

****

SAQ 3.6a

Pick out from the following list of reagents:

(*i*) an acid giving an anion that is difficult to eliminate completely from the matrix by volatilisation;

(*ii*) an acid that cannot be used in glass vessels;

(*iii*) a flux material that can be removed from the matrix by volatilisation;

(*iv*) a reagent that is unsuitable for trace analysis because of difficulties in obtaining it pure.

B_2O_3	H_3PO_4
HCl	H_2SO_4
$HClO_4$	$K_2S_2O_7$
HF	Na_2CO_3
HNO_3	NaOH

Response

(*i*) The acid that gives an anion that is difficult to eliminate by volatilisation is HF, because of the high stability of certain fluoro-complexes. The procedure normally adopted is to evaporate to dryness several times with H_2SO_4, but this can be laborious.

(*ii*) All the acids named can be used safely in glass vessels except HF, which attacks silica and silicates with the formation of volatile SiF_4.

(*iii*) The one flux material that can be eliminated from the matrix by volatilisation is B_2O_3. Reaction of B_2O_3 with an excess of methanol produces $B(OCH_3)_3$, which can be distilled off from the matrix.

(*iv*) The reagent from the list given that is least easily obtained pure is NaOH. KOH (not on the list) is likewise difficult to purify. Both these fluxes can therefore give problems if used in trace analysis.

All the answers to this question are to be found in Section 3.6. If you had difficulty with this question re-read the Section and see if you can find the answers therein.

****

SAQ 3.6b

In each of the following dissolution or opening-out procedures an error is expected because of loss of a component or contamination of the sample. Identify this error in each procedure.

(*i*) Dissolution of a chrome steel by using a mixture of H_2SO_4 and HCl in a glass vessel.

$\longrightarrow$

SAQ 3.6b (cont.)

> (*ii*) Dissolution of an alumina sample by using HF and H_2SO_4 in a pyrex glass vessel. Trace refractories to be determined.
>
> (*iii*) Dissolution of $CaCO_3$ in HCl. The sample is taken to dryness. Traces of Hg^{2+} are to be determined. A glass vessel is used for the process.
>
> (*iv*) Dissolution of a mixture of sulphides, phosphates, and silicates by 'fuming to dryness' with H_2SO_4 and dissolving the residue in dilute acid. A platinum crucible is used. The anions are to be determined.

Response

(*i*) This is a straightforward acid dissolution procedure which should occur safely in glass vessels. But if you look back to Fig. 3.5b, amongst the list of potential losses through volatility, you will find that in the presence of H_2SO_4 and Cl^-, Cr can be lost. In fact volatile CrO_2Cl_2 is formed under these conditions.

(*ii*) Another acid dissolution, again in a glass vessel. However, here HF is involved, which can cause loss of Si and B and other refractory elements as volatile fluorides as well as attack on the glass, leading potentially to highly meaningless values for several refractory elements, especially Si.

(*iii*) The simple problem here is that in the solid state $HgCl_2$, formed in the dissolution of $CaCO_3$ in HCl is volatile and can be lost from the sample unless great care is taken not to heat the sample excessively.

(*iv*) Sulphides on treatment with acid may lose sulphur as H_2S

(see Fig. 3.6c). The table also indicates that H_3PO_4, formed by double decomposition of the phosphates with an acid, is liable to be lost. The silicate is unlikely to be dissolved under these conditions, so in total, none of the anions here are likely to be quantitatively taken into solution. An alkaline fusion is likely to be more successful here.

****

SAQ 3.6c

Here is a list of dissolution and opening-out procedures. Select the one that you think is most suitable for use in each of the analytical situations below. Each procedure may be used once, more than once, or not at all.

The procedures are as follows.

(i) Dissolution in $2:1$ HNO_3–HCl. The residue is then treated with HF.

(ii) Dissolution in $3:1$ HCl–HNO_3.

(iii) Dissolution in $1:1$ aqueous HNO_3.

(iv) Fusion with $K_2S_2O_7$. The residue is taken up in H_2O.

(v) Fusion with NaOH. The residue is then taken up in dilute HCl containing a little H_2O_2.

(vi) Treatment with a mixture of HF and H_2SO_4 in a bomb.

The analytical situations are as follows.   ⟶

SAQ 3.6c (cont.)

(A) Dissolution of a mixture of ignited metal oxides, including Al_2O_3, BeO, Fe_2O_3, and Cr_2O_3. The metals are to be determined.

(B) Dissolution of a sample of cement (assumed to be basically a calcium aluminosilicate) for determination of the major constitutents.

(C) Dissolution of a stainless steel containing small amounts of some refractory elements, for analysis of alloying elements.

(D) Dissolution of a brass sample for determination of the copper : zinc ratio.

(E) Dissolution of dolomite (basically a mixture of $CaCO_3$ and $MgCO_3$).

(F) Dissolution of a gold/copper alloy.

Response

Here is the best match.

$A - (iv)$

$B - (v)$

$C - (i)$

$D - (iii)$

$E - (iii)$

$F - (ii)$

Congratulations if your match agrees with the above! That indicates that you now know something about dissolution of inorganic samples.

Here is a quick run-down on how we came to make the above selections. Remember that where alternatives exist we shall select the method that is quickest, simplest, and make use of the least toxic reagents and least dangerous conditions.

A – (iv).

Perhaps a number of these oxides might dissolve in acid, but the refractory ones are acid resistant. Fusion, which can be quicker in action than exhaustive acid treatment, offers the best possibility, using $K_2S_2O_7$ to give soluble metal sulphates. The NaOH fusion might produce some complex oxyanions, but is more effective on acidic oxides (eg SiO_2).

B – (v).

I have chosen the NaOH fusion here because this should result in the formation of soluble sodium silicates and aluminates, with the calcium being taken up in the HCl. Treatment with HF and H_2SO_4 in a bomb could also be used, but is more complex, hazardous, and likely to be slower.

C – (i).

The stainless steel is effectively oxidised with the powerful oxidising acid mixture (HNO_3–HCl), leaving the refractory elements which are then solubilised by the HF. Working this way, the amount of HF can be kept to a minimum.

D – (iii).

Copper and zinc are two metals that are easily dissolved in acid. Nitric acid must be used for the copper, although the zinc would have dissolved in HCl. Any of the acid mixtures would have been effective here, but for safety and convenience, there is no need to use any conditions more drastic than necessary.

E – (*iii*).

The carbonates of these two electropositive elements should dissolve in almost any acid or acid mixture to give the metal salts plus CO_2. So for the same reasons as in D, we use the mildest acid conditions offered.

F – (*ii*).

Here with this alloy, the copper can be dissolved in nitric acid, but then the gold would be precipitated as the free metal. Very powerful oxidising conditions are required. These are provided by use of aqua regia, a $3:1$ HCl–HNO_3 mixture, which is the only common acid mixture available for oxidising and dissolving such electronegative metals as gold. The nitric-hydrochloric acid mixture containing HF is less suitable as HF is not required here.

****

SAQ 4.1a   In order to determine a specific element or elements in an organic or biological sample, you are most likely:

(*i*)   to decompose the sample by hydrolysis to release the element into solution in a simple inorganic form;

(*ii*) to dissolve the sample in a suitable organic solvent, ensuring that no decomposition occurs;

(*iii*) to decompose the sample completely by oxidation, thus converting the element into a suitable simple inorganic form; $\longrightarrow$

SAQ 4.1a
(cont.)

> (*iv*) to convert the sample (by chemical means
>   if necessary) into a water-soluble form and
>   take it into aqueous solution.
>
> Select the most appropriate response to the
> above.

Response

(*iii*) is the most appropriate response. Well done if this was your choice. You understand that far and away the most common way of releasing any element from an organic or biological matrix is exhaustively to oxidise the matrix, so that all the elements involved are converted into simple inorganic chemical forms.

Let us now see why the other responses were not so appropriate.

Response (*i*) might be appropriate for the extraction of a complexed metal-ion, by using acid for the extraction of the metal from its complexed form into aqueous solution as a simple inorganic ion, but is not of general applicability.

Response (*ii*) would be suitable for the determination of individual organic compounds in a complex organic or biological medium, for example by chromatography, but would be of no help in elemental analysis where the compounds *must* be decomposed to *release* the elements.

Response (*iv*) is not specific enough. Certainly oxidation to a simple inorganic form (which is what is required) often amounts to the same thing, as converting by chemical means into a water-soluble form (as stated in the response). But the emphasis in the response is not on the need for chemical decomposition. Furthermore, often the result is not an aqueous solution.

****

SAQ 4.2a   Having read through the description of the combustion tube apparatus for determining C and H in organic compounds, answer the following questions without referring to the text.

(*i*)   In what chemical forms are the carbon and hydrogen when they are absorbed in absorption tubes for weighing?

(*ii*)   The sample is oxidised in three stages. What are these?

(*iii*)   How can you be sure that N, S, Cl, Br and I do not interfere in the analysis?

(*iv*)   What is the function of the guard tube?

(*v*)   What would happen if you reversed the order of the two absorption tubes?

Response

(*i*) CO_2 and H_2O. These are normal combustion products formed when any organic compound is oxidised completely.

(*ii*)   The first stage is initial burning off from the porcelain boat, the second is oxidation in the presence of the heated platinum-gauze catalyst, and the third is oxidation by heated CuO.

(*iii*)   In the combustion tube we have a section after the oxidation is complete, where the gases pass over heated PbO_2 plus $PbCrO_4$ and Ag. These reagents between them remove any oxidation products of those elements which might interfere. (N_2 gas may also be produced. This is not absorbed, but does not interfere.)

(*iv*) The guard tube, containing a mixture of both absorbants (a desiccant for H_2O and 'Ascarite' for CO_2) prevents any H_2O vapour or CO_2 creeping back up the tube from the atmosphere and giving high results.

(*v*) If we reversed the order of the tubes, we should absorb the CO_2 correctly in the 'Ascarite', but KOH is also a desiccant and would absorb at least some of the H_2O vapour. Thus the weight increase for the 'Ascarite' tube would be too high. Correspondingly the weight increase for the desiccant tube would be low. Thus we would get a high carbon result and low hydrogen result.

Well done if your answers corresponded to the above in broad principle. You understand the basic idea of the combustion-tube technique sufficiently to see how it can be modified for other elements. If you were unable to answer the questions correctly, go through the description of the combustion-tube technique for C and H analysis and try the questions once again.

****

SAQ 4.2b

On pp. 471–471 are given block diagrams for six combustion-tube analyses. However, the individual stages of the analyses have been muddled up (ie in each column the vertical order of the blocks has been randomised). Furthermore, some inappropriate or erroneous blocks have been included.

Select one block (numbered 1 to 6) from each of the four columns A to D to produce viable procedures for the determination of the following elements by the combustion-tube technique: carbon and hydrogen, nitrogen, sulphur, chlorine, oxygen.   ⟶

SAQ 4.2b (cont.)

Each block may be used once, more than once, or not at all.

For instance, if you think that element 'X' can be analysed by:

A – heating the sample in stream of O_2;

B – passing vapours over heated Cu;

C – absorbing products in aqueous KI;

D – titrating the solution with standard acid;

then your answer should appear as:

$$A = 3, \quad B = 6, \quad C = 3, \quad D = 2$$

	A	B	C	D
1	Heat sample with powdered carbon in a stream of O_2	Pass over heated Pt then over heated Cu then over heated PbO_2 and $PbCrO_4$ + Ag	Absorb in an aqueous solution of Na_2CO_3 + Na_2SO_3	Measure increase in weight of tubes
2	Heat sample with CuO in a stream of CO_2	Pass over heated CuO	Absorb all other gases in aqueous KOH	Titrate solution with standard acid
3	Heat sample in a stream of O_2	Pass over heated Pt gauze	Absorb product in aqueous KI	Titrate with standard alkali

	A	B	C	D
4	Heat sample with powdered carbon in a stream of H_2	Pass over heated I_2O_5	Absorb in desiccant and Ascarite tubes in that sequence	Determine gravimetrically with $AgNO_3$
5	Heat sample in a stream of CO_2	Pass over hot Pt gauze then over hot CuO then over hot PbO_2 and $PbCrO_4 + Ag$	Absorb in tubes packed with Na_2CO_3 and Na_2SO_3	Titrate with standard $Na_2S_2O_3$ solution
6	Heat sample with powdered Cu in a stream of H_2	Pass over heated Cu	Absorb in aqueous H_2O_2	Measure volume of remaining gas in gas burette

Response

I have prepared a grid for you to check your answers. If you have a perfect match, then congratulations! You obviously have a good knowledge of combustion techniques and you need go no further with the response. If your answers do not match up, then following the grid are some notes taking you through the various analyses, which you should read.

Elements	A	B	C	D
C and H	3	5	4	1
N	2	6	2	6
S	3	3	6	3
Cl	3	3	1	4
O	4	4	3	5

Notes

Carbon and Hydrogen

We need first of all to oxidise the sample which is best done by heating it in an oxygen atmosphere and sweeping the partially oxidised products over a catalyst (hot Pt – still in the presence of O_2), and over an auxiliary oxidising agent (hot CuO) to complete the oxidation. We then need to remove other oxidation products (nitrogen oxides, halogens, and S). Hence our choice of A = 3 and B = 5. The H and C are now in the form of H_2O and CO_2 which can be absorbed in desiccant and Ascarite (KOH) tubes (C = 4) which are weighed and the increase in weight measured (D = 1).

Nitrogen

Here we require an oxidising environment for the sample but in a stream of easily removed gas. Hence we choose CuO as the oxidant and an atmosphere of CO_2 (A = 2). The evolved gases may contain oxides of nitrogen which must be reduced to N_2 so the gases must next pass through a reducing environment. This is provided only by hot Cu in column B. Hence B = 6. We should measure the volume of N_2 produced, so all other gases must be removed (C = 2) and the remaining N_2 measured in a gas burette (D = 6).

Sulphur and chlorine

We will take these two together as at least the first part of the analysis is the same for both. Once again, as with C and H an oxidising environment is required (supplied by A = 3). We don't want to remove the volatile products from S and Cl so we certainly don't want B = 1 or B = 5. Instead we just pass the gases over a catalyst (hot Pt) to aid the oxidation (B = 3) and then collect the gases in suitable absorbing solutions (aqueous H_2O_2 for SO_2 and SO_3, C = 6, and aqueous Na_2CO_3 + Na_2SO_3 for Cl products, C = 1). The final analyses are alkalimetric titration of H_2SO_4 (D = 3 for S) and gravimetry with $AgNO_3$ (D = 4 for Cl).

Oxygen

Here we start with a reducing environment to convert O_2 into CO. This is only satisfactorily provided by A = 4, since we need an excess of C to guarantee conversion of O_2 into CO. The CO reacts with I_2O_5 (B = 4) to produce I_2 vapour which is then absorbed in KI solution (C = 3) and titrated with $Na_2S_2O_3$ (D = 5).

****

SAQ 4.3a

In each of the following dry-ashing procedures a loss is likely to occur. Explain why, and how the loss might be prevented.

(*i*)   Ashing of a foodstuff with a high salt content in a porcelain crucible at 600 °C. Lead to be determined.

(*ii*) Ashing of plant material in a platinum crucible at 500 °C with H_2SO_4 as ashing aid. Gold to be determined.

(*iii*) Ashing of organic matter in a platinum crucible at 500 °C. Fluorine to be determined.

Response

(*i*)   The problem here is that $PbCl_2$ is formed from the lead and the Cl^- in the salt. The $PbCl_2$ is partially volatile at this relatively high ashing temperature. Reduce the ashing temperature and add H_2SO_4 as an ashing aid to keep the Pb in the involatile $PbSO_4$ form.

(*ii*) Gold is a noble metal and will be present in the ash as the metal. As such, it can alloy with the platinum crucible. So the simple answer here is to use a porcelain crucible.

(*iii*) Most non-metals are lost unless the ash is alkaline when metal salts of the non-metal acids are formed. It appears that the ash here is insufficiently alkaline and so an ashing aid consisting of an alkali-metal hydroxide or carbonate, for example, is called for. If, say, Na_2CO_3 is used, then the fluorine is retained as involatile NaF.

****

SAQ 4.4a State two advantages and two disadvantages that low-temperature ashing has compared with conventional dry ashing.

Response

I can suggest to you in fact *three* advantages:

(*i*) losses through volatility are reduced;

(*ii*) reactions between analyte and container material are reduced;

(*iii*) the technique does not rely on ashing aids, so there are no dangers of sample contamination.

There are also *three* disadvantages that spring to mind:

(*i*) ashing times can be very long;

(*ii*) unless the sample is broken up, the ash may prevent penetration of the excited oxygen throughout the sample;

(*iii*) the apparatus is more complex and expensive than that required for simple dry ashing.

****

SAQ 4.5a

Below are described some malfunctions of the
oxygen flask described to you by a fellow analyst who asks for your advice. What would your
advice be? Also give the reasoning behind any
advice that you give.

(*i*)   'All my results are lower than expected.
The oxidation proceeds smoothly without
problems although I do notice some black
particles in the flask after oxidation.'

(*ii*) 'I am trying to determine phosphorus in
a sample. My results are low and unpredictable. I always allow the flask to stand
for five minutes after the oxidation before
I open the flask to allow the pressure to be
reduced to about atmospheric.'

(*iii*) 'After years of use of my oxygen flask, the
platinum basket eventually disintegrated
and fell off. Because of the high cost of
platinum, I coiled the wire (also made
of platinum) connecting the basket to the
stopper to make a temporary basket. My
redesigned flask is shown below.

$\longrightarrow$

SAQ 4.5a (cont.)	'Regrettably, however, I have had a series of explosions with flasks using this stopper.'
(*iv*)	'I have been trying to determine sulphur by using the oxygen flask. I have used a small sample and a large flask and am happy that oxidation is as complete as can be. I assumed that the sulphur is oxidised to sulphur trioxide which I have absorbed into water to give sulphuric acid which I can titrate. I have agitated the flask for up to an hour although the initial mist has dispersed after ten minutes. My results have been low. On testing the solution, I find that oxidation has produced a mixture of sulphur trioxide and sulphur dioxide.'

Response

If *you* were the unfortunate analyst with all these problems, this is how *I* would advise you.

(*i*) 'The black particles indicate soot or particles of carbon, which means that your oxidation is incomplete. The most likely cause for this is that you have not allowed a sufficient volume of oxygen to oxidise all the sample plus filter paper. Try using less filter paper, less sample, or a flask of larger volume.'

(*ii*) 'I wonder if you are allowing sufficient time for the absorbing solution to absorb all the P_2O_5 produced in the oxidation. I notice you allow the flask to stand for only five minutes. I suggest you agitate the flask rather than leaving it still and that you allow at least 10 minutes. Also check that any mist in the flask has dispersed and note that the pressure should drop to well below atmospheric, if oxidation products are absorbed into the solution.'

(*iii*) 'This has happened to me. To begin with, I was mystified but then it occurred to me that the fierce burning was occurring in the narrow neck of the flask with the heat being concentrated onto the glass near the seal with the ground-glass joint, where the strain in the glass was at its highest. It was this that was causing the explosions. Placing the platinum coil on a glass stalk bringing it about two thirds the way into the flask solved the problem until we purchased a new basket.'

(*iv*) 'Well, you seem to have done all you can to ensure complete oxidation and absorption of vapour, so we must assume that the sulphur is oxidised to a mixture of SO_2 and SO_3 as a matter of course. We observed this to be so in the combustion tube method, but there we found that SO_2 is absorbed into H_2O_2 with the production of H_2SO_4. So the simple answer is to do the same thing here. Use a mixture of H_2O and H_2O_2 as absorbing solution, when the SO_2 is oxidised to H_2SO_4 in the solution phase rather than by the oxygen in the flask.'

****

SAQ 4.6a	Which of the following statements are *true* and which are *false*?

(*i*)   The following acids are those most commonly used, either on their own or in mixtures, for wet-ashing organic or biological samples:

HCl, HNO_3, H_2SO_4, $HClO_4$, HF, H_3PO_4.

(*ii*) $HClO_4$ is extremely dangerous to use on its own in wet-ashing procedures, although the hazards involved are much reduced

$\longrightarrow$

SAQ 4.6a **(cont.)**	when it is used in conjunction with other acids.
	(*iii*) The long, narrow neck of the Kjeldahl flask not only acts as an air condenser, reducing losses through volatility, but also helps to reduce losses through spray and foaming if the wet-ashing reaction becomes vigorous.
	(*iv*) Wet-ashing by using the Kjeldahl-flask technique is really best suited for non-metallic and metalloidal elements such as the halogens, P, S, As, Se, Sb, Te, B, and just a few metals such as Hg.
	(*v*) Very special care must be taken when samples are being wet-ashed with H_2SO_4 as part of the Kjeldahl nitrogen determination, to avoid losses of N as NH_3.
	(*vi*) The Carius-tube method is suitable for wet-ashing in the determination of all the halogens, S, P, Se, As, Hg, and other elements.

Response

(*i*)   *False*. The list of acids that I have given you here is the one that related to dissolution of inorganic samples, when we used the acid's oxidising and complexing properties. For wet-ashing, we simply require acids with pronounced oxidising properties. This basically narrows the list down to HNO_3, H_2SO_4, and $HClO_4$ as normal choices.

(*ii*)   *True*. $HClO_4$ when concentrated and hot can react explosively with organic matter, especially when the latter is easily oxidised. However, by mixing the acid with other acids or removing the easily oxidised portion of the sample by wet-ashing with another acid first, you can very substantially reduce the risk of explosion. But nevertheless, always treat $HClO_4$ with the greatest respect, whatever the conditions for wet-ashing that you use.

(*iii*) *True*. We mentioned in the text the role of the long neck as an air condenser to prevent loss of certain volatiles (although it won't help to prevent loss of gases, for instance). If the solution spits, or a spray is produced, or some frothing, then indeed the likelihood of loss is reduced by the presence of the long, narrow, neck of the flask.

(*iv*) *False*. In fact these are just the elements that could prove troublesome in a Kjeldahl flask since they are typical of the elements that form volatile derivatives and thus are liable to be wholly or partially lost from the flask. If you want to determine any of these elements by a wet-ashing procedure, you must either take special precautions to prevent loss from the flask or alternatively use the much less convenient Carius-tube technique.

(*v*) *False*. During the wet-ashing procedure, the nitrogen is converted, not into NH_3, but into the NH_4^+ ion since the acid concentration is very high. This is a ionic species and is therefore not volatile. However, after the wet-ashing, the sample solution is made alkaline and the NH_4^+ is converted into NH_3. This is then steam distilled and collected quantitatively in standard acid. You do need to be careful that no loss of NH_3 occurs at this stage.

(*vi*) *False*. The list is almost correct, but not quite all the halogens can be suitably handled by this means. We forgot that we must exclude fluorine from the list, because it is converted into HF by the wet-ashing procedure and the HF promptly reacts with the glass wall of the tube to form volatile SiF_4, an unsuitable chemical form for the analysis of fluorine.

****

SAQ 4.8a	Comment in an informed and critical way on the available ashing techniques for each of the following problems, and then recommend one particular ashing technique for each problem.

(*i*) Determination of traces of transition and heavy metals in a variety of samples of dried plant material collected from the environment.

(*ii*) Determination of phosphorus in a sample of polyvinyl chloride plastic (the phosphorus is present as an organic phosphate plasticiser and the amounts present are quite high).

(*iii*) Determination of fluorine in a newly prepared organic compound.

(*iv*) Determination of nitrogen in a sample of a soya-bean product (soya beans being a valuable source of protein).

(*v*) Determination of lead, cadmium and mercury in a sample of tuna fish in brine.

Response

If you had difficulty with these, then the responses to the exercises in Section 4.8 should be of help to you. Consult these, if you did not do so previously, and try again. Also note that, in most cases, there is no single correct answer with regard to the technique that you recommend for any application, since more than one technique could well work quite satisfactorily. The actual choice of technique in practice might well depend on personal preference and available skills and equipment in addition to the actual quality of results obtained.

(*i*)   In principle, this could be achieved by dry ashing, low-temperature ashing, oxygen-flask, or wet-ashing techniques (the latter in a Kjeldahl flask). Dry ashing would be a particularly simple and easy method and capable of handling a fairly large sample (important, to reduce sampling errors). Because the metals may be unevenly distributed in the material, it is necessary to take a sufficiently large amount of material to get a representative sample. But we have to guard against losses through volatility. Perhaps we could use H_2SO_4 as an ashing aid to overcome this problem. We could also use low temperature ashing when such losses are less likely, but the equipment is not generally available. The oxygen-flask method would not suffer from such losses but can handle only a small sample, thus sampling errors might occur. Wet-ashing with $HNO_3/HClO_4$ suffers from none of the problems above, however, and would be a fairly sure way of carrying out the analysis. I would recommend either dry-ashing with precautions taken to avoid volatility losses, or wet-ashing with $HNO_3/HClO_4$.

(*ii*)   This should be possible in an oxygen flask or by wet-ashing either in a Kjeldahl flask or a Carius tube. We would use the Carius tube to overcome losses through volatility that might occur with the Kjeldahl flask, but this is a bomb technique requiring a certain degree of skill and carrying a risk of explosions. We would rather try to use the more convenient Kjeldahl flask and guard against losses of P (as H_3PO_4) through volatility by using acids with not too high a boiling-point ($HNO_3 + HClO_4$) and not allowing the acids to boil out of the flask. In other words we should be sure to use the neck of the flask truly as a reflux condenser. The ashing might be quite time-consuming, but with these precautions, the loss of H_3PO_4 should be eliminated. The use of the oxygen flask, on the other hand, is rapid and not prone to volatility losses. In practice both wet-ashing and oxygen-flask techniques will work, but the latter will be faster in action.

(*iii*) Methods that could be considered for the ashing of an organic compound include the combustion tube method, dry ashing with an alkaline ashing aid present (to prevent release of HF), the oxygen-flask method, peroxide fusions, and reductive extraction with metallic Na or K. Assuming that we wish to carry out the analysis with the minimum of apparatus and specialised expertise, we then rule out the combustion tube. We can use the oxygen flask, but we need a silica or PTFE flask to minimise reaction with the vessel. The sodium peroxide fusion involves the use of bomb techniques and is thus rather inconvenient. The reductive extractions are rather specialised, but do work under appropriate conditions. The dry-ashing procedure is quite successful (except that high blank values from contamination have to be taken into consideration). There is no clear winner here, but dry-ashing (with an alkaline ashing aid) or the oxygen-flask technique would probably be the best choices.

(*iv*) There are really only two choices here: a combustion-tube method or the Kjeldahl nitrogen method. The latter is applicable only to amine or amide nitrogen. However, the nitrogen in soya beans is present as protein nitrogen (since soya beans are a valuable source of protein) and proteins are essentially polyamides. Hence the Kjeldahl method will be applicable. In practical terms it is easier to carry out than the combustion-tube method and needs no special apparatus. It is the method I would recommend, and in fact protein nitrogen determinations are the commonest application of the Kjeldahl nitrogen method.

(*v*) In principle dry ashing, low-temperature ashing, the oxygen-flask method and wet-ashing in a Kjeldahl flask or in a Carius tube can be considered. We must, however, immediately eliminate dry-ashing and low-temperature ashing at least for mercury because of volatility problems. Dry-ashing of lead in the presence of chloride from the salt in the brine can lead to problems due to the volatility of lead chloride. Mercury is also liable to be lost in wet-ashing procedures unless they are carried out in a sealed system, such as a Carius tube, or unless conditions are carefully controlled to eliminate the loss (ie make sure that the vapours are efficiently refluxed in the air condenser, or even use an auxiliary condenser). The Carius tube should be quite effective, but is rather inconvenient. There are no problems with the oxygen flask provided the small sample needed does not cause sampling errors. In practice, with suitable precautions, any of these three methods could be your choice.

****

SAQ 5.1a

Here is an outline procedure for the identification of an unknown drug in a sample of urine by using thin-layer chromatography (tlc). The individual stages of the procedure have been muddled up. Sort them out into their correct order. Which stage(s) involve(s) (a) chemical change(s) in the analyte?

(i)   Place the tlc plate in a chromatography tank and develop the chromatogram in a suitable solvent.

(ii)   Thus obtain the identity of the unknown drug.

(iii) Compare these for the unknown with those for the standards.

(iv)   Extract the drug from the sample with an organic solvent.

(v)   Apply the extract to a tlc plate, along with standards (samples of drugs suspected of being present).

(vi)   Measure the R_f values and note the colours of all spots on the chromatogram.

$\longrightarrow$

SAQ 5.1a **(cont.)**	(*vii*) Buffer the sample to a pH at which the drug is in an extractable form. (*viii*) Spray the plate with a solution of a reagent to convert the separated drugs into visible forms.

Response

Here is the analysis, sorted out into its correct order.

(*vii*) Buffer the sample to a pH at which the drug is in an extractable form.

(*iv*) Extract the drug from the sample with an organic solvent.

(*v*) Apply the extract to a tlc plate, along with standards (samples of drugs suspected of being present).

(*i*) Place the tlc plate in a chromatography tank, and develop the chromatogram with a suitable solvent.

(*viii*) Spray the plate with a solution of a reagent to convert the separated drugs into visible forms.

(*vi*) Measure the R_f values and note the colours of all spots on the chromatogram.

(*iii*) Compare those for the unknown with those for the standards.

(*ii*) Thus obtain the identity of the unknown drug.

You should have been able to cope with this satisfactorily, if you have a good understanding of the analytical sequence. The actual stages, although not corresponding exactly to the list in Section 1.1, show the general principles of sample pretreatment, separation, chemical conversion, identification, and drawing an inference, in that order.

The analyte is chemically converted into a visible (ie coloured derivative) in stage (*viii*), although stage (*vii*) also involves an acid/base conversion (conversion from ionised or protonated form by means of a change in pH).

****

SAQ 5.11a   Which of the following statements are *true* and which are *false*? Give reasons for your answers.

(*i*)   The concept of identification of an un-known compound by comparison of melt-ing-points with those of known compounds is applicable only to crystalline solids that are capable of being converted into crys-talline solid derivatives (since at least two melting-points are required to identify the compound).

(*ii*)   Universal spray reagents that are capable of reacting with any organic material are invaluable in paper chromatography for the general screening of chromatograms.

(*iii*) Polymers are substances consisting of large macromolecules. They therefore have zero volatility and when heated are degraded thermally. Therefore, gas chromatography is a technique that has no relevance to their analysis.

(*iv*)   hplc of non-uv absorbing compounds by

$\longrightarrow$

SAQ 5.11a
(cont.)

using a uv-photometric detector is possible only if we can prepare uv-absorbing derivatives of the compounds that can be separated under the same conditions as for the original compounds.

Response

All four statements are in fact false! Let us see why this is so by looking at each statement in turn.

(*i*)   The situation described here, where both the compound and its derivative must be crystalline solids capable of giving well-defined melting points is certainly more likely to produce an unambiguous result. But this does not preclude the usefulness of the technique in other situations. In fact, liquid samples can be characterised by melting-point determinations if solid crystalline derivatives can be prepared from them.

(*ii*)   This would be true for thin-layer chromatography which can be carried out on layers of an inorganic stationary phase. But paper itself is organic in origin, and so a totally universal spray reagent for organic materials would react with the paper as well, and so be of no use in paper chromatography.

(*iii*) You might well have been forgiven if you thought that this was true. But we can use chemical reactions to convert the polymer into volatile derivatives. In fact, the clue here is in the thermal degradation that occurs on heating the polymer. What we can do is to 'pyrolyse' (ie strongly heat) the polymer, when it is degraded thermally (breaks up) into smaller molecules (including monomers). These can be analysed by gas chromatography, and often their analysis provides much useful information about the polymer.

(*iv*) It is quite unnecessary to have to stick to the original chromatographic conditions when separating the derivatives. Why not simply prepare the uv-absorbing derivatives and then look for the best conditions for separating these? The conditions that were relevant for the original compounds now have no relevance to the analysis of the derivatives.

****

SAQ 5.11b

> Explain how you could distinguish between the members of the following pairs of compounds non-instrumentally and non-chromatographically:
>
> (*i*) C_6H_5OH and $C_6H_5OCH_3$;
>
> (*ii*) $HOCH_2CH_2OH$ and $HOCH_2CH_2OCH_2CH_2OH$.

Response

(*i*) These two compounds contain different functional groups. One is a phenol, the other is not. Therefore a simple chemical test for a phenol should suffice to tell them apart. Treat each compound in solution with a solution of $FeCl_3$. Only the phenol (C_6H_5OH) will give an intense purple colour.

(*ii*) These two compounds contain the same functional groups ($-OH$) but in different proportions. The first compound contains a greater proportion of $-OH$ groups per total number of atoms than the second. We should therefore be able to distinguish them if we were able to determine the proportion of $-OH$ groups in each compound, which we can do by titration.

We treat the sample with acetic anhydride, hydrolyse the excess of acetic anhydride and titrate the resulting acetic acid with standard KOH.

****

SAQ 5.11c

Consider the following list of analytical techniques and then group them in four columns under the numbers (*i*) to (*iv*) as follows:

(*i*)   those techniques that are applicable to organic compounds without derivatisation (normally);

(*ii*) those techniques that are applicable to organic compounds either with or without derivatisation;

(*iii*) those techniques that are applicable to organic compounds only after derivatisation;

(*iv*) those techniques not applicable to organic compounds at all.

(*a*) Atomic absorption spectroscopy.

(*b*) Thin layer chromatography.

(*c*) Spectrofluorimetry.

(*d*) Autoradiography.

(*e*) Nuclear magnetic resonance spectroscopy. ⟶

SAQ 5.11c
(cont.)

(*f*) Spectrophotometry.

(*g*) Infra-red spectroscopy.

Response

Here is my grouping which I have done by letters:

(*i*)	(*ii*)	(*iii*)	(*iv*)
(*e*)	(*b*)	(*d*)	(*a*)
(*g*)	(*c*)		
(*e*)			

Here are a few comments if you went wrong with this one:

(*a*) Atomic absorption is a technique specifically applicable to metal ions and quite irrelevant here.

(*b*) Thin-layer chromatography often, but not always, requires coloured derivatives of the separated species to be produced by spraying them with a suitable reagent.

(*c*) and (*f*). These techniques require the production of derivatives: coloured or uv-absorbing (spectrophotometry), or fluorescent (spectrofluorimetry). (But only if the original compounds do not possess these properties.)

(*d*) Autoradiography refers to the detection of radioactively labelled derivatives, say on a chromatogram, by photographic means. Derivatives containing a radioactive isotope must therefore be produced.

(*e*) and (*g*). Both infra-red and nuclear magnetic resonance spectroscopy reveal so much significant structural information about organic compounds that derivatisation is not usually needed.

****

SAQ 6.1a Consider the following analytical techniques. One is particularly associated with the determination of metal ions in solution in complexed form, one with the determination of metal ions in solution as free ions, and one is not particularly associated with the determination of metal ions at all. Which is which?

(*i*) Ultra-violet and visible spectrophotometry.

(*ii*) Infra-red spectroscopy.

(*iii*) Atomic absorption spectroscopy.

Response

(*i*) Ultra-violet and visible spectrophotometry are associated with the determination of metal ions in solution in complexed form (as well as with the determination of other species).

(*ii*) Infra-red spectroscopy is associated with the analysis of organic species. It has no application to the determination of metal ions in solution.

(*iii*) Atomic absorption spectroscopy is particularly associated with the determination of metal ions in solution as free ions.

****

SAQ 6.1b

Here are *six* descriptions of the properties of certain ligands and of the resulting metal complexes, together with *six* analytical techniques. Match the descriptions to the techniques.

The descriptions

(*i*)   Reacts selectively with certain types of metal ion to form water-soluble, non-absorbing complexes thus preventing the reaction of such metal ions with other ligands.

(*ii*) Reacts with metal ions to form complexes that are very much more soluble in organic solvents than in water.

(*iii*) Reacts selectively with metal ions to form water-insoluble complexes.

(*iv*) Reacts non-selectively with a wide range of metal ions to form stable, strongly coloured complexes.

(*v*)   Reacts selectively with certain metal ions only (ideally under specific conditions with a single metal ion) to form stable, strongly coloured complexes.

(*vi*) Reacts with metal ions to form negatively charged complexes.

The techniques

(*a*)   Spectrophotometric determination of metal ions in solution. $\longrightarrow$

SAQ 6.1b **(cont.)**	(b) Gravimetric determination of metal ions in solution. (c) Detection of metal ions separated by paper or thin layer chromatography. (d) Separation of metal ions by ion-exchange. (e) Masking of interfering metal ions in an analysis. (f) Solvent extraction of metal ions.

Response ·

Here is the correct match:

Descriptions	–	Techniques
(i)	–	(e)
(ii)	–	(f)
(iii)	–	(b)
(iv)	–	(c)
(v)	–	(a)
(vi)	–	(d)

Well done if you got a perfect match! If you did not, then here are a few points to check.

If you got techniques (b) and (f) the wrong way round, then note the emphasis on solubility in organic solvents rather than water for solvent extraction, and the simple requirement that the complex should be insoluble in water in gravimetry.

You might very well have got techniques (a) and (c) the wrong way round. If you did, can I emphasise that when you are using spectrophotometry in metal-ion determination, you are most likely to be determining a single, individual metal ion - hence the need for selectivity? However in chromatography, you are separating the metal ions in a mixture. Here you want the ligand to react with as many metal ions as possible.

Note for technique (e), the need to produce a complex with no particular properties that will cause it to interfere in other analyses. Here we are simply trying to prevent the metal ion from forming other complexes.

Finally for technique (d), we use complexation reactions to convert cations into anions, enabling separation by ion exchange to occur (in fact *via* the selectivity of such reactions).

****

SAQ 6.2a	Here are some molecular and ionic species. Which of these can act as ligands towards metal ions and which cannot?

    (i) NH_3,

    (ii) F^-,

    (iii) CH_4,

    (iv) H_2O,

    (v) PO_4^{3-},

    (vi) $CH_2{=}CH_2$,

    (vii) $CHCl_3$, $\longrightarrow$

SAQ 6.2a
(cont.)

> (*viii*) C_2H_5SH,
>
> (*ix*) $Si(CH_3)_4$,
>
> (*x*) $B(CH_3)_3$.

Response

The following species can act as ligands towards metal ions:

$$NH_3, F^-, H_2O, PO_4^{3-}, CH_2{=}CH_2, C_2H_5SH.$$

The following cannot:

$$CH_4, CHCl_3, Si(CH_3)_4, B(CH_3)_3.$$

This is how we come to the above conclusions. NH_3, H_2O, PO_4^{3-}, and C_2H_5SH are all simple molecules or anions containing the atoms, N, O or S, which have lone-pair electrons on them available for coordination (yes – H_2O is a ligand and our so-called 'free' metal ions in aqueous are themselves complexes with H_2O molecules as ligands!). F^- is an anion that can complex strongly with many metal anions in its own right. $CH_2{=}CH_2$ is an alkene and the π-electrons can complex with certain metal ions (eg Ag^+) to form π-complexes.

CH_4 contains no lone-pair electrons, neither does $Si(CH_3)_4$. $B(CH_3)_3$ is actually an electron-deficient molecule and so would act as a Lewis acid. $CHCl_3$ has atoms with lone-pair electrons, but we noted that in general the halogen atoms only show an inclination to complex formation when in the form of halide ions.

****

SAQ 6.2b

Below are a number of organic ligands. Which are likely to be capable of forming chelate complexes and which are not? (Answer *yes* or *no* for each ligand).

(*i*)

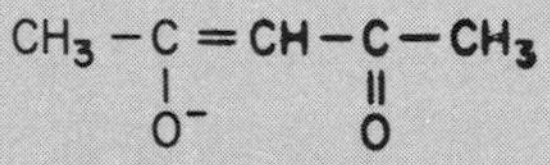

$$CH_3-C=CH-C-CH_3$$
$$\quad\; | \qquad\qquad \|$$
$$\quad O^- \qquad\quad O$$

(*ii*)

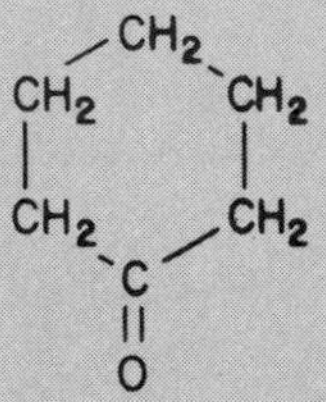

(*iii*)

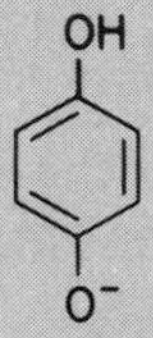

(*iv*)

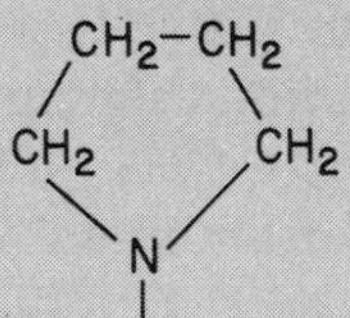

Response

(*i*)   *Yes*. This is the anion of the enol form of acetylacetone. Note the two oxygen atoms strategically placed to coordinate with a single metal ion to produce a favourable 6-membered ring structure.

(*ii*) *No*. The ligand contains only one coordinating atom (one oxygen atom) and so a ring structure will not be formed on coordination with a metal ion. The existence of a ring structure within the ligand itself does not in itself produce a chelate. This must be done by forming a ring on coordination. (Basically the metal ion must be part of the ring).

(*iii*) *No*, even though this ligand has two oxygen atoms. Here, the coordinating atoms are at opposite ends of a rigid planar molecule and it would be very difficult to 'manipulate' the molecule so that both oxygen atoms could coordinate simultaneously with the same metal ion.

By the same token

(structure: benzene ring with O⁻ group and OH group in meta positions)

would not form stable chelates, but

(structure: benzene ring with O⁻ and OH groups in ortho positions)

would.

(*iv*) *Yes*. In practice by coordination through the N atom and the negatively charged S atom to form 4-membered chelate rings. Although we noted that the optimum size for a chelate ring was 5 or 6 atoms, this does not preclude other ring sizes. In fact, the pyrollidine dithiocarbamate anion (which is what this is) forms stable chelates with a very wide range of metal cations.

****

SAQ 6.2c	The compound, 8-hydroxyquinoline has the structure below.

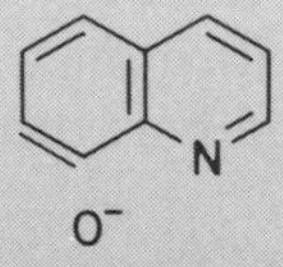

It reacts with Al^{3+} ions to form a chelate complex with a stoicheiometry of $1:3$ that is neutral (ie is not charged).

(*i*) What is the actual molecular species that reacts with the Al^{3+} ion (ie what is L?)

(*ii*) Write expressions for the stepwise stability constants and the overall stability constant for the ML_3 complex.

(*iii*) At low pH, formation of the complex is suppressed. What other species are formed instead of ML_3?

(*iv*) Finally draw the structure of the neutral ML_3 complex.

Response

(*i*) We note that the Al^{3+} ion has a charge of $+3$, that the complex is neutral and that three ligands are involved per metal ion. Thus the ligands must carry a charge of -1 each. A negatively charged ligand is easily visualised from the molecule shown by assuming that the phenolic $-OH$ group ionises as below:

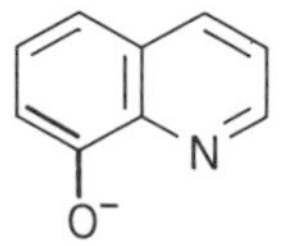

This is the actual ligand involved in the formation of the complex.

(*ii*) We can do this quite easily if we remember how we did it for the Cu^{2+}-1-aminoethane system by using the following symbols:

— M for Al^{3+},

— L for the negatively charged anion [the true ligand as we saw in response (*a*)],

— $K_1 \rightarrow K_3$ for the stepwise stability constants,

— β for the overall stability constant,

— square brackets for molar concentrations,

Then

$$K_1 = \frac{[ML]}{[M][L]}$$

$$K_2 = \frac{[ML_2]}{[ML][L]}$$

$$K_3 = \frac{[ML_3]}{[ML_2][L]}$$

$$\beta = K_1 . K_2 . K_3 = \frac{[ML] [ML_2] [ML_3]}{[M][L] [ML][L] [ML_2][L]}$$

$$= \frac{[ML_3]}{[M][L]^3}$$

(*iii*) At low pH we encounter high concentrations of hydrogen ions which can react with the negatively charged ligand in competition with the Al^{3+} ions. First, the neutral molecule can be regenerated.

Secondly, at even lower values of pH, there is the possibility of producing a protonated (positively charged) cation.

(*iv*) The structure of the complex. We know that it has a stoicheiometry of $1:3$ (ie ML_3). We have established that the ligand is the negatively charged anion of the original molecule with which we were provided. We also know that the complex is a chelate. So, we need to look for at least two points for coordination in the ligand suitably placed to form a stable chelate ring, and we immediately find these in the N atom and the negatively charged O atom. These will give ideal 5-membered chelate rings.

The Al^{3+} ion is 6-coordinate. In other words it likes to have 6 coordinating atoms around it in an octahedral arrangement (4 at the 4 corners of a square, one above and one below – all equivalent as in Fig. 6.2c)

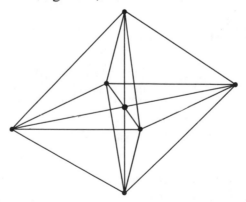

Fig. 6.2c. *Octahedral arrangement of six atoms around a single central atom*

All this fits perfectly, and here is the structure:

The arrangement of the atoms around the central Al^{3+} is octahedral.

* *

SAQ 6.2d Which of the following statements are *true* and which are *false*? Explain your choice in each case.

(*i*)   Amongst metal cations there is a tendency towards increasing 'hard' character as the charge on the ion increases.

(*ii*) Amongst both transition-metal cations and coordinating non-metallic elements there is an increasing trend towards 'soft' character as you progress to the right of the Periodic Table.

(*iii*) Amongst the coordinating non-metallic elements there is a marked increase in 'soft' character as you move down any group in the Periodic Table, but amongst metallic elements the trend is in the opposite direction. $\longrightarrow$

SAQ 6.2d   (*iv*) Acetylacetone would be a better chelating
(cont.)   agent for Al^{3+} than it would be for Pb^{2+}.

(*v*) Dithizone would be a good chelating agent
for Pb^{2+}, whilst EDTA would be a good
masking agent for interfering transition-
metal ions.

Acetylacetone   is $CH_3 \overset{\|}{\underset{O}{C}} CH_2 \overset{\|}{\underset{O}{C}} CH_3$

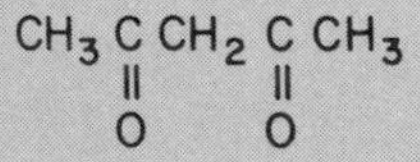

Dithizone   is   [benzene ring] NHNH

[benzene ring] N=N   C=S

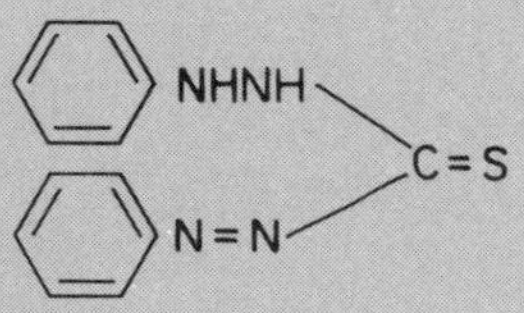

EDTA anion is $^-O_2CCH_2$ $CH_2CO_2^-$

$N\ CH_2CH_2\ N$

$^-O_2CCH_2$ $CH_2CO_2^-$

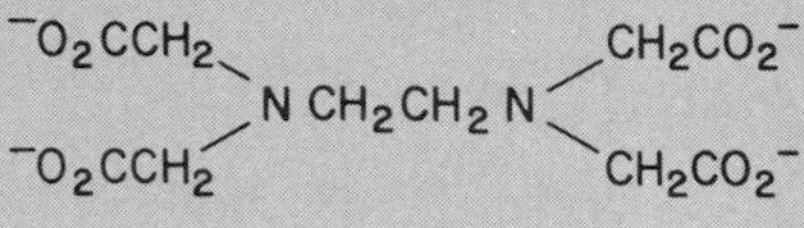

Response

(*i*)   In general, this is *true* (although it is possible to point to odd
individual exceptions). Thus, we find the 'hardest' character
amongst some of the high-valence refractory metal ions and
the 'softest' character amongst the low-valence heavy-metal
ions (the alkali metals, on the other hand are not particularly
'soft' in character).

(*ii*)   This is *false*. The statement is basically true for the transition
metal ions as you go, say, from Sc^{3+} to Zn^{2+}, but for the non-
metallic elements, exactly the opposite is true. For example,
as you go from N to O to F, 'hard' character increases.

(*iii*) This is *false*. Again for non-metallic elements, the statement is quite true (eg F → Cl and O → S), but there is no evidence to suggest the opposite for metal ions. If anything, the same trend towards increasing 'softness' occurs as you move down a group.

(*iv*) This is *true*. Acetylacetone is a reasonably 'hard' ligand, due to its O atoms, while Al^{3+} is a reasonably 'hard' metal ion. Pb^{2+} though, is typical of a 'soft' metal ion (belonging to the so-called 'heavy metals').

(*v*) This is *true*, at least in principle. Note that dithizone contains a 'soft' S atom and is a 'soft' ligand. It forms very stable complexes with heavy-metal ions. EDTA and transition-metal ions are well-matched with neutral 'hard-soft' character.

****

SAQ 6.2e

Comment on the trends in chelating properties that you might expect to observe within the following two series of chelating ligands.

(i) (*a*)   (*b*)   (*c*)

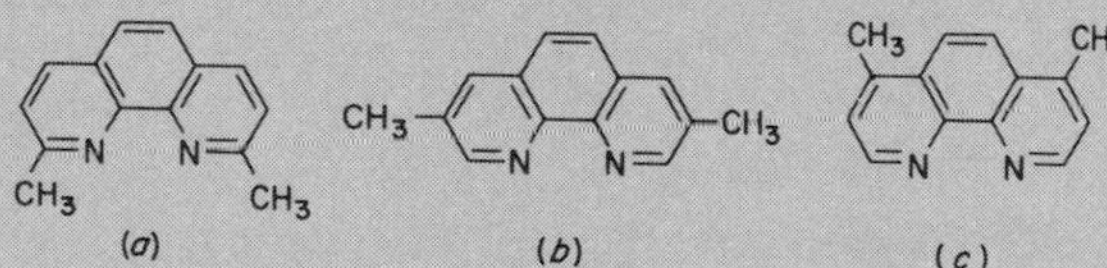

(ii) $H_2NCH_2CH_2NH_2$ $H_2NCH_2CH_2CH_2NH_2$

      (*a*)   (*b*)

   $H_2NCH_2CH_2CH_2CH_2NH_2$

      (*c*)

Response

(*i*) I would expect molecule (*a*) to form less stable chelates that either molecule (*b*) or (*c*) and to be more selective in its reactions, due to steric hindrance from the methyl groups, which are close to the coordinating N atoms in (*a*). In (*b*) and (*c*) the methyl groups are well removed from the chelating part of the molecule and are unlikely to hinder chelate formation.

(*ii*) These three molecules respectively form 5-, 6- and 7-membered chelate rings with metal ions. I would thus expect molecules (*a*) and (*b*) to form fairly stable chelates (5- and 6-membered rings) but molecule (*c*) to form somewhat less stable chelates because of the 7-membered ring.

****

SAQ 6.2f   Explain the following two observations:

(*i*)   Acetylacetone appears to form stable complexes with a stoicheiometry of 1:3 with Al^{3+} and Fe^{3+} but not with Cr^{3+}.

(*ii*) You measure the stability constant of a particular metal complex as a function of metal ion concentration. You observe the value to be independent of concentration until the concentration is of the order of 10^{-3} mol dm^{-3}. At higher concentrations you observe a dependence of the stability constant on concentration.

Response

(*i*) The obvious conclusion to come to here is simply that the stability constants for the Al^{3+} and the Fe^{3+} complex is large and for the Cr^{3+} complex it is small. But this is in fact not so. There are no great differences in the thermodynamic stabilities of the complexes, but there are in their rates of formation. The Cr^{3+} ion is an inert ion and reacts only very slowly. So the apparent failure of the Cr^{3+} ion to form a complex is kinetic in origin. Theoretically, given sufficient time, Cr^{3+} will also form an acetylacetone complex of stability comparable to that of the other ions.

(*ii*) We are observing the effect of activities coming into play at the higher concentrations and we should not be ignoring the need to use activities rather than concentrations, at concentrations above about 10^{-3} mol dm^{-3}, in the expression for the stability constant. If we do include activity coefficients (to convert concentrations into activities) then we should find that our stability constant is indeed truly concentration independent.

****

SAQ 7.1a	Which of the following involve separations within the meaning of the term as explained in Section 7.1?
	(*i*) Paper chromatography of transition-metal ions. The metal ions are resolved by development with a mixture of acetone (propanone) and hydrochloric acid, and detected and identified by reaction with various selective organic reagents.
	(*ii*) Removal of the interference by iron(III) in the spectrophotometric determination $\longrightarrow$

SAQ 7.1a
(cont.)

of dichromate ion by complexing the former with phosphoric acid.

(*iii*) Spectrographic analysis of a powder for the metals present. Each metal is represented by a line or lines on a photographic plate spatially resolved from each other in the spectrum.

(*iv*) Removal of nickel ions from a solution containing nickel and cobalt ions by precipitation as an organic complex.

(*v*) Gas chromatography of a petroleum fraction. The chromatogram consists of a series of peaks produced as each component of the petroleum fraction passes through a detector after a length of time which is characteristic of that component.

Response

(*i*), (*iv*) and (*v*) represent true separations whereas (*ii*) and (*iii*) do not. Let us briefly look at each one in turn.

(*i*) Separation has occurred here because after the chromatogram has been developed, different types of metal ion are found as spots (or bands or rings) at different locations on the paper. You can identify different metal ions *via* different colour reactions with various reagents only as a consequence of this physical separation.

(*ii*) No separation has occurred here, because the iron(III) ions are still present in the dichromate solution, even if their presence is no longer noticed when we measure the absorbance of the dichromate ions. In other measurements their presence may well be observed.

(*iii*) No separation of matter has occurred here. The metal atoms emit their characteristic spectra in the presence of each other. The spectrograph separates radiation according to wavelength (ie this is *energy* separation).

(*iv*) Clearly the nickel ions are converted into a solid form and filtered off from the solution, leaving the cobalt ions in solution. This is a simple example of physical separation.

(*v*) Physical separation has occurred here because each component takes a different length of time to reach the detector. Although the analytical result is a series of peaks on chart paper (cf peaks in a spectrum), if we were to replace the detector with some trapping device we could (in principle) finish up with pure samples of each component of the mixture.

****

SAQ 7.2a

Six analytical situations follow in which a specific problem has been identified and has yet to be solved. For each situation, explain whether incorporation of a separation stage might help to solve the problem and, if so, how, in relation to the eight areas of application of separation to analysis described in Section 7.2.

(*i*) Analysis of a liver extract for a drug metabolite by liquid chromatography. Unfortunately, the chemical composition of the extract is so complex as to make identification and quantification of the peak very difficult. $\longrightarrow$

SAQ 7.2a
(cont.)

(*ii*) Analysis of a 5 μl sample of child's blood for lead. Unfortunately, the method available has an absolute limit of detection (expressed in ng) that is not low enough for the analysis to be possible because of the small sample size.

(*iii*) Colorimetric determination of the copper content of tap water as its 1-(2-pyridylazo)-2-naphthol complex. Unfortunately the complex is precipitated from aqueous solution. It is however, soluble in trichloromethane.

(*iv*) Titration of cobalt in the presence of nickel by using EDTA as titrant. Unfortunately the minimum values of pH for the titration of the two ions are too close together for the titration to distinguish between the ions.

(*v*) Determination of traces of a toxic metal present in drinking water. Unfortunately we can't find a technique with a limit of detection (expressed in concentration units) that is low enough for us to carry out the analysis.

(*vi*) Total analysis of an alloy sample (ie determination of each individual metal present to give a total equal to 100% within experimental error). The trouble is, we don't know which metals are present.

Response

(*i*)   *Yes*. Here we have a matrix that is too complex for liquid chromatography. So a preliminary separation to *simplify the matrix* may overcome the problem (see Section 7.2.6).

(*ii*) *No*. Limits of detection may be increased by preconcentration only if it is the concentration that is too low. We can't use preconcentration to generate material that wasn't there in the first place. Here it is the total amount of material that is the limiting factor. We must either obtain a larger sample or, if this is not possible, look for a more sensitive measuring technique (see Section 7.2.2).

(*iii*) *Yes*. Transfer the complex to a trichloromethane phase by solvent extraction and make the colorimetric measurements thereon. This is an example of transfer of the analyte to a more suitable phase for analysis (see Section 7.2.3).

(*iv*) *Yes*. Here we need physically to remove the nickel from the solution by a separation procedure before we titrate the cobalt (for example, by precipitation with dimethylglyoxime (butane-2,3-dione dioxime)). This is an example of separating an interferent from an analyte (see Section 7.2.1).

(*v*)   *Yes*. This is a simple and straightforward example of the application of preconcentration. The limit of detection is unsatisfactory in terms of concentration but not in terms of the total amount of analyte, provided we can take a larger sample. So we preconcentrate the analyte into a smaller volume and we hope that the increased concentration will be sufficient for us to complete the analysis satisfactorily (see Section 7.2.2).

(*vi*) *Maybe*. You could carry out a sequential separation of all the metals (as ions in solution) and identify each individually. But nowadays there are very efficient selective atomic spectroscopic techniques (such as atomic emission or X-ray fluorescence spectroscopy) which will allow you to detect and determine all the metals in the presence of one another without separation.

Although in several of these responses, we have concluded that separation may be helpful to the analysis, that is not to say that there might not be other ways of solving the problem, not involving separation. For instance in (*iv*), the interference might be overcome either by masking the nickel, or by use of a different, more selective, analytical technique such as atomic absorption. At this stage we also make no claims that the approach by using separation is better or worse than any alternative approach.

****

SAQ 7.3a   Sulphate ions in water may be determined gravimetrically as barium sulphate. This analysis may be considered as consisting of:

(*i*)   a separation step, involving isolation of the sulphate ions in the form of pure, insoluble barium sulphate,

(*ii*) a measurement step, involving the weighing of the separated barium sulphate.

Describe the stages of the separation part of this analysis (ie (*i*) above) under the headings of:

chemical conversion,

distribution between two phases, (name the two phases),

physical separation of the phases.

Which two of these stages occur simultaneously and which sequentially?

Response

Compare your answers with the following and check that they agree in principle.

(*a*) *Chemical conversion.* The first part of the procedure involves adding a solution of barium chloride to the sample (perhaps after pH adjustments or other preliminary steps) in order to convert the sulphate into insoluble barium sulphate.

(*b*) *Distribution between two phases.* The barium sulphate is precipitated, creating a second phase (pure solid) in equilibrium with the liquid solution phase. The distribution then involves exchange of barium and sulphate ions between the two phases, the equilibrium constant now being the solubility product.

(*c*) *Physical separation of the phases.* This is simply the filtering, washing, and drying part of the operation, the result of which is to produce barium sulphate in a pure solid form, suitable for weighing.

The two stages that effectively occur simultaneously are the first two: chemical conversion and distribution. In reality in this separation, they more or less amount to a single process that can be described by the equation below.

$$Ba^{2+}_{(soln)} + SO^{2-}_{4(soln)} \rightleftharpoons BaSO_{4(solid)}$$

If your answer did not agree with this or you were not able to produce an answer at all, reread Section 7.3. to clarify your mind further on the stages of a separation.

****

SAQ 7.4a   Here are brief descriptions of four separations.
                Classify each as one of the following:

                — a batch separation,
                — a multiple batch separation,
                — a continuous separation,
                — separation involving a trapping technique.

                (*i*)   Preparation of de-ionised water. Water is
                        continuously passed through a mixed-bed
                        ion-exchange resin. Positive and negative
                        ions are quantitatively replaced by H^+
                        and OH^- respectively. When the cartridge
                        containing the ion-exchange resin is ex-
                        hausted, ie contains an excess of ions re-
                        moved from the water, it is replaced with
                        a new one.

                (*ii*) Iodine is quantitatively removed from an
                        aqueous solution by extraction with suc-
                        cessive portions of tetrachloromethane. As
                        soon as a portion of tetrachloromethane
                        obtained from the extraction shows no pur-
                        ple colour due to extracted iodine, the ex-
                        traction is assumed to be complete.

                (*iii*) The simple sugars in a mixture are sepa-
                        rated by placing spots of solution along the
                        edge of a sheet of adsorbent paper, dipping
                        this in a mixture of propan-1-ol, ethyl ac-
                        etate (ethyl ethanoate), and water, allowing
                        the solvent to rise by capillary action along
                        most of the length of the paper, carrying
                        the spots with it, drying it, spraying it with
                        aniline phthalate (a special colour-forming
                        reagent for sugars) and observing that each
                        sugar has travelled a different characteris-
                        tic distance along the paper. $\longrightarrow$

SAQ 7.4a **(cont.)**	(*iv*) Separation of copper ions from a mixture of other ions in solution. Under a particular set of chemical conditions copper ions and no other react with α-benzoin oxime to form an insoluble precipitate which can subsequently be collected by filtration.

Response

(*i*) This is a *trapping technique*. There is a continuous flow of 'sample' through the cartridge with the sole purpose of trapping all ions on the resin. There is no subsequent recovery stage here simply because what we need is the de-ionised water, but in principle you could recover the ions in a preconcentrated form from the spent cartridge.

(*ii*) This is a *multiple-batch separation*. Iodine is strongly, but not quantitatively, extracted from water into tetrachloromethane, so we repeat the extraction with fresh volumes of solvent until no further iodine is extracted.

(*iii*) This is an example of separation by paper chromatography. The stationary phase is water adsorbed onto the paper, the mobile phase, the mixed solvents, and *continuous* separation of the sugars occurs between the eluent rising by capillary action through the paper and the adsorbed water.

(*iv*) This is a simple *batch separation* involving precipitation. Its high degree of selectivity arises from the fact that only one ion (the copper ion) reacts to form an insoluble complex with α-benzoin oxime. There is no difficulty with this separation (at least in principle) once we have found a selective chemical reaction for the copper ions.

I hope that you did not have too many difficulties with this SAQ, which provides you with four fairly specific examples of the four types of separation that we have discussed. If you did, then revise

Section 7.4. Don't worry too much if the chromatography seems a little unclear to you. There are separate Units of *ACOL* devoted to this subject. Here I just want you to understand the idea of continuous separation (ie one phase flowing continuously over another with continuous distribution of substance between them, to put it in the simplest terms).

****

SAQ 7.5a

Here are *four* statements, only *one* of which is completely and unambiguously true. Which one is this and what are the errors in the other three statements?

(*i*) The distribution coefficient is a constant given by the ratio of the concentrations of a substance distributed between two phases at thermodynamic equilibrium. It is a function of temperature and pressure but is independent of concentration.

(*ii*) Iron(III) reacts with acetylacetone (pentane-2,4-dione) to give a 1:3 (metal to ligand) complex. Chromium(III) apparently does not react. From this fact we can conclude that the complex formed between iron(III) and acetylacetone has high thermodynamic stability whereas that between chromium(III) and acetylacetone does not.

(*iii*) Chromatography, amongst separation techniques in general, offers the greatest amount of separating power. This is because the separation factors between various species are usually larger for chromatographic techniques than for other separation techniques. $\longrightarrow$

SAQ 7.5a (cont.)	(*iv*) The value of the distribution ratio for a particular analyte distributed between two phases is dependent, and very markedly so, on the position of any chemical equilibria in which that analyte is involved.

Response

The only completely true statement is statement *four*. The distribution ratio is defined in terms of the concentration of an analyte without regard to its chemical form. If the relative concentrations of different chemical forms are being affected by the different positions of relevant chemical equilibria, then, since the different chemical forms will in general have different K_D values, the position of the distribution equilibrium for the analyte, taking into consideration all chemical forms will vary, and this will be reflected in a changing value for the distribution ratio.

Statement *one* is almost true and would have been, had we expressed our distribution coefficient in terms of activities rather than concentrations. In practice it may well be true over certain more-or-less limited concentration ranges, although there are situations where distribution coefficients are concentration dependent. Ideally, especially in chromatography, it is best to have K_D independent of concentration, and this is most likely to be achieved at low concentrations.

Statement *two* falls down because we did not consider the kinetic aspects of the situation. Chromium(III) is electronically inert, which means that its complexes are formed only extremely slowly, giving an apparent indication of lack of stability. We cannot therefore conclude that the complex is unstable. In practice, as it happens the complex is thermodynamically quite stable.

Statement *three* is false in that the separating power of chromatography stems from its ability to separate substances with quite small separation factors between them, whereas other, less powerful techniques require large separation factors between substances. Separation factor is a function of the distribution, not the way the separation is carried out.

****

SAQ 7.6a   For each of the following descriptions of a separation technique, give the name of the technique to which the description most aptly applies. A different separation technique should be selected for each description. Try to do this question from memory. If you don't feel confident to do this immediately, then revise Section 7.6 which contains all the information that you need for the question. Then attempt the question.

Descriptions

(*i*)   A chromatographic technique used for polymers, in which molecular size is the most important factor.

(*ii*)   A technique in which different rates of migration of different charged species in the presence of an electric field leads to separation.

(*iii*)   A chromatographic technique which has largely supplanted fractional distillation for the separation of mixtures of volatile organic compounds.

(*iv*)   A technique involving a membrane permeable to small molecules only. $\longrightarrow$

**SAQ 7.6a
(cont.)**

(*v*)   A preconcentration technique involving a solid-solution equilibrium. The solid phase is not pure, which can be a problem in a related well known classical quantitative method.

(*vi*)   Two liquid phases are involved in this batch technique, and solubility is the over-riding factor determining the separations possible.

(*vii*)   Paper chromatography is an example of this type of chromatography, which is related to (*vi*).

(*viii*) A preconcentration technique suitable for organic vapours in the atmosphere, involving reversible processes occurring on the surface of an 'active' solid with a large surface-area-to-mass ratio.

(*ix*)   A technique used largely in inorganic analysis but also for organic analysis, in which charged species are preconcentrated or separated chromatographically.

(*x*)   Metallic cations are separated by this technique, according to their position in the electrochemical series.

Response

The best matches are as follows:

(*i*)   Exclusion Chromatography,
(*ii*)   Electrophoresis,
(*iii*)   Gas–Liquid Chromatography,
(*iv*)   Dialysis,
(*v*)   Coprecipitation,
(*vi*)   Solvent Extraction,
(*vii*)   Partition chromatography,
(*viii*) Adsorption/Desorption,
(*ix*)   Ion Exchange and Ion-Exchange Chromatography,
(*x*)   Electrodeposition.

Congratulations if you got a perfect match! You have a good appreciation of the various types of separation technique available to the analyst, and are in a good position now to study some of the more important of these in greater detail.

If you got, say, between seven and nine correct, then just re-read in Section 7.6 the paragraphs describing the techniques that you missed. Then when you are happy about all the answers proceed to the next section.

Six or less? I suggest that you re-read Section 7.6 in its entirety and then attempt the question again. This time you should have improved your score to seven or more. Then just check once again any that you missed and proceed to the next section.

****

SAQ 8.1a Consider each of the following five solvent systems. For each system decide whether or not it would be useful for solvent extraction, and write down a sentence or two of explanation for your decision.

	Aqueous phase	*Organic phase*
(*i*)	Decimolar aqueous NaCl	Trichloromethane
(*ii*)	De-ionised water (unbuffered)	Acetonitrile
(*iii*)	50% aqueous methanol	Cyclohexane
(*iv*)	Water buffered to pH 3	4-Methylpentane-2-one
(*v*)	Molar aqueous ammonia	Ethanoic acid

Response

The correct responses are:

(*i*) *Useful*. Trichloromethane is commonly used for extraction.

(*ii*) *Not useful*. Acetonitrile is miscible with water in all proportions and therefore we should not obtain our two-phase system.

(*iii*) *Useful*. Methanol and water are both immiscible with cyclohexane whereas methanol and water are completely miscible with each other. This should give a two-phase system.

(*iv*) *Useful*. 4-Methylpentane-2-one (methyl isobutyl ketone, or MIBK for short) is a ketone of low solubility in water and is used extensively for extraction.

(*v*) *Not useful*. Not only are water and ethanoic acid (acetic acid) miscible in all proportions, but the acid will react with the ammonia.

****

SAQ 8.2a

Which of the following separations would you expect to be achieved by a solvent extraction procedure (none, one, more than one, or all might be suitable)?

(*i*) Preconcentration of traces of fluoride ion in water.

(*ii*) Separation of cobalt ions from nickel ions in aqueous solution.

(*iii*) Preconcentration of traces of a halogenated hydrocarbon (eg a pesticide) present in a biological fluid and removed from the bulk of the matrix.

(*iv*) Resolution of a group of amino acids into individual components on the basis of acid/base properties.

Response

Your responses should be as follows.

(*i*) *No*. You would be likely to consider solvent extraction in this situation. Remember that solvent extraction is most useful for metal ions and organic species. Anions are not normally the subject of solvent extraction. Here you might consider ion exchange, or distillation after conversion of the F^- into volatile SiF_4.

(*ii*) *Yes*, provided you could find an organic reagent that would form a water-insoluble complex with one of the ions. Such a reagent is dimethylglyoxime, which, under appropriate conditions, reacts selectively with nickel ions to form a complex containing almost all the nickel present.

(*iii*) *Yes*. This is a good example of solvent extraction being used for preliminary sample clean-up and preconcentration.

(*iv*) *No*. Although solvent extraction may be used to separate the components of organic mixtures by way of acid-base properties, this is only a rather crude fractionation process. Resolution of a complex mixture such as this requires the power of chromatography which does the job well. (This separation is actually possible by discontinuous countercurrent extraction, but this is such a slow, complex and expensive procedure compared with chromatography as not to be used nowadays).

****

SAQ 8.3a	Place the following operations in their correct sequence to produce a procedure for preconcentrating traces of metal ions from water into 4-methylpentane-2-one (MIBK). The preconcentration factor is 100 and the metal ions are extracted as their pyrrolidine-dithiocarbamate complexes. The complexes are formed in aqueous solution between pH 3 and pH 10. MIBK is less dense than water. (*N.B.* There is more than one acceptable order). ⟶

SAQ 8.3a
(cont.)

(*i*)   Add 10 cm^3 of MIBK.

(*ii*)   Evaporate the sample by gentle heating to 1/5th of its original volume.

(*iii*) Allow the phases to separate.

(*iv*)   Shake gently for 2 to 3 minutes.

(*v*)   Add a few cm^3 of a solution of ammonium pyrrolidine dithiocarbamate.

(*vi*)   Draw off the lower phase.

(*vii*) Draw off the upper phase.

(*viii*) Acidify this sample.

(*ix*)   Retain this phase.

(*x*)   Take a 1 dm^3 sample of water.

(*xi*)   Discard this phase.

(*xii*) Transfer to a separating funnel.

(*xiii*) Add buffer to give a pH of about 5.

(*xiv*) Cool the solution.

Response

The correct sequence of operations is as follows.

(*x*)   Take a 1 dm^3 sample of water.

(*viii*) Acidify this sample.

(*ii*) Evaporate the sample by gentle heating to 1/5th of its original volume.

(*xiv*) Cool the solution.

> (*xii*)   Transfer to a separating funnel.
>
> (*xiii*) Add buffer to give a pH of about 5.
>
> * (*v*)   Add a few cm^3 of a solution of ammonium pyrrolidine dithiocarbamate.
>
> (*i*)   Add 10 cm^3 of MIBK.

(*iv*) Shake gently for 2 to 3 minutes.

(*iii*) Allow the phases to separate.

(*vi*) Draw off the lower phase.

(*xi*) Discard this phase.

(*vii*) Draw off the upper phase.

(*ix*) Retain this phase.

If you got the above order (or one in which operations (*xii*), (*xiii*), (*v*) and (*i*) are rearranged), well done, you are now beginning to understand how solvent extraction can be applied in a real analytical situation. If you got any other order, check the procedure given in Section 8.3 'Practical Procedures in Solvent Extraction' once again, and then check the following additional points.

(*a*) We acidify the sample before evaporation to prevent adsorption of metal ions onto the surface of the container. We subsequently adjust the solution to pH 5 to ensure complete complex forma-

---

* The exact order in which operations (*xii*), (*xiii*), (*v*) and (*i*) are carried out is not critical. If you got these in a different order [eg (*xii*), (*v*), (*xiii*), (*i*)] you may consider your response correct.

tion, since we are told the complexes are formed between pH 3 and pH 10.

(*b*) We evaporate the original 1 dm^3 water sample to 200 cm^3, not the MIBK extract, since we know that this is already 10 cm^3 and anyway we could not have shaken 1 dm^3 of water with 10 cm^3 of MIBK. (*NB* An alternative approach might have been to shake the 1 dm^3 of water with 50 cm^3 of MIBK and evaporate the latter to 10 cm^3. The pros and cons of these two approaches are discussed later).

(*c*) Unlike the procedure in Section 8.3, here the organic phase is the less dense phase, and so the first phase drawn off will be the aqueous phase which we discard and the second is the organic phase, which we retain.

****

SAQ 8.3b	Indicate which of the following statements are *true* (T) and which are *false* (F).
	(*i*) In order to carry out a solvent extraction successfully, it is necessary to shake the two solutions together as vigorously as possible to disperse the two phases into minute droplets to ensure complete mixing. T / F
	(*ii*) Equilibrium in solvent extraction is normally established quite quickly, say, after shaking the liquids for 2–3 minutes. T / F
	(*iii*) After equilibration, the different substances present will have concentrated themselves entirely into one phase or the $\longrightarrow$

SAQ 8.3b (cont.)	other, depending on their relative solubilities.
	T / F
	(*iv*) For solvent extraction to be effective, there is no particular need for the relative volumes of the phases to be equal or even nearly equal.
	T / F

Response

The correct responses are as follows.

(*i*)   *False*. Too rigorous shaking could lead to emulsion formation and this makes separation of the two solvents into two distinct layers difficult. This problem varies with different organic solvents and different aqueous conditions. Extraction into trichloromethane from an alkaline medium is a case where emulsion formation might be troublesome.

(*ii*) *True*. Normally, two to three minutes are all that is required to establish equilibrium in straightforward situations.

(*iii*) *False*. Were it always true! In many practical situations in fact, certain compounds may well be concentrated almost entirely into one phase or the other, but other compounds may rather annoyingly distribute themselves in such a way that they are partly dissolved in one phase, partly in the other, leading to incomplete extraction.

(*iv*) *True*. When we want to use solvent extraction to preconcentrate a species we expect a large difference in the volumes of the two phases. But remember that there are nevertheless practical limits imposed by the miscibility of the two solvents and practical handling difficulties. So, we cannot expect a larger phase volume ratio than 20 to 50 (depending on the choice of organic solvent) for a single extraction step.

****

SAQ 8.4a

Match the following types of compound on a one-to-one basis with the following statements which all refer to extraction from an aqueous phase.

Compounds

(*i*)   Non-polar organic compounds.

(*ii*) Organic compounds of medium to high polarity.

(*iii*) Strongly ionised inorganic electrolytes.

(*iv*) Weakly ionised organic electrolytes.

Statements

(*a*) Cannot normally be extracted.

(*b*) Can be extracted efficiently into non-polar organic solvents of high immiscibility with water.

(*c*) Can possibly be extracted into polar organic solvents provided ionisation is suppressed.

(*d*) Can be extracted into organic solvents of medium to high polarity.

Response

The correct match is below.

Compounds	Statements	
(*i*)	(*b*)	Non-polar organic compounds can be extracted efficiently into non-polar organic solvents of high immiscibility with water.
(*ii*)	(*d*)	Organic compounds of medium to high polarity can be extracted into solvents of medium to high polarity.
(*iii*)	(*a*)	Strongly ionised inorganic electrolytes cannot normally be extracted.
(*iv*)	(*c*)	Weakly ionised organic electrolytes can possibly be extracted into polar organic solvents provided ionisation is suppressed.

If you failed to get the correct match, here are the points for you to check once more.

(*i*)   Non-polar solutes are easily extracted into non-polar solvents.

(*ii*)   More highly polar (but non-ionised) solutes are extracted into more polar solvents.

(*iii*)   Normally ions are not extracted from aqueous media.

(*iv*) Weakly ionised compounds can, when under appropriate conditions, have their ionisation suppressed when they may be extracted as polar neutral molecules.

****

SAQ 8.4b

Insert phrases selected from the list below to fill the numbered gaps in the following. Each phrase in the list may be used once, more than once, or not at all.

A urine sample was to be analysed for a drug with basic properties. It was necessary to obtain an extract of the ... 1 ... in a somewhat simplified matrix in an organic solvent for analysis. An aliquot of the sample was treated with ... 2 ... until ... 3 ... was observed. The sample was then shaken with a suitable volume of ... 4 ... and the ... 5 ... was discarded. The ... 6 ... was then treated with ... 7 ... until ... 8 ... was observed. It was then shaken with ... 9 ... and the ... 10 ... was retained for analysis.

The Phrases

'aqueous phase',
'an alkaline reaction',
'dilute aqueous sodium hydroxide',
'propanone',
'a neutral reaction',
'drug',
'urine sample',
'an acidic reaction',
'diethyl ether',
'organic phase',
'dilute aqueous hydrochloric acid'.

Response

(1) 'drug'
Our final extract should ideally contain the drug and nothing else. In practice of course it will contain other compounds with similar acid/base and solubility properties.

(2) 'dilute aqueous hydrochloric acid'

(3) 'an acidic reaction'
Perhaps you might have chosen 'dilute aqueous sodium hydroxide solution' and 'an alkaline reaction' here, but had you done so you would have had problems with responses to (6) and (7) where it becomes clear that we are further treating the aqueous phase. So the drug must have been retained in the aqueous phase in its protonated form. In fact, here we are removing any extractable neutral or acidic interferents.

(4) 'diethyl ether'
The only other possible response would be propanone but this is quite inappropriate as it is completely miscible with water.

(5) 'organic phase'

(6) 'aqueous phase'
At this stage the drug is in the aqueous phase in its protonated form with a selection of unwanted substances in the organic phase.

(7) 'dilute aqueous sodium hydroxide'

(8) 'an alkaline reaction'
We are now converting the drug into its uncharged form prior to its extraction.

(9) 'diethyl ether'
The reason is the same as for (4).

(10) 'organic phase'
We now have the drug finally extracted into the organic phase, hope-

fully in the presence of a far less complex matrix than in the original urine sample.

Thus the final complete procedure should read as follows:

'A urine sample was to be analysed for a drug with basic properties. It was necessary to obtain an extract of the *drug* in a somewhat simplified matrix in an organic solvent for analysis. An aliquot of the sample was treated with *dilute aqueous hydrochloric acid* until *an acidic reaction* was observed. The sample was than shaken with a suitable volume of *diethyl ether* and the *organic phase* was discarded. The *aqueous phase* was then treated with *dilute aqueous sodium hydroxide* until *an alkaline reaction* was observed. It was then shaken with *diethyl ether* and the *organic phase* was retained for analysis.'

****

SAQ 8.4c   Devise a scheme in block diagram form to carry out a broad, rough separation of a complex mixture of organic compounds into the following fractions (quantitative concentration of individual compounds into a single fraction is not required – rather the composition of each fraction should roughly correspond to the description below).

(*i*)   Neutral (non-acidic or non-basic) extractable compounds.

(*ii*)   Acidic extractable compounds.

(*iii*)   Basic extractable compounds.

(*iv*)   Non-extractable (ie water soluble) compounds.   ⟶

SAQ 8.4c
(cont.)

(NB 'extractable' means extractable from the aqueous into the organic phase).

Use the following devices to present your block diagram.

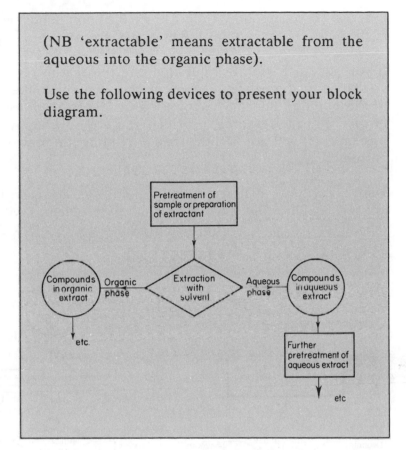

Response

The basic principles that we shall utilise for this fractionation are as follows.

Fraction (*i*) compounds (neutral extractable compounds) will be extracted into the organic phase at any pH.

Fraction (*ii*) compounds (acidic extractable compounds) will *not* be extracted into the organic phase at *high* pH.

Fraction (*iii*) compounds (basic extractable compounds) will *not* be extracted into the organic phase at *low* pH.

Fraction (*iv*) compounds (non-extractable compounds) will remain in the aqueous phase at all values of pH.

If you got stuck with this question and the reason for this is that you failed to appreciate the above points, you might like to go back to the question and have another go at it before proceeding any further with the response, provided you are now happy about the principles. If you are not, then try re-reading the previous Section on solubility.

Here is my block diagram. I would point out, however, that there are two equally valid ways of solving this problem. I will describe the alternative route after presenting you with my block diagram.

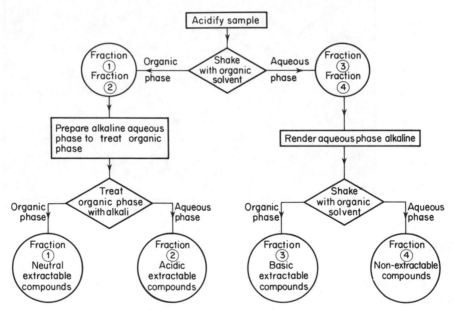

The alternative route would simply be the reverse of the above. Start by making the sample basic, and then shake the organic phase with acid, and adjust the aqueous extract to acid pH before shaking it once again with organic solvent.

We haven't mentioned the organic solvent that we would use in this extraction scheme yet. The actual choice is not important to the problem, but common choices might be: hexane (non-polar), dichloromethane (moderately polar), or diethyl ether or trichloromethane (somewhat more polar).

****

SAQ 8.4d

Copper ions in an aqueous solution of $CuCl_2$ are not extractable into trichloromethane as such. Which *one* of the following might be an effective way of extracting Cu^{2+} into an organic solvent?

(*i*) Replace the trichloromethane with another organic solvent in which $CuCl_2$ is highly soluble.

(*ii*) Convert the Cu^{2+} ions into a neutral chelate complex with an organic ligand such as 8-hydroxyquinoline.

(*iii*) Render the Cu^{2+} ions insoluble in water by the addition of alkali, and thus by inference soluble in an organic solvent.

(*iv*) Convert the Cu^{2+} ions into a highly stable and bulky organic complex form by using the non-selective organic ligand, EDTA (ethylenediaminetetra-acetic acid).

Response

The correct answer is (*ii*). Well done if this is the response that you chose.

If you chose (*i*) I would ask you the question 'what solvent?' $CuCl_2$ is soluble in solvents such as ethanol but these are miscible with water. No, we do have to produce some form of comp!ɔxed Cu ions to achieve the extraction.

We shan't get very far with (*iii*) either because it is *not* correct to infer that the product must be soluble in organic solvents. In fact, copper hydroxide is itself insoluble in all solvents (although it does, of course, form water-soluble complex ions with, for instance, ammonia).

Approach (*iv*) might at first seem promising, but there is no indication that the ion is neutral, extractable, or capable of forming ion-pairs. In fact, EDTA complexes are all highly water-soluble and EDTA is useful for preventing the extraction, say, of an interfering ion.

****

SAQ 8.5a	A sample of iodine weighing 0.560 g is dissolved in 100 cm^3 of water. This solution is shaken with 20.0 cm^3 of tetrachloromethane. Analysis of the aqueous phase shows that it contains 0.280 g dm^{-3} of iodine. What is the value of K_D for the distribution of iodine between water and tetrachloromethane?

Response

K_D for iodine between water and tetrachloromethane is 95.

This is how we arrived at this result.

The concentration of iodine in water of 0.280 g dm^{-3} means that 0.028 g were left in the 100 cm^3 of aqueous phase.

Thus $0.560 - 0.028$ g $= 0.532$ g were extracted into 20.0 cm^3 of tetrachloromethane.

Thus the concentration of iodine in the tetrachloromethane phase

$$= \frac{0.532 \times 1000}{20.0} \text{ g dm}^{-3} = 26.6 \text{ g dm}^{-3}$$

Now, $K_D = \dfrac{\text{concentration of iodine in tetrachloromethane}}{\text{concentration of iodine in water}}$

$$= \frac{26.6 \text{ g dm}^{-3}}{0.28 \text{ g dm}^{-3}} = \mathbf{95}.$$

****

SAQ 8.5b

Only *one* of the following five statements is completely true. Select this statement and identify *all* the errors in the other four statements.

(*i*) The ratio of the total masses of a substance, X, in a particular chemical form dissolved in two immiscible solvents at equilibrium is given by the distribution coefficient only when the volumes of the two solvents are equal.

(*ii*) The distribution coefficient is strongly dependent on pH for weak acids and bases but is effectively independent of pH for neutral substances.

(*iii*) 10.0 cm^3 of water containing 1.00 g of substance X is shaken with 10.0 cm^3 of trichloromethane. The organic layer is found to contain 0.100 g of substance X.

$\longrightarrow$

SAQ 8.5b
(cont.)

Thus it follows that K_D for this system is exactly 0.100.

(*iv*) The distribution coefficient, K_D, is defined as the ratio of the activities of substance X distributed between two immiscible solvents at equilibrium, and is a constant provided temperature and pressure remain constant and provided X is not involved in any chemical equilibria with other species.

(*v*) The total range of possible values of K_D goes from infinity for a substance completely retained in the aqueous phase, to 1 for a substance whose concentration in both phases is equal, to 0 for a substance completely retained in the organic phase.

Response

(*i*) Well done if you selected this as the only completely true statement.

(*ii*) This would be true had we been talking about distribution ratio, D, which refers to a substance in all its chemical forms. The distribution coefficient, K_D, refers only to one chemical form of a substance, such as either an ionised or an unionised form. It does not, therefore, depend on pH, even if the total extractability of the substance does.

(*iii*) If you thought that this was correct, then perhaps you argued along the lines that we have 1.00 g of X in the aqueous phase, 0.100 g of X in the organic phase and thus:

$$K_D = \frac{0.100 \times 10}{1.00 \times 10} = 0.1$$

(remembering, quite correctly, that K_D is defined as the ratio of concentrations of solute in organic and aqueous phases). But you forgot to take into account the fact that when 0.100 g of X was transferred to the organic phase, the amount in the aqueous phase was reduced to 0.900 g. Thus, in fact:

$$K_D = \frac{0.100 \times 10}{0.900 \times 10} = 0.1111 \text{ (to 4 significant figures).}$$

(*iv*) There are two errors here. First of all, remember that K_D was defined in terms of concentrations not activities. This was for the sake of convenience, even though the true thermodynamic constant should indeed have been expressed in activities. Secondly, there is no need to specify that X is not involved in any chemical equilibria with other species. We simply have to understand that if X exists in different chemical forms as a result of participating in chemical equilibria then each chemical form of X will be characterised by its own unique value of K_D.

(*v*) This would be totally correct if we simply interchanged the words 'aqueous' and 'organic'. If you thought that this was correct, then perhaps you need to remind yourself that K_D was defined as:

$$\frac{\text{concentration of substance in organic phase}}{\text{concentration of substance in aqueous phase}}$$

at equilibrium. This results in large values of K_D for species extracted efficiently from an aqueous into an organic phase. This is the way that extractions are most commonly carried out.

****

SAQ 8.5c   The neutral, unionised form of an organic acid has a value of K_D of 150 for extraction from water into trichloromethane. At pH 5 the value of D for the acid taking into account ionised (anionic) and unionised (neutral) forms is 12. Thus the percentage extraction of the acid, taking into account both ionised and unionised forms from 100 cm^3 of water at pH 5 into 25 cm^3 of trichloromethane is (to two significant figures):

(*i*)   97%

(*ii*)   92%

(*iii*) 98%

(*iv*)   75%

Response

(*i*)   97%.

This is *not* the correct answer, but is the answer that is obtained if K_D is used in the equation for E instead of D. The distribution ratio, D, must be used if we wish to find out how much of the total acid in all ionic forms is extracted. You have actually correctly established that 97% of the unionised neutral form of the acid has been extracted.

(*ii*) 92%.

I am afraid that if you got this answer, you probably used the wrong equation for E. Did you perhaps use

$$E = 100 \, D/(D + 1)?$$

This is the simplified form of the equation and is valid only when the volumes of the phases are equal. Try again using the correct equation.

$$E = \frac{100\ D}{(D\ +\ x/y)}$$

(*iii*) 98%.

Sorry, I think that you have got the volumes of the phases interchanged in the equation for E. This is easily done. Check that in the equation below:

$$E = \frac{100\ D}{(D\ +\ x/y)}$$

x is the volume of the aqueous phase and y is the volume of the organic phase. If you find it confusing remembering which way round these should be, try this line of thought. If the ratio x/y is increased, ie the aqueous phase volume is increased at the expense of the organic phase volume, then we would expect less substance to be extracted. As x/y is in the denominator of the expression for E, E decreases as x/y increases as expected.

(*iv*) 75%.

Well done if this was the answer you selected as it is indeed the correct answer. The equation you used was:

$$E = \frac{100\ D}{D\ +\ x/y}$$

where D is the distribution ratio $= 12$.

x is the volume of the aqueous phase $= 100\ cm^3$

y is the volume of the trichloromethane phase $= 25.0\ cm^3$

$$\therefore \quad E = \frac{100 \times 12}{12\ +\ 100/25.0} = \frac{1200}{16} = 75\%$$

****

SAQ 8.5d   Test the hypothesis that a more efficient total extraction is obtained by dividing the organic solvent into a number of aliquots, shaking the aqueous phase successively with these aliquots and combining them afterwards. Use the following hypothetical data.

Volume of aqueous phase = 100 cm^3
Total volume of organic phase = 100 cm^3
Distribution ratio = 10
Extraction carried out:

(*i*) in one stage,

(*ii*) in 2 to 4 stages, dividing the 100 cm^3 of organic solvent into 2 to 4 equal sized aliquots.

Response

This is a simple application of the equation:

$$W_n = W_o \left(\frac{x}{Dy/n + x} \right)^n$$

From the question, we have:

$D = 10$

$x = y = 100$ cm^3

n is varied from 1 to 4.

We are not given a value of W_o, but then we don't need an absolute value for W_1 if we are testing the theory. Any value will do. Let's use:

$$W_O = 1.00 \text{ g}$$

Thus:

For $n = 1$ $\quad W_1 = \dfrac{100}{10.0 \times 100 + 100} = 0.0909 \text{ g}$

For $n = 2$ $\quad W_2 = \left(\dfrac{100}{10.0 \times 50 + 100}\right)^2 = 0.0278 \text{ g}$

For $n = 3$ $\quad W_3 = \left(\dfrac{100}{10.0 \times 33.3 + 100}\right)^3 = 0.0123 \text{ g}$

For $n = 4$ $\quad W_4 = \left(\dfrac{100}{10.0 \times 25 + 100}\right)^4 = 0.0067 \text{ g}$

Thus the efficiency of extraction is indeed improved by carrying out the extraction with $4 \times 25 \text{ cm}^3$ aliquots of organic solvent rather than with just one 100 cm^3 aliquot. The improvement is better than 13.5 times. So at least in this situation the hypothesis that a more efficient extraction is obtained by dividing the organic extract into a number of aliquots, extracting solutes from the aqueous phase successively with these aliquots and combining them afterwards is well proven.

You can test the hypothesis with other figures and you will always see an improvement, but I do warn you that the improvement is more significant with a more efficient extraction to begin with (ie the improvements to be gained are greater the greater the value of D to start with).

****

SAQ 8.5e　　State whether the following statements, which all relate to *continuous extraction*, are *true* (T) or *false* (F).

(*i*)　For a solute to be extracted efficiently by continuous extraction, it must be sufficiently volatile to be volatilised along with the extracting solvent.

T / F

(*ii*)　All other factors being equal, the xylenes boiling at about 140 °C are less useful than diethyl ether boiling at 35 °C, as extraction solvents.

T / F

(*iii*)　Continuous extraction is invaluable for increasing selectivity in extraction systems.

T / F

(*iv*)　Continuous extraction is more versatile if the refluxing liquid is the organic solvent and the solutes are extracted *from* the aqueous phase.

T / F

Response

(*i*)　*False*. The solutes should not volatilise in the refluxing solvent at all, otherwise they would be returned to the solvent whence they originally came, thus defeating the point of the exercise.

(*ii*)　*True*. Setting aside questions as to whether D or K_D values are higher with the xylenes than with diethyl ether, the higher boiling-points of the xylenes mean that any thermally labile compounds that are extracted *might* decompose in the refluxing solvent.

(*iii*) *False*. Continuous extraction is used when extraction efficiency, rather than selectivity, is poor. There is a technique which can be used to increase selectivity in extraction systems known as *discontinuous countercurrent extraction*. The apparatus for this is complex and bulky, and separations are long-winded and expensive in use of solvents. *Continuous countercurrent extraction* is, in effect, the same as *partition chromatography* which is the way such separations would be carried out nowadays. Discontinuous countercurrent extraction is really an idealised form of partition chromatography. If you want to read more about discontinuous countercurrent extraction, then there are some excellent accounts on the subject (see reference 3 for example).

(*iv*) *True*. Extractions are more versatile this way round because we so often wish to buffer or add other reagents, electrolytes, etc, to the aqueous phase. Continuous extraction in which water is the refluxing solvent would mean extraction with pure water or, at the most, with water and a volatile acid or alkali that forms a constant boiling mixture – a much more limited choice.

****

SAQ 8.5f

On the basis of the K_D and K_A values Fig. 8.5d, which of the following values (*i*) to (*vi*) is most appropriate for the separation factor, β, between benzoic acid and our hypothetical acid for extraction from water at pH 9 into diethyl ether?

(*i*) 163

(*ii*) 13 100 $\longrightarrow$

SAQ 8.5f
(cont.)

(*iii*) 88.0

(*iv*) 14 300

(*v*) 1.04

(vi) 18 200

Response

The correct answer is (*iv*) – 14 300.

Well done if this is the answer that you obtained.

Let us go through the calculation.

$$D = \frac{K_D}{1 + K_A/[H^+]}$$

For benzoic acid:

$$D_1 = \frac{720}{1 + 6.30 \times 10^{-5} \times 10^9} = 0.0114$$

For the hypothetical acid:

$$D_2 = \frac{750}{1 + 3.60 \times 10^{-9} \times 10^9} = 163$$

Then $\beta = D_2/D_1 = 163/0.0114 = 14\ 300 = 1.43 \times 10^4$

Let us now see how the other wrong answers might have been arrived at.

(*i*)   *163*. This is the value of D for our hypothetical acid. Perhaps you obtained this answer by subtracting the D values rather than dividing them. You may have been prompted to do this because you remembered subtracting E values to obtain the differences listed in Fig. 8.5d but, remember β is a ratio, not a difference.

(*ii*) *13 100*. Although this is close to the correct answer I am afraid it is wrong. If you got this answer, check that you did not get either the K_A values for the two acids the wrong way round in your calculations of D_2 or the K_D values. Check very carefully how you put the numbers into the equations.

(*iii*) *88.0*. This is the answer you would have obtained if you calculated the ratio of the percentage extractions (E_2/E_1). Remember β is the ratio of the distribution ratios.

(*iv*) *1.04*. You should have guessed that this answer was not a sensible one because it is so close to unity. If this was the correct answer, we would need a good quality capillary gas-chromatography column for the separation. However, you might have arrived at this answer if you had calculated β by using distribution coefficients (K_D values) instead of distribution ratios (D values). The latter must be used here because we have a chemical equilibrium resulting in the acids existing in two forms (neutral and anionic) in the aqueous phase.

This answer might also have been obtained if you used pH instead of $[H^+]$ (remember $pH = -\log_{10}[H^+]$) or if you multiplied K_A by $[H^+]$ in the equation for D, instead of dividing by it.

(*v*) *18 200* If you got this answer, then it is most likely that you got K_A and K_D confused in the equation for D, ie, you put the K_A value into the numerator. Check this point, and re-do the calculation. This time you should get the right answer.

****

SAQ 8.6a

For each of the following, indicate whether the statement is *true* (T) or *false* (F).

(*i*) It is virtually unknown for a metal ion to be extractable from water into a water-immiscible organic solvent as a free hydrated cation.

T / F

(*ii*) A good method for the extraction of metal ions from water into organic solvents is to convert them, by using suitable organic ligands, into neutral, coordinatively saturated chelate complexes containing the minimum of extraneous polar groups.

T / F

(*iii*) Metal ions may often be extracted into organic solvents as charged species, provided that this charged species has been somehow incorporated into an ion-association complex. Complex formation and solvation by the extracting solvent are important features of such a system.

T / F

Response

All three statements are *true*.

Statement (*i*) tells us that in almost every known case we have got to change the form of the metal ion from the free hydrated state by one means or another before extraction. It is just not feasible to attempt to extract a metal ion in its free hydrated state. In no way does it satisfy the requirements for an extractable species as described in Section 8.4.

Statement (*ii*) describes one good way of doing this. Here, a suitable organic reagent is used to convert the metal ion into a neutral (ie uncharged) chelate complex so that all the co-ordination sites around the metal ion are occupied by ligand atoms. Thus, all co-ordinated water molecules are eliminated. The complex when formed, should normally be free of extraneous polar groups that would otherwise reduce its solubility in the extraction solvent.

Statement (*iii*) describes an alternative approach to extraction of metal ions that may, at first, be a little more difficult to understand. Up to now we have always said that charged species are not extractable. It is, however, possible in appropriate circumstances to extract a charged species if it can be combined with a suitable counter-ion to form an electrostatically bound ion-association complex. This complex is, of course, neutral. The basic requirement of such a system is that one or other (or both) of the ions should have a large, bulky organic character. This can be achieved either by complex formation or by solvation by the extracting solvent.

If you are still uncertain about any of these points then try re-reading Section 8.4.

****

SAQ 8.6b

> If we work at about pH 10, magnesium(II) is only poorly extracted into trichloromethane as a neutral chelate complex with 8-hydroxyquinoline. If butylamine is used in place of 8-hydroxyquinoline, then no extraction of magnesium(II) occurs at all. However, if a mixture of the two reagents is used (8-hydroxyquinoline plus butylamine) then the extraction of magnesium(II) suddenly becomes highly efficient. $\longrightarrow$

| SAQ 8.6b (cont.) | Can you offer explanations for these three observations from what you have just learnt about neutral chelate extraction systems? Magnesium(II) has a coordination number of six and so exists in aqueous solution as the $Mg(H_2O)_6^{2+}$ ion. |

Response

We'll consider the three cases separately, ie the use of 8-hydroxyquinoline alone, butylamine alone, and then of a mixture of the two. If you were hopelessly stuck on this question, perhaps you might like to study the first response and then go back and have another go at the other two. If you are still stuck, read the second response and try the third part. If it is still not clear to you, then the third response should finally complete the picture.

(*i*)   *Use of 8-hydroxyquinoline alone.* This reagent reacts as a bidentate ligand in the singly charged anionic state (have a look back as SAQ 6.2c). Thus two ligand molecules need to react with the metal ion in order to produce a neutral chelate. This means, however, that only *four* co-ordination sites around the metal ion are occupied by ligand atoms and so two water molecules remain. This reduces the solubility of the complex in the organic solvent, because an extractable neutral-chelate complex should contain *only* organic ligand molecules around the metal ion. Perhaps at a lower pH, where the concentration of free ligand is higher, the complex $ML_2 \cdot HL$ might be formed to some extent. This satisfies our requirements and so should be extractable, but is likely to be of low stability.

(*ii*) *Use of butylamine alone.* Butylamine does not ionise significantly but can complex with metal atoms. There is therefore the possiblity of forming ML_6^{2+}, but I am sure that you have immediately noticed that the neutral ligand molecule does not neutralise the charge on the metal ion. Thus the complex formed is not neutral nor is it chelated and so is not extractable, even though all the water molecules have been displaced and the metal ion is surrounded by organic ligands.

(*iii*) *Use of the two reagents together*. When the two reagents are used together, not only the complexes already discussed, but also *mixed ligand* complexes may be formed. Now one of these is $Mg(Ox)_2(BA)_2$, where Ox is the anion of 8-hydroxyquinoline and BA is a neutral butylamine molecule. This particular complex satisfies our requirements for efficient extraction, since the Ox ligands neutralise the charge and displace some of the water molecules whereas the BA ligands displace the remaining water without affecting the charge. Thus, this complex is formed virtually exclusively since the equilibria are forced in the direction of its formation, because it is immediately removed from the aqueous phase by the extraction.

This is an example of 'synergism' whereby the two ligands 'help' each other so that their combined effect far exceeds the sum of their individual effects. Other examples are known of the use of synergism in this way in solvent extraction.

****

SAQ 8.6c   Below are *three* examples of attempted extraction of a metal ion from an aqueous phase into an organic phase. Only *one* of these truly corresponds to a neutral-chelate extraction that is likely to be successful. The other three might be expected to fail as neutral chelate extractions or else correspond to extraction systems other than neutral chelate. Pick out the one potentially successful neutral-chelate extraction system and explain why the others fail as neutral-chelate ex-

$\longrightarrow$

SAQ 8.6c **(cont.)**	traction systems. Assume that the metal ion has a coordination number of six.

    (*i*)   Extraction of high concentrations of iron(III) from a 6*M*-hydrochloric acid medium into diethyl ether.

    (*ii*) Extraction of aluminium(III) as its 8-hydroxyquinoline-5-sulphonic acid complex from an aqueous medium, pH 9, into trichloromethane.

    (*iii*) Extraction of traces of nickel(II) from an aqueous medium, pH 5, into trichloromethane (as its 1-(2-pyridylazo)-2-naphthol (PAN) complex.

Response

Only (*iii*) is likely to work truly effectively as a neutral-chelate extraction system. The anionic form of PAN acts as a tridentate ligand, co-ordinating through the pyridine N atom, the azo-group and the $-O^-$ (ionised phenolic) group. Therefore, if two PAN anions co-ordinate with one nickel cation, the charges are neutralised, all co-ordination sites around the nickel ion are filled, and a neutral-chelate compound suitable for extraction is formed (formula below).

Note the lack of any exposed polar groups in the complex, the complete neutralisation of all charges and the filling of all six co-ordination sites round the nickel ion. Also note that the extraction is being performed at a non-extreme value of pH and that traces of nickel are involved, both conditions normally required for successful neutral chelate extraction.

Now let us look at the other systems and see why they fail to correspond to true neutral-chelate extraction systems.

(*i*) This fails to correspond to a neutral chelate extraction system on at least three counts. First, there is no chelate-forming organic reagent present; secondly, the extraction takes place from a strongly acidic medium where chelates are not normally formed; and thirdly, a high concentration of metal ion is involved. In fact this is a useful and successful extraction system, but the mechanism is quite different. We shall look at the mechanism of this extraction later on.

(*ii*) Here, the inclusion of sulphonic acid groups (completely ionised at all relevant values of pH) in the 8-hydroxyquinoline derivative, means that the ML_3 complex we might expect to be formed will carry a triple negative charge. The formation of a *charged* chelate does not satisfy our requirements for *neutral*-chelate extraction, and therefore no extraction will occur.

$$*********************************$$

SAQ 8.6d

In the following paragraph, for each group of three words or expressions in italics, *delete* the least suitable two in order to produce a description of the *best* procedure for the preconcentration of lead from a 1 dm^3 water sample into 10 cm^3 of organic solvent, before analysis by atomic absorption. $\longrightarrow$

SAQ 8.6d (cont.)	'The sample is first rendered *acidic/neutral/ alkaline* and then evaporated from 1 dm³ to a volume of *600 cm³/250 cm³/30 cm³*. It is then buffered to *pH 1/pH 5/pH 12* and an excess of *EDTA/dimethylglyoxime/ammonium pyrroli- dine dithiocarbamate* is added to complex the lead. Finally the complex is extracted into 10 ml of *MNAK/benzene/MIBK*'.

Response

The responses indicated as being correct are those which should *not* have been deleted, ie those that will remain as part of the best procedure.

(*i*)

(*a*) *Acidic. Correct response!* The acid is there to prevent adsorp- tion or precipitation of the lead or other cations during the evaporation.

(*b*) *Neutral*. At neutral pH values, there is a danger of adsorption of lead onto the walls of the container during evaporation. This can be controlled by making the solution acidic.

(*c*) *Alkaline*. This is definitely unadvisable! Not only will adsorp- tion be a problem, but cations including lead might be precipi- tated as hydroxides. Overcome both problems by acidifying the sample.

(*ii*)

(*a*) *600 cm³*. You are right to keep the time-consuming evaporation stage to a minimum, but you have left too much for the solvent extraction to do. A single-stage solvent extraction cannot give a preconcentration factor of more than 50. Here you need a factor of 60.

(*b*) *250 cm³. Correct response!* This represents a nice compromise between keeping the time-consuming exaporation stage to a minimum and not leaving the solvent extraction with too big a preconcentration factor (25 in this case).

(*c*) *30 cm³.* You have evaporated the solution far more than was necessary, presumably taking a long time, and leaving the relatively quick extraction stage to deal with a preconcentration factor of only 5.

(*iii*)

(*a*) *pH 1.* I am afraid that you are unlikely to get very much extraction at this low pH because protonation of the reagent will supress complex formation.

(*b*) *pH 5. Correct response!* The complex is formed between pH 3 and pH 10.

(*c*) *pH 12.* Complex formation may well be incomplete, I am afraid, at this high pH because of competing reactions involving hydroxide formation.

(*iv*)

(*a*) *EDTA.* Not a good choice, I am afraid, because metal-EDTA complexes, although quite stable usually, are always charged and so non-extractable.

(*b*) *Dimethylglyoxime.* We of course require a universal reagent or at least one that reacts efficiently with lead. Dimethylglyoxime is well known as a specific reagent for nickel and palladium only. No extraction with lead therefore!

(*c*) *Ammonium pyrrolidine dithiocarbamate. Correct response!* This reagent reacts with a wide range of metal ions over a fairly wide pH range and is well known as a complexing agent for use in solvent extraction preconcentration procedures. Its complexes are of the extractable neutral chelate type.

(*v*)

(*a*) *MNAK. Correct response!*. Although this solvent is less widely available and more expensive than MIBK, its lower solubility in water means that we can carry out solvent extractions at higher phase-volume ratios. Our selected ratio of 25 might imply problems due to the limited but significant solubility of MIBK in water.

(*b*) *Benzene*. This might be a suitable solvent from an extraction point of view, but we would not recommend it because of its high toxicity compared with the ketones.

(*c*) *MIBK*. This is the commonest solvent to use in this type of extraction, but because of the high preconcentration factor involved, its solubility in water might cause problems. Had you happened to answer (*c*) in (*ii*), then this would indeed be the best solvent to use.

Thus the best procedure (from the options in front of us) for the preconcentration would be:

'The sample is first rendered *acidic*, and then evaporated from 1 dm^3 to a volume of *250 cm^3*. It is then buffered to *pH 5* and an excess of *ammonium pyrrolidine dithiocarbamate* is added to complex the lead. Finally, the complex is extracted into 10 cm^3 of *MNAK*'.

****

SAQ 8.6e

Explain the following observation which, although at first a little surprising, is in fact true.

'In choosing between two organic solvents for extracting neutral chelate complexes of metal ions where *n* (the stoichiometric factor) is equal to 2 or more, the solvent giving rise to the

$\longrightarrow$

SAQ 8.6e (cont.)

smaller value of K_{DM} (distribution coefficient of ML_n), ie the poorer solvent for the extraction of the complex, is actually the solvent producing the higher value of D (the distribution ratio for the metal ion) ie, the better solvent for the overall extraction'.

If you find that you haven't a clue about the explanation to this, don't worry! The response is arranged as a series of 'clues' to the explanation. Look at the response clue by clue and after considering each clue, see if you can come up with the explanation. Perhaps you won't need many of the clues but if you do, at the end you may have a better understanding of some of the factors influencing the overall extraction.

To get you started, I would recommend that you look at Eq. 8.27, which relates D to K_{DM}. Keep it in front of you during your deliberations.

Response

Here are your clues one by one. If you think you have the explanation then you can work straight through them to check your argument. Otherwise, read them one by one, and after considering each one in turn, return to the question and have another go at it. A complete explanation, ie answer to the question, is formed by reading all the clues together in sequence.

Clue 1. The complex is not the only species present that can be extracted.

Clue 2. The neutral form of the ligand can also be extracted, ie, K_{DL} and K_{DM} both appear in the equation for D.

Clue 3. If we change the solvent, we shall affect not only the value of K_{DM} but also the value of K_{DL}.

Clue 4. If we choose a 'better' solvent for the extraction of the complex, then we select a solvent giving an increased value of K_{DM}, but since the effect of changing solvent is a general one, it is reasonable to expect it also to increase the value of K_{DL} and, what is more, to about the same extent. Thus, the ratio $K_{DM} : K_{DL}$ should be approximately independent of the solvent (or again in the simplest language a 'good' solvent for the complex is also a 'good' solvent for the ligand).

Clue 5. A 'better' solvent will thus reduce the concentration of free ligand, HL in the aqueous phase, and therefore also the concentration of L^-, the anionic form of the ligand that reacts with M^{n+} to give ML_n. (*NB* K_{DL} appears in the denominator in Eq. 8.27).

Clue 6. Reducing the concentration of L^- reduces the degree of formation of ML_n and this reduces the extent of extraction of M^{n+}. But this effect competes with the more efficient extraction of ML_n. Which effect wins?

Clue 7. The reduction in the concentration of L^- is the more significant effect because of the value of n; ie, K_{DL} appears in Eq. 8.27 raised to the power of n whereas K_{DM} does not. In fact, *all* the terms involving the ligand appear in the equation raised to the power of n. The fact that one metal ion reacts with *more* than one ligand molecule means that factors affecting the ligand will be more significant than those affecting the metal ion or the complex. Therefore, a better extraction solvent removes ligand from the aqueous phase preventing (or reducing) formation of the complex and therefore preventing (or reducing) extraction of the metal ion. This is more significant than the improved extraction of the metal-ion complex whenever the value of $n > 1$.

****

SAQ 8.6f | Under a particular set of conditions (of pH, etc.), a metal ion in equilibrium with various complexed (chelate) forms is known to have a distribution ratio, D of 60.5 between trichloromethane and water. How many extractions, each involving 25.0 cm^3 of trichloromethane, are required to remove >99.9% of the metal ion from 100.0 cm^3 of water?

The answer to this question can be obtained quite quickly and easily if you can recall from memory an equation that we actually derived earlier in this Unit. However, if you cannot recall this equation, or there is any likelihood of your recalling it incorrectly, *do not* refer back to the text, but instead attempt the problem by working from basic principles, which in this case means essentially working from the definition of D.

Response

The equation that can be used quite simply for working out the answer to this question is Eq. 8.20. All we have to do is to put suitable values into the equation and solve it for successive values of n until we get a value of $W_n < 0.001 \times W_0$.

However, I shall assume that in fact you could *not* remember the equation and that you did *not* refer back to the text. It is quite possible to solve this type of problem working simply from the definition of D (which I hope you can by now remember).

Let us assume that we start with a mass W_0 of metal ion (in all chemical forms) in the 100 cm^3 of water. After n extractions we are left with W_n of metal ion in the aqueous phase (ie W_1, W_2 etc, after

1, 2, etc extractions). We need to find the lowest value of n for which $W_n < 0.001 \times W_0$ (for 99.9% removal of the metal ion).

Remember that:

$$D = \frac{\text{total concentration of metal ion in the organic phase}}{\text{total concentration of metal ion in the aqueous phase}}$$

Therefore, after one extraction;

$$D = \frac{(W_0 - W_1)}{25.0} \times \frac{100.0}{W_1} = 60.5$$

$$\therefore \quad 100 \, W_0 - 100 \, W_1 = 60.5 \times 25.0 \times W_1$$

$$\therefore \quad W_1 = \frac{100 \, W_0}{100 + (60.5 \times 25.0)} = \frac{100 \, W_0}{1612.5} = 0.0620 \, W_0 \text{ g}$$

This is $>0.001 \, W_0$, and therefore we need a second extraction.

After the second extraction:

$$D = \frac{(0.0620 \, W_0 - W_2)}{25.0} \times \frac{100.0}{W_2} = 60.5$$

$$\therefore \quad 6.20 \, W_0 - 100 \, W_2 = 1512.5 \, W_2$$

$$\therefore \quad W_2 = \frac{6.20 \, W_0}{1612.5} = 0.00385 \, W_0$$

This is still $>0.001 \, W_0$, so let's try a third extraction.

After the third extraction:

$$D = \frac{(0.00385 \, W_0 - W_3)}{25.0} \times \frac{100.0}{W_3} = 60.5$$

$$\therefore \quad 0.385 \, W_0 - 100 \, W_3 = 1512.5 \, W_3$$

$$\therefore \quad W_3 = \frac{0.385 \, W_0}{1612.5} = 0.000239 \, W_0$$

This is now $<0.001\ W_0$, and therefore we have achieved better than 99.9 % extraction by using just *three* extractions. So the answer to the question is *three*.

NB If you did use Eq. 8.20 then you should have proceeded as follows:

$$W_n = W_0 \left(\frac{x}{Dy + x}\right)^n$$

$$W_1 = W_0 \left(\frac{100}{60.5 \times 25.0 + 100}\right) = 0.0620\ W_0\ (\text{ie} >0.001\ W_0)$$

$$W_2 = W_0 \left(\frac{100}{60.5 \times 25.0 + 100}\right)^2 = 0.00385\ W_0\ (\text{ie} >0.001\ W_0)$$

$$W_3 = W_0 \left(\frac{100}{60.5 \times 25.0 + 100}\right)^3 = 0.000239\ W_0\ (\text{ie} <0.001\ W_0)$$

****

SAQ 8.6g Two divalent metal ions, A and B, which are present at low and approximately equal concentrations in water are to be separated by solvent extraction by using selective pH control. Neutral, extractable chelates, AL_2 and BL_2 are to be formed with the anion of the ligand, HL and the extraction is to be performed by using equal volumes of phases. The ligand is present initially in the organic phase at a concentration of 0.100 mol dm^{-3}. The following data relating to equilibrium constants are provided

Acid dissociation
constant for the
ligand, HL $= 10^{-5.7}$ mol dm^{-3}

$\longrightarrow$

SAQ 8.6g
(cont.)

Distribution coefficient
for the neutral ligands,
HL $= 10^{3.2}$

Overall stability
constant for the
complex, AL_2 $= 10^{6.9} \; (mol \; dm^{-3})^{-2}$

Overall stability
constant for the
complex BL_2 $= 10^{10.7} \; (mol \; dm^{-3})^{-2}$

Distribution coefficient
for the complex AL_2 $= 10^{4.8}$

Distribution coefficient
for the complex BL_2 $= 10^{5.2}$

Which of the following statements is the only
one that is completely true?

(i)   The $pH_{0.5}$ values for the ions A and B are
         4.05 and 1.95 respectively. These pH val-
         ues represent the limits of the pH range
         within which the separation can reasonably
         be considered as quantitative.

(ii)   The difference in $pH_{0.5}$ for the two ions (Δ
         $pH_{0.5}$) is 2.10. Ion A is the more extractable
         ion and is therefore the one found in the
         organic phase after extraction at the opti-
         mum pH.

(iii) The optimum pH for the separation is 3.0.
         This value was obtained by averaging the
         values of $pH_{0.5}$ for each ion. This is a gen-
         eral method for calculating the optimum

                                                      $\longrightarrow$

SAQ 8.6g **(cont.)**	pH for the separation of two metal ions in neutral chelate extraction systems. (*iv*) The separation factor between the two ions at the optimum pH is 16×10^3 (to 2 significant figures). This represents somewhat less than 1% cross contamination between the ions.

Response

(*i*) *Not true*. The $pH_{0.5}$ values have indeed been correctly calculated as 4.05 and 1.95, but we are being over-optimistic in assuming that separation is quantitative at all pH values between these limits. At the $pH_{0.5}$ value itself the ion in question is only 50% extracted (remember, this is the definition of $pH_{0.5}$), hardly appropriate for a quantitative separation! Have a look at Fig. 8.6b. From it, I hope that you can see that if an ion is to be virtually quantitatively extracted (or *not* extracted) then for $n = 2$ we ought really to be working at at least one pH unit away from the $pH_{0.5}$ value. This means that here extraction can be considered as quantitative *only* between about pH 2.95 and pH 3.05, in other words the adjustment of pH is going to be quite critical.

Let us now check through the calculation of $pH_{0.5}$ for the two ions. This can be done by entering the appropriate values for the equilibrium constants and other values into Eq. 8.27a.

$$K' = \frac{\beta_M K_{DM} K_A^n [HL]^n}{K_{DL}^n}$$

to obtain a value for K', and then to use this value in Eq. 8.30:

$$pH_{0.5} = \frac{-\log K'}{n}$$

to obtain $pH_{0.5}$.

For ion A:

$$K' = 10^{6.9} \times 10^{4.8} \times 10^{(-5.7) \times 2} \times 10^{-2} \times 10^{(-3.2) \times 2}$$

$$= 10^{-8.1}$$

$$\therefore \quad pH_{0.5} = \frac{8.1}{2} = 4.05$$

For ion B:

$$K' = 10^{10.7} \times 10^{5.2} \times 10^{(-5.7) \times 2} \times 10^{-2} \times 10^{(-3.2) \times 2}$$

$$= 10^{-3.9}$$

$$\therefore \quad pH_{0.5} = \frac{3.9}{2} = 1.95$$

(*ii*) *Not true*. We have calculated $\Delta pH_{0.5}$ correctly. This is simply the difference in $pH_{0.5}$ values for the two ions, ie: $\Delta pH_{0.5} = 4.05 - 1.95 = 2.10$

However, we made the wrong choice when we said that ion A would be the ion that was extracted at the optimum pH. The optimum pH (3.0) is *below* the $pH_{0.5}$ value for ion A (pH 4.05) and above the $pH_{0.5}$ value for ion B (pH 1.95). Now, if you look at any of the pH extraction curves for neutral-chelate extraction systems *up to* this point in the text, you will see that we must be working at pH values *above* $pH_{0.5}$ for the ion to be extracted and that at pH values below $pH_{0.5}$, the ion is not extracted. Thus, it is ion B that is extracted while ion A remains in the aqueous phase.

(*iii*) *Not true*. If the two pH-extraction curves have the same slope, ie the complexes have the same value for n, then the point on the pH scale (*x*-axis) where the difference in percentage extraction for the two ions is greatest, which corresponds to our optimum pH, (remember what we said in the exercise on p. 260), will be exactly midway between the two $pH_{0.5}$ values. This corresponds to the mean value of the two individual $pH_{0.5}$

values, *viz* 3.0. Look at Fig. 8.6b to see a pictorial justification for this argument, although in *that* particular example there is a fair amount of leeway on either side of the optimum pH (not so in this question).

So, in this example we have indeed calculated the optimum pH correctly. But it is *not true* to say that we can always calculate the optimum pH simply by averaging $pH_{0.5}$ values in this way. First of all we had to assume equal volumes of phases. Secondly, if the charges on the ions are unequal, then the slopes of the extraction curves will be different, and calculation of the optimum pH becomes more involved. This is perhaps best done graphically from the pH-extraction curves (see Fig. 8.6c).

(*iv*) *True!* Well done if you picked this one! The separation factor can be calculated from a rearranged form of Eq. 8.32:

$$\beta = 10^{n \times \Delta pH_{0.5}} = 10^{2 \times 2.1}$$

$$= 10^{4.2} = 16 \times 10^3 \text{ (to 2 significant figures)}$$

This figure is somewhat in excess of 10 000 which we said earlier (p. 259) represents approximately 1% cross contamination. Thus, here, cross contamination will be somewhat less than 1%.

****

SAQ 8.6h   Given the following information and by using the block diagram approach introduced in SAQ 8.4c, devise a multistage solvent extraction scheme for the complete separation of the following ions which were initially present at low concentrations in an aqueous sample. $\longrightarrow$

The aqueous sample initially contained

Fe^{2+}, Fe^{3+}, Mn^{2+}, Co^{2+}, Ni^{2+}, Cu^{2+}, Pb^{2+}.

Here is the information for your work.

(*i*)   Neocuproine (2,9-dimethyl-1,10-phenan-throline) is a selective reagent for the chelated ion-association extraction of Cu^+ from a neutral aqueous phase into tri-chloromethane.

(*ii*)   Dithizone reacts with most transition-metal and heavy-metal ions to form ex-tractable complexes. Some $pH_{0.5}$ values for extraction into tetrachloromethane were given in Fig. 8.6d. Here is some more in-formation specifically on the metal ions present in our sample (obtained from ref-erence 16):

Fe^{2+} – extracted quantitatively between pH 7 and 9,

Fe^{3+} – not extracted (oxidises complex),

Mn^{2+} – extracted at about pH 10 only (the complex is unstable),

Co^{2+} – quantitatively extracted into te-trachloromethane between pH 5.5 and 8.5 or into trichlorometh-ane around pH 8,

Ni^{2+} – extracted into tetrachloromethane at pH 6–9 or into trichloromethane at pH 8–11, $\longrightarrow$

SAQ 8.6h (cont.)

Cu^{2+} – extracted into trichloromethane or tetrachloromethane at pH 1–4,

Pb^{2+} – extracted into trichloromethane at pH 8.5–11.5 or tetrachloromethane at pH 8.0–10.0.

(*iii*) In the presence of 6 *M*-hydrochloric acid the only ion of those in our sample to be extractable into diethyl ether is Fe^{3+}.

(*iv*) Cyanide ion reacts with most transition-metal ions but not with main-group metal ions to form very stable, water-soluble complexes. The metal ions may subsequently be decomplexed by boiling them with acid when volatile HCN is driven off, *provided that* adequate precautions are taken to avoid inhaling the *highly toxic* HCN.

(*v*) In a slightly alkaline medium, and in the presence of masking agents, dimethylglyoxime is a selective chelating agent for the extraction of nickel. Of the ions in our sample likely to interfere, Fe^{3+} is masked with tartaric acid, Co^{2+} either with CN^- plus H_2O_2 or formaldehyde, or with NH_3, Cu^{2+} with thiosulphate and Mn^{2+} with hydroxylamine hydrochloride.

(*vi*) In acid solution, ammonium peroxydisulphate is a powerful oxidising agent, oxidising Fe^{2+} to Fe^{3+}, Cr^{3+} to Cr^{4+} and Mn^{2+} to Mn^{7+}. Nitric acid and dichromate ion are milder oxidising agents, oxidising Fe^{2+} to Fe^{3+} only. $\longrightarrow$

**SAQ 8.6h
(cont.)**

(*vii*) The Jones reductor (a zinc/mercury amalgam) reduces Fe^{3+} to Fe^{2+} and Cu^{2+} to Cu metal, whereas the Walden reductor (silver/silver chloride) reduces Fe^{3+} to Fe^{2+} and Cu^{2+} to Cu^{+}. Both reductors work on a column principle and the reduced solutions so obtained are free from excess of reducing agent. The Walden reductor works best in the presence of hydrochloric acid.

Response

There is no single 'correct' answer to this problem and, no doubt, there are a number of workable schemes that could be produced 'on paper'. To know which would be the most satisfactory in practice, you would have to try them out in the laboratory. The object of this exercise, though, is for you to gain experience by manipulating the data to propose a reasonable multistage extraction scheme.

I have divided this response into two halves. The first consists of a list of comments and pieces of advice as well as warnings about 'pitfalls' and the like, which you can read through. Then I have produced a possible scheme. If your scheme doesn't correspond to mine, don't worry, yours may still be viable. Use the responses as follows.

(*a*) If you haven't a clue how to start (and I hope you have!) start reading the comments, and these, I hope, may help you to start. Look back at the response to SAQ 8.4c to see how we tackled the problem there. If you are still stuck then look at the start of my scheme, copy out the first stage and then try to continue. Finally you can work through my scheme, see how I approached the problem and try to work out a different scheme that also works.

(*b*) If you have got some way but got stuck, or if you are not happy about your scheme, then read the comments and check through your work and have another go at completing it.

(*c*) If you have a scheme that you are reasonably happy with, then go straight to my scheme and see if they substantially correspond. If they do, then you may congratulate yourself on a successful response, but if there are differences, still check through the comments to see whether you have created any problems for yourself. You may well not have done so. If you have, then you can amend your scheme accordingly. In any event reading the comments should be useful.

Comments

(*i*) The basic idea of the exercise is to combine the information given, or as much of it as you need, to produce a sequence of events, including six extractions. Each extraction should be selective towards one or more ions of those present, but remember that as you proceed the mixture will get increasingly simple and so the demands on selectivity will lessen.

(*ii*) Optimum values of pH, acidity, etc, for one extraction may well have to be altered for the next stage of the separation. Large changes may result in high concentrations of electrolyte which in turn will mean a more complex, unpredictable matrix. So try to keep these to a minimum, although some adjustments will, of course, be necessary.

(*iii*) Addition of masking agents to make one stage of the separation more selective may mean that a subsequent stage will be interfered with. So, be careful if you use a masking agent at any point other than the end of the sequence, that the masked ion is subsequently released. Note that the removal of CN^- as HCN although theoretically possible is practically hazardous because of the toxicity of HCN.

(*iv*) Use selective extractions, when the matrix is complex, early on (eg ether extraction and use of neocuproine) and the less selec-

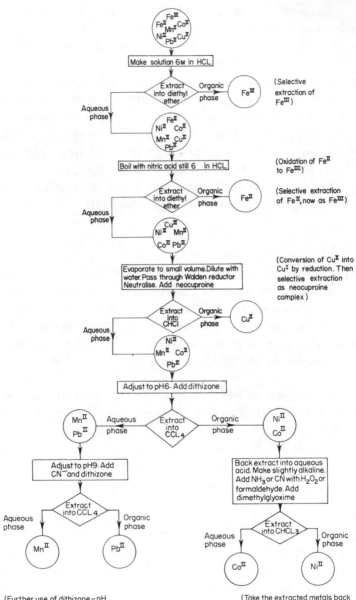

Make solution 6M in HCL

Extract into diethyl ether → Organic phase → Fe^III   (Selective extraction of Fe^III)

Aqueous phase

Ni^II Fe^II Co^II Mn^II Cu^II Pb^II

Boil with nitric acid still 6 in HCL   (Oxidation of Fe^II to Fe^III)

Extract into diethyl ether → Organic phase → Fe^II   (Selective extraction of Fe^II, now as Fe^III)

Aqueous phase

Ni^II Cu^II Mn^II Co^II Pb^II

Evaporate to small volume. Dilute with water. Pass through Walden reductor. Neutralise. Add neocuproine   (Conversion of Cu^II into Cu^I by reduction. Then selective extraction as neocuproine complex)

Extract into CHCl → Organic phase → Cu^II

Aqueous phase

Ni^II Mn^II Co^II Pb^II

Adjust to pH6. Add dithizone

Mn^II Pb^II ← Aqueous phase ← Extract into CCL₄ → Organic phase → Ni^II Co^II

Adjust to pH9. Add CN⁻ and dithizone

Extract into CCL₄

Aqueous phase   Organic phase

Mn^II   Pb^II

(Further use of dithizone – pH raised to enable extraction of Pb^II, at the same time masking Mn^II with CN⁻)

Back extract into aqueous acid. Make slightly alkaline. Add NH₃ or CN with H₂O₂ or formaldehyde. Add dimethylglyoxime

Extract into CHCL₃

Aqueous phase   Organic phase

Co^II   Ni^II

(Take the extracted metals back into the aqueous phase. Mask the Co^II with NH₃ or CN⁻ (in which case the H₂O₂ or formaldehyde demasks Ni^II from the CN complex) and then extract the Ni^II with dimethylglyoxime)

tive extractions (eg with dithizone) later on, when the matrix is less complex.

(v) Be careful about the order in which you carry out any oxidations or reductions. You should be able to use the gentler, more selective procedure at each stage, but be sure to remove the Fe^{3+} by extraction into ether before attempting any oxidation or reduction, so that you can separate the Fe^{2+} from the Fe^{3+}.

(vi) You can get rid of a substantial proportion of an acid matrix by evaporating the solution (in a fume cupboard!) to a small volume.

My proposed scheme for the separation is shown opposite.

****

SAQ 8.6i

Match the following extractions with the following general extraction systems on a one-to-one basis.

Extractions

(i) Extraction of uranium(VI) into trichloromethane in the presence of 8-hydroxyquinoline(HOx). The $UO_2(Ox)_3$ ion is formed and the tetrabutylammonium cation is present as counter-ion. ⟶

**SAQ 8.6i
(cont.)**

(*ii*) Extraction of the permanganate anion into trichloromethane in the presence of the tetraphenylarsonium cation as counter-ion.

(*iii*) Extraction of gold(III) from $3M$-hydrochloric acid into diethyl ether.

(*iv*) Extraction of nickel(II) into trichloromethane at neutral to weakly acidic pH in the presence of 1-(2-pyridylazo)-2-naphthol.

(*v*) Extraction of lanthanum(III) into dibutyl hydrogen phosphate (a liquid).

(*vi*) Extraction of uranium(VI) into tri-iso-octylamine hydrochloride (a liquid) from a sulphuric acid medium, in which the species $UO_2(SO_4)_2^{2-}$ is formed.

General Extraction Systems

(*a*) Neutral-chelate extraction system.
(*b*) Ion-association extraction system involving no chelation or solvation.
(*c*) Liquid-cation exchanger extraction system.
(*d*) Liquid-anion exchanger extraction system.
(*e*) Chelate-ion association extraction system.
(*f*) Oxonium extraction system.

Response

The correct match is as follows.

Extractions	General Extraction Systems
(i)	(e)
(ii)	(b)
(iii)	(f)
(iv)	(a)
(v)	(c)
(vi)	(d)

Here are a few words of explanation about each extraction, should you not have obtained this match.

Extraction (i). Here the UO_2^{2+} cation reacts with the Ox^- anion to form an ML_3 species, because the co-ordination number of UO_2^{2+} is six. This therefore carries a negative charge of one, and the tetrabutylammonium cation thus acts as counter-ion to form an ion pair.

Extraction (ii). Here we have an example of extraction involving simple ion association between a complex metallic anion and a bulky organic cation. No chelation or solvation is involved.

Extraction (iii). This is a classic oxonium system in which the ion association complex formed is of the type $(C_2H_5O)_2H^+ . AuCl_4^-$ or of a structure similar to this (it is difficult to be precise about the structure of such species).

Extraction (iv). This is a simple neutral chelate extraction system, operating under mild conditions of pH. The extracted species is $Ni(PAN)_2$ where PAN^- is the anionic form of the ligand.

Extraction (*v*). This is a so-called liquid cation-exchanger system in which lanthanum(III) ions pass into the organic phase and hydrogen ions into the aqueous phase. The lanthanum(III) ions are then chelated by the dibutyl hydrogen phosphate which simultaneously acts as extracting solvent.

Extraction (*vi*). This is a so-called liquid-anion exchanger system. Chloride ions pass into the aqueous phase from the organic phase, allowing the complex $UO_2(SO_4)_2^{2-}$ anion to pass in the other direction to form an ion-association complex in the organic phase.

****

SAQ 8.7a

Suggest outline procedures for carrying out the following separations and preconcentrations by solvent extraction. Write a paragraph or two about each, detailing the procedure you would adopt, and explain why you feel that the approach you have selected is a suitable one. Take your time over this question and don't try to do it from memory. In fact, I would encourage you to consult the references if possible and to do a little literature 'research' into the problems, although you can produce satisfactory answers from the material I have given you previously. In this respect, you might find reference 16 particularly helpful.

Do please also note that there are no single 'correct' answers to this SAQ and there may well be several quite acceptable approaches to any of these problems. So, if your response does not correspond with mine and you have researched it from a reference, don't worry, your method may well be appropriate. If you can think of

$\longrightarrow$

SAQ 8.7a (cont.)

more than one adequate approach, then discuss them both (or all of them!) In my responses, I have endeavoured to outline the general approach to the problem and then give one or two practical solutions together with comments about them.

(*i*) Separation of nickel(II) and cobalt(II) ions present at low concentrations in an aqueous solution.

(*ii*) Preconcentration by a factor of at least 100 of traces of cadmium(II), mercury(II) and lead(II) ions in a river water sample.

(*iii*) Separation of iron(III) and chromium(III) ions in an acidic medium.

(*iv*) Preconcentration of traces of halogenated pesticides (weakly polar organic compounds) in a water sample by a factor of about 1000, before gas chromatographic analysis.

Response

(*i*) *Nickel(II) and Cobalt(II)* Here we have two late transition-metal ions of low valency present at low concentrations and without any extremes of pH being involved –ideal conditions for a *neutral-chelate extraction*. However, the extraction system chosen must be selective towards one of the two ions. If you chose one of the less selective reagents then you will have to be very particular about conditions of pH and the use of masking agents in order to satisfy yourself about the selectivity of the extraction. Perhaps a more certain approach would be to look for a *specific chelating agent* for one of the two ions. Even so, masking agents may be required to ensure selectivity.

There is such a reagent for nickel(II). In weakly alkaline media, *dimethylglyoxime* forms extractable complexes *only with nickel*. So the separation could be carried out by adding dimethylglyoxime to the sample, adjusting the solution to mildly alkaline conditions with ammonia and extracting the nickel complex into trichloromethane. Ammonia acts as a masking agent towards cobalt. Alternatively CN^- may be added to mask the cobalt, but we then have to add H_2O_2 or formaldehyde to *demask* the $Ni(CN)_4^{2-}$ complex. On the other hand, α-nitroso-β-naphthol is a specific chelating agent for cobalt. This is a bidentate ligand that oxidizes cobalt(II) to cobalt(III) and then forms an ML_3 complex. Cobalt(III) has an inert electronic configuration and is stable only in a complexed form. So the complex has high kinetic stability (see Section 7.5.4). Under the right conditions, therefore, the reagent is selective for the extraction of cobalt (see reference 16).

(*ii*)   *Cadmium, Mercury and Lead*. The problem here is that a preconcentration factor of 100 is too much to achieve with a single-stage extraction alone. Two suitable approaches are possible. We could either combine solvent extraction with an evaporation step as we did in SAQ 8.6d, in which case the preconcentrated ions will end up in the organic phase. Alternatively we could carry out a two-stage extraction, in which case the ions will be first preconcentrated into an organic phase but then back extracted with further preconcentration into the aqueous phase. Let's look further at these two approaches.

Evaporation plus preconcentration. One question that we can ask here is whether it is better to evaporate the aqueous phase before extraction or the organic phase after extraction. Evaporation of the aqueous phase has the advantage that, as long as the sample is acidified, we should not suffer any losses of analyte (metal ions) either through volatilisation or adsorption, nor yet should any toxic fumes be produced. However, rather large volumes of water may have to be evaporated. Evaporation of the organic solvent should therefore be quicker and easier. On the other hand, some neutral metal chelates have a discernible volatility (some can be analysed by gas chromatography) and the evaporation of organic solvents can be costly and somewhat hazardous.

So, choose a non-selective chelating agent (dithizone, 8-hydroxyquinoline, or a dithiocarbamate would be a suitable choice), a nearly neutral pH for which the extraction is efficient rather than selective, an extraction solvent with a low solubility in water (4-methylpentane-2-one or, better, heptan-2-one, are good choices if the subsequent analysis is to be by flame spectroscopy) and carry out the extraction with a water-to-solvent-phase ratio of 25:1. Obtain the other 4-fold preconcentration either by evaporating the acidified water sample before extraction or (probably less likely in practice) the organic phase after extraction.

Two-stage solvent extraction. You can use the same chelating agent and pH conditions, but here the choice of extraction solvent is less critical. So, use the less expensive 4-methylpentane-2-one. Carry out an extraction into the organic phase with an water-to-solvent-phase ratio of 10:1. Then back extract into an acidic aqueous medium, now with a solvent-to-aqueous-phase ratio of 10:1, giving you 100:1 overall. The former approach gives the ions in an organic medium, the latter in an aqueous medium. In practice, such preconcentrations are often used in conjunction with atomic absorption spectroscopy where it is often beneficial to have your sample in an organic medium. So the first approach is perhaps the commoner.

(*iii*) *Iron(III) and Chromium(III) in Acid.* There is one fairly obvious approach to adopt here, and that is the *oxonium* extraction system, whereby the sample is made $6M$ in hydrochloric acid and shake with diethyl ether. Iron(III) is extracted as a chloro-complex, the only other ions to be quantitatively extracted being Au^{3+}, Tl^{3+}, Ga^{3+}, Sb^{5+} and Ti^{3+}. $Chromium^{3+}$ has an inert electronic configuration and forms complexes only extremely slowly. It is not, therefore, extracted under these conditions. Neutral-chelate extractions won't work from acidic media.

(*iv*) *Halogenated Pesticides*. Here, we have a huge preconcentration factor to achieve, so once again, we need to tackle the extraction in stages. Fortunately for us, the halogenated pesticides have low solubility in water but high solubility in nonpolar organic solvents. So, we should get a good recovery if we shake, say, one dm^3 of water with two successive 20 cm^3 portions of hexane and combine the solutions to obtain a 40 cm^3 organic phase. This two-stage procedure will ensure a more complete recovery. Now hexane is a fairly volatile solvent, so we can evaporate our organic phase down to 1 cm^3 quite easily. If there is any question of losses of pesticides through evaporation, then you should consider using a more volatile solvent for the extraction (eg pentane, dichloromethane, or diethyl ether).

****

SAQ 8.7b   We have seen that solvent extraction is a useful means in chemical analysis for removing certain substances quantitatively from an aqueous phase into an organic phase or *vice versa*. We have seen this applies especially for preconcentration in trace analysis and for the separation of two or more interfering substances. However, in order to get the best out of a solvent extraction it is very important to ensure that all the experimental conditions for the extraction have been carefully optimised. This applies not only to the chemical composition of each phase, but also to the actual physical way in which the extraction is carried out.

I want you, therefore, to make two lists of chemical and physical factors that must be optimised either generally or in specific situations in order to get the best out of a solvent extraction procedure. The two lists should be under the following headings: $\longrightarrow$

SAQ 8.7b
(cont.)

> (*i*)   *Factors affecting the overall efficiency of extraction* (ie factors affecting the magnitude of *E* for solutes of interest).
>
> (*ii*) *Factors affecting the selectivity of an extraction* (ie factors affecting the magnitude of ΔE for two particular solutes).
>
> Put your lists side by side and see if you can correlate points in each list. You won't always be able to do this, but I think you will find it instructive to note when this is possible and when it is not.
>
> Try initially to prepare this list from memory without referring to the text, but if this proves impossible then refer to the text and previous SAQ's.

Response

I have prepared two lists in the way I suggested in the question. Briefly glance at them without studying them in detail and check that you were on the right course. If for any reason you were not, then have another go without consulting this response any further. If you feel that you were completely stuck, then look at one or two of the items on my list and see if they give you some idea of the kind of points that we are looking for. Then try to produce some of your own, if necessary by referring to the text. If you have produced a list that you are fairly happy with, then compare it with mine and see to what extent we agree. The final list (yours or mine) should be an excellent aid to revision.

Factors Affecting Overall Efficiency in Extraction

Factors Affecting Selectivity in Extraction

(*a*) *Choice of Extraction Solvent*. This will have an important effect on K_D and *D* (and therefore *E*)

(The choice of solvent is not very significant in improving extraction selectivity because the ratio of K_D values (ie β)

through such fundamental properties as polarity and dielectric constant (related to polarity). Also in preconcentration, the lower the solubility of the extraction solvent in water, the larger the phase-volume ratio, and thus the preconcentration factor, that can be supported.

(b) *Choice of Optimum pH*. So many species that we are likely to extract show pH-dependent extraction behaviour (remember all those pH-extraction curves!) that it is generally necessary to be sure to choose a pH where D (rather than K_D) is a maximum.

(c) *Choice of Complexing Agent (for metal ions)*. To obtain an efficient extraction of a metal ion, it is necessary to choose a complexing agent that produces a species that is:

(i) highly stable,
(ii) highly extractable.

Occasionally, the use of two or more reagents, is necessary to get the best extraction efficiency.

is normally fairly constant with change ibn extraction solvent.)

(a) *pH Control*. Because of the pH dependence of so many extraction systems selectivity may be enhanced by studying the relevant pH-extraction curves with the object of choosing a pH where the difference in E between two species is maximal.

(b) *Selective Complexing Agents (for metal ions)*. It is often profitable to look for complexing agents that react with metal ions selectively or even specifically as regards the formation of extractable complexes (eg nickel and dimethylglyoxime).

(c) *Concentration of Reagents*. Since extraction efficiency is often significantly affected by reagent concentration, so selectivity can be in certain cases. For example, in oxonium systems, reducing the acid concentration significantly increases the selectivity of extraction.

(*d*) *Concentration of Reagents*. This is generally important. For instance, extraction in oxonium systems may vary from 0 to almost 100% over quite a small range of hydrochloric acid concentrations.

(*e*) *Salting-Out Agents*. These are neutral inorganic electrolytes (added to the aqueous phase in high concentrations) that are especially useful in ion-association extraction systems for increasing extractability into the organic phase. They may also be useful in other types of extraction system.

(*f*) *Multiple Extractions and Continuous Extraction Techniques*. By carrying out an extraction with several small volumes of extraction solvent, you can get much greater efficiency in the overall extraction than by using a single extraction. Continuous extraction carries this to a kind of 'ultimate' conclusion whereby we have a continuous stream of the extraction

(*d*) *Masking Agents*. These are additional complexing agents that form water-soluble complexes with interfering ions, thus preventing the extraction of such ions and so increasing selectivity, particularly in neutral-chelate extraction systems.

(*e*) *Back Washing, Back Extractions, and Multistage Extractions*. Back washing involves re-equilibrating the organic phase containing the extracted species with a fresh aqueous phase containing all relevant reagents, buffers, etc, but no analyte. Back extraction involves a similar re-equilibration but with an aqueous phase of altered composition. A multi-stage extraction would involve a sequence of equilibrations involving different aqueous and organic phases. All these techniques are effective in increasing selectivity and achieving more complex separations.

phase running through the solution to be extracted.

(*g*) *Kinetic Effects in Complex Formation.* These are not often significant in increasing extraction efficiency, but α-nitroso-β-naphthol reacts with cobalt(II) to form a cobalt(III) ML$_3$ complex that is kinetically inert and so resistant to dissociation by acid. Under such conditions it should be selectively extracted.

(*f*) *Kinetic Effects in Complex Formation.* Some metal ions are extracted only extremely slowly because they possess inert electronic configurations, resulting in slow complex-forming reactions. The commonest ion affected in this way is chromium(III). Such ions can be selectively retained in the aqueous phase.

****

SAQ 9.1a

The ion exchange process is briefly described in the paragraph below. Various words and phrases, however, have been omitted. Fill in the missing words or phrases in the paragraph from the list below. Each word or phrase in the list may be used once, more than once, or not at all.

'Ion exchange consists of equilibration between a(n) ... 1 ... and a(n) ... 2 ... material which has ... 3 ... chemically bonded to it, which is ... 4 ... in water or other solvents, and which has ... 5 ... associated with it. Ions in the ... 6 ... of the ... 7 ... sign as the ... 8 ... may exchange ... 9 ... with the ... 10 .... The commonest materials used for this purpose analytically

$\longrightarrow$

SAQ 9.1a (cont.)

are ... 11 ... onto which ... 12 ... have been chemically bonded. They come in two complementary forms: ... 13 ... containing negatively charged ... 14 ... and ... 15 ... containing positively charged ... 16 ... bonded to them. The ... 17 ... is most commonly aqueous. The technique is generally suitable for the separation and preconcentration of ... 18 ... in solution'.

List of words and phrases

inorganic materials
highly polar molecules
ions
same
solid polymeric
organic
anion exchange resins
cation exchange resins
electrolyte solution
functional groups
irreversibly

inorganic groups
organic polymers
complexing
reversibly
opposite
aqueous
insoluble
counter-ions
soluble
chelating

NB There is more than one correct answer.

Response

Here are the best matches:

1. electrolyte solution,
2. solid polymeric,
3. ionogenic groups,
4. insoluble,
5. counter-ions,
6. electrolyte solution,
7. same } or { opposite,
8. counter-ions } { ionogenic groups,

 9. reversibly,
10. counter-ions,
11. organic polymers,
12. ionogenic groups,
13. cation-exchange resins,
14. ionogenic groups,
15. anion-exchange resins,
16. ionogenic groups,
17. electrolyte solution,
18. ions.

Here is the full paragraph once again in its complete and correct form. If you had difficulty in completing this question, reading this paragraph should help. Otherwise, re-read Section 9.1 and try the question again.

'Ion exchange consists of equilibration between an electrolyte solution and a solid polymeric material which has ionogenic groups chemically bonded to it, which is insoluble in water or other solvents and which has counter-ions associated with it. Ions in the electrolyte solution of the same sign as the counter-ions may exchange reversibly with the counter-ions. The commonest materials used for this purpose analytically are organic polymers onto which ionogenic groups have been chemically bonded. They come in two complementary forms: cation-exchange resins containing negatively charged ionogenic groups and anion-exchange resins containing positively charged ionogenic groups bonded to them. The electrolyte solution is most commonly aqueous. The technique is generally suitable for the separation and preconcentration of ions in solution.'

****

SAQ 9.2a

Three separations based on ion-exchange follow. Describe each separation as batch mode, trapping technique, or ion-exchange chromatography and indicate under what circumstances each might be useful (ie what do you think might have been the objective of the experiment?). Can you suggest how each experiment might have been finished?

(*i*)   A known volume of solution containing a known concentration of a metal cation was shaken in a bottle together with a known amount of cation-exchange resin in the H^+ form. Then the solution was separated from the resin by filtration.

(*ii*)   A column was packed with anion-exchange resin in the nitrate form and a solution of nitrate ions was continually passed through the column. When equilibrium was established a small volume of solution containing a mixture of chloride, bromide and iodide ions was placed on top of the column and the flow of nitrate-ion solution through the column was continued.

(*iii*)   A large volume of water containing a variety of trace metal ions was passed through a small column containing a cation-exchange resin in the H^+ form. This resulted in the complete retention on the column of all the metal ions and the release into the water of an equivalent amount of hydrogen ions.

Response

(*i*)   This experiment is a simple example of a batch-mode separation. Compare it with the general description of batch-mode separations either in Section 7.5.1 or more specifically to ion-exchange, in Section 9.2.1. It will probably have been carried out in order to measure the equilibrium constant for the exchange of hydrogen ions with this specific cation (the selectivity coefficient) as part of a study of the fundamental properties of the resin. The experiment would be finished by carrying out an analysis on an aliquot of the solution after ion-exchange, for H^+ and metal-ion concentration. Knowing initial concentrations and the number of ionogenic groups per gram of resin (determined by measuring the total amount of counter-ion released on changing the resin form) we have the information required to calculate the selectivity coefficient.

(*ii*)   This is an example of ion-exchange chromatography (albeit a rather simple one in practice). Compare it with more general descriptions give in Section 7.5.3 and 9.2.3. The objective of this experiment is to separate the three halide ions quantitatively. To finish the experiment, the halide ions will exchange reversibly with the nitrate ions and will start to move as bands down the column at different rates (in fact the chloride moves fastest, the iodide slowest). If you collect aliquots of solution from the bottom of the column and titrate each with silver nitrate to determine the halide ions, you should find halide ions present in three distinct bands corresponding to chloride, bromide, and iodide (assuming that the separation has been successful).

(*iii*)   This is an example of the use of a trapping technique. Literally the metal ions have been trapped on the column. Compare it with more general descriptions given in Section 7.5.4 and Section 9.2.2.

There may be a number of reasons why this experiment was initiated. We are looking ahead a bit here, but briefly, we might have been preconcentrating the metal ions, in which case we need to recover them from the column by eluting them with a small volume of strong acid (to reverse the reaction); we could have been in the process of removing all ionic material from the water (preparing de-ionised water) in which case the anions could be removed by passing the water through a column of anion-exchange resin in the hydroxide ion form; finally we might be interested in the total metal-ion concentration expressed as hydrogen ion equivalents, in which case we could titrate the liberated hydrogen ions with standard alkali.

****

SAQ 9.3a   Answer the following questions about the desirable properties of ion-exchange resins.

(*i*)   What sort of selectivity/affinity for ions would you look for in a resin to be used for general preconcentration purposes?

(*ii*) You are looking for an anion exchange resin suitable for separating anions of strong acids from those of weak acids. What *particular* property would you look for in your chosen resin?

(*iii*) A new ion-exchange resin has been prepared with an exchange capacity of over 500 micro-equivalents per gram of dry resin. What comment would you make about this figure?

(*iv*) Solutions resulting from ion-exchange separations by using PSDVB resins appear to show a very small absorbance in the ultraviolet around 250 to 300 nm, which cannot
⟶

SAQ 9.3a (cont.)	be explained by any of the ions present. What is the cause of this absorbance? (NB This assumes some knowledge of uv spectroscopy. If you have no such knowledge, start to read the response).

Response

(*i*) The clue to the answer here is in the question, in which we asked for a resin for 'general preconcentration purposes'. We want a resin, therefore, with low selectivity but with high general affinity for all ions of one type (cations or anions). If you had said that you needed a resin with a high affinity for hydrogen ions (for cations) because of the need to remove the ions subsequently, then that is also a useful point to make.

(*ii*) Our resin should be selective in its affinity for anions of strong acids and should not retain those of weak acids. Such properties are shown by a weakly basic anion-exchanger which, remember, can be used only at low pH values. So we would look for an anion-exchange resin with weakly basic cationic ionogenic groups.

(*iii*) Note that the exchange capacity has been quoted in micro-equivalents per gram of dry resin, whereas the range of exchange capacities that I quoted to you was in units of milli-equivalents per gram of dry resin. So we need to convert our figure of 500 micro-equivalents g^{-1} into 0.5 meq g^{-1}. Comparison of this with our typical range of 2 to 10 meq g^{-1} shows us that this resin has a rather poor exchange capacity. This does not mean to say, of course, that the resin is of no value. It may well have other useful properties.

(*iv*) If you know nothing about uv spectroscopy then let me point out that the observation described might well lead to the conclusion that an impurity has entered the solution, possibly an aromatic compound. Now, can you suggest where this impurity might have come from? The most likely explanation of this impurity is that there are traces of residual monomer (or perhaps oligomer or degradation products) being washed off the column. This is a problem when solutions of high purity are being obtained from such columns. Careful washing of the resin may help, but it is usually difficult to eliminate all traces of such impurities.

****

SAQ 9.3b State whether each of the following statements is *true* or *false*:

(*i*) Since ion-exchange resins are invariably solid materials of extremely low solubility with a semi-rigid structure, exchange processes must of necessity be restricted to the surfaces of the resin particles only.

(*ii*) For a cation-exchange resin, cations are the only species present in the resin phase (besides the resin itself), since the latter contains anions as ionogenic groups.

(*iii*) A PSDVB resin containing 20% of divinylbenzene would show a greater affinity towards small ions than would one containing 8% of divinylbenzene.

(*iv*) Ion-exchange is possible between two liquid phases without the presence of a solid ion-exchange material.

Response

(*i*)   *False*. Organic resins such as PSDVB resins, which are of such tremendous importance in ion-exchange, have a semi-rigid structure permitting water molecules to permeate the structure. The resin swells due to the imbibed water, and becomes porous. Ions of both sign may enter the pores and thus permeate the structure. This permits ion-exchange to occur at sites within the resin beads, rather than just on the surface.

There are at least two situations, however, in which ion-exchange is limited just to the surface of the ion-exchange particles. The first is, when ions are involved that are too large to enter the pores of the resin. The second is with some ion-exchange materials that are prepared from a rigid non-porous material (eg silica). Here, ionogenic groups are present only on the surface of the particles.

(*ii*)   *False*. We have already seen in the response to (*i*) that water molecules will be present, but do not forget that the resin contains ionogenic groups and counter-ions and so is electrically neutral. Therefore ions of both sign will penetrate the resin, and thus be present in the resin phase, and must do so if electrical neutrality is to be maintained.

(*iii*)   *False*. This statement would be true if we used the word 'selectivity' in place of 'affinity'. Let's just check what is happening. Increasing the quantity of divinylbenzene increases the degree of cross-linking in the resin. This generally decreases the resin's affinity for ions, because of the smaller pores involved (penetration of ions into the resin is harder if the pores are smaller). However, this reduction in affinity is more marked for larger ions than for smaller ions because the pore size has a more critical effect on larger ions. Thus, the resin becomes more selective towards the smaller ions.

(*iv*)   *True*. Well done if you spotted this one! Do you remember the liquid ion-exchangers that we met in the section dealing with solvent extraction (Section 8.6)? These are basically solvent-extraction systems, but in which ions of like sign are exchanged reversibly across a liquid boundary, there being a large ion of opposite sign (equivalent to our ionogenic groups) in the organic solvent (equivalent to our organic polymer). Re-read the section on liquid ion-exchangers now that you know something about conventional ion-exchange and think a bit more about the analogy.

****

SAQ 9.3c

Here are *five* specifications for an ion-exchange material, together with *five* ion-exchange materials. Match the materials to the specifications.

Specifications

An ion-exchange material suitable for:

(i)   strong retention of transition-metal cations,

(ii)   removal of all metallic and organic cations from solution,

(iii)   high-performance liquid chromatography,

(iv)   general work in the separation and preconcentration of anionic species,

(v)   selective retention of more highly basic organic cations.

Ion-exchange materials

(a)   Silica containing chemically bound ionogenic groups on the particle surfaces.

$\longrightarrow$

SAQ 9.3c
(cont.)

(*b*)   Sulphonated polystyrene divinylbenzene (PSDVB) resin in the hydrogen-ion form.

(*c*)   PSDVB resin containing quaternary ammonium groups.

(*d*)   Polyacrylic acid-divinylbenzene resin.

(*e*)   PSDVB resin containing iminodiacetic acid groups.

Response

The correct match is:

(*i*) $\rightleftharpoons$ (*e*)
(*ii*) $\rightleftharpoons$ (*b*)
(*iii*) $\rightleftharpoons$ (*a*)
(*iv*) $\rightleftharpoons$ (*c*)
(*v*) $\rightleftharpoons$ (*d*)

Well done if you got a perfect match! Here are some general comments if you did not quite manage a perfect match, or if you need some more help in this area

(*i*)   The iminodiacetic acid groups in this material (*e*) are strong chelating groups (cf EDTA). Hence the resin will strongly retain metal cations of the type likely to form chelates with EDTA and similar ligands. At the correct pH, the transition-metal cations are such ions.

(*ii*) Sulphonated PSDVB resins are general-purpose cation-exchange resins usable over a wide range of pH. In the hydrogen-ion form such a resin will non-selectively replace all cations in solution with H^+.

(*iii*) Silica-based materials will withstand higher pressures than will organic resins. This is the main reason why they are perhaps the best choice for hplc, since high pressures are involved.

(*iv*) Material (*c*) (PSDVB resin with quaternary ammonium groups) is a general-purpose anion-exchange resin suitable for a wide range of applications in anion-exchange.

(*v*) Material (*d*) (polyacrylic-acid-divinylbenzene resin) is a weak cation-exchange resin, since the $-CO_2H$ groups of the acrylic acid will be significantly ionised only at high pH. Such materials show selectivity towards strongly basic materials only and do not retain weakly basic materials.

****

SAQ 9.3d Imagine that you are trying to select the most appropriate ion-exchange material (resin or otherwise) for a particular purpose and you are looking at specifications of commercially available materials in chemical manufacturers' catalogues. What properties of the material would you look for in the specification? Make a list containing at least *six* important properties, and more if you can.

Response

Here is my list which actually contains nine properties, that could be described in the specification. This is not to say, incidentally, that if you do look into catalogues that you will find each property always specified. Although most should be included in a reasonable specification, some may overlap to a certain extent, and so could be considered as alternatives.

(*i*)   Whether cation- or anion-exchanger. Obviously of fundamental importance.

(*ii*)   Chemical nature of ionogenic groups. This tells you whether the material is a strong or weak exchanger, a chelating exchanger, etc.

(*iii*)   Nature of the polymeric portion. Whether it is a PSDVB resin, whether it is a silica based, etc.

(*iv*)   Degree of cross-linking. This affects properties in a number of ways as we have already seen. For PSDVB resins, it is expressed as the percentage of divinylbenzene used.

(*v*)   Mesh-size of particles. Small particles are needed for fast equilibration and for ion-exchange chromatography. Large particles are physically easier to handle.

(*vi*)   The counter-ions supplied. In other words, the way in which material is supplied.

(*vii*)   Whether supplied dry or not. If not supplied dry, the resin will already be swollen. In that case, in what medium is it supplied? This helps in resin pretreatment.

(*viii*) The exchange capacity. This would have been determined for a particular ion under a particular set of conditions and probably features only in a very full specification.

(*ix*)   Whether ionogenic groups are on the surface of the particles or dispersed throughout them. Probably this is alternative to (*viii*) or deducible from other properties. For example, silica-based materials exchange on the surface only, whereas PSDVB resins are porous as a rule. But you may come across other materials where this has to be specified.

****

SAQ 9.4a	Select the *single true statement* from the following:

(*i*) Selectivity coefficients in ion exchange are true thermodynamic constants, independent of concentration.

(*ii*) The selectivity coefficient in ion exchange is the equivalent of the separation factor in solvent extraction.

(*iii*) The value of the distribution coefficient for an ion in ion exchange is independent of the concentration of that ion, but depends on the concentration of the ion with which it is exchanging, under all normal conditions.

(*iv*) In ion-exchange chromatography, variations of the concentration of the eluting ion in the external phase will not affect its concentration in the resin phase.

(*v*) Increasing the concentration of electrolyte in the external phase will lead to an increase in concentration of electrolyte in the resin phase.

Response

Item (v) is the single true statement. Congratulations if this was your choice, commiserations if it was not! This increase in electrolyte concentration in the resin phase arises through an attempt by the system to remove the perturbation (ie change in concentration in the external phase). The driving force is the difference in activity between the two phases, the mechanism the permeation by ions of the resins pores, and the basic principle is Le Chatelier's principle.

Let us now see what is wrong with each of the other statements.

(*i*)   Selectivity coefficients are defined in terms of concentrations, but the true thermodynamic constant must be defined in terms of activities. Under the conditions of ion exchange, where electrolyte concentrations may be high and variable, activity effects become very significant, and as a result of this, large changes in selectivity coefficients can be observed as the concentration of either or both ions changes.

(*ii*)   This is wrong because the selectivity coefficient is the fundamental constant from which we derive the distribution coefficient. The separation factor is then defined as the ratio of the distribution coefficients for two ions competing for exchange in the presence of a background electrolyte. The values may well depend on the nature and composition of that electrolyte.

(*iii*) This is nearly correct, but we have forgotten to consider activity effects. If we can assume that activity coefficients do not vary during the ion-exchange process, then the statement is true. This will be so in practice if the concentration of the exchanging ion is small and that of the ion with which it is exchanging is high. We must therefore make this stipulation instead of saying 'under all normal conditions'. These conditions are realised in practice in ion-exchange chromatography.

(*iv*)   This is the opposite to (*v*) and so we must assume that it is wrong for much the same reasons that we found (*v*) to be correct. However, perhaps I can just elaborate a bit further. It is only the penetration of electrolyte into the resin that makes this statement false, because all exchange sites (ie ionogenic groups) on the resin will be occupied by eluting ions. No further ions can be incorporated by ion exchange and this line of argument may have led you to believe this statement to be true. For an ion-exchange material that is non-porous (eg surface-coated silica) this statement would be true.

****

SAQ 9.4b Answer the following questions, which all refer to exchange of cations on a sulphonated PSDVB resin in the H^+ form. Give concise explanations for each of your answers.

(*i*) Which of the following orders will the selectivity coefficients for the alkali metals follow:

$$Li^+ < Na^+ < K^+ < Rb^+ < Cs^+$$

or $Cs^+ < Rb^+ < K^+ < Na^+ < Li^+$?

(*ii*) Which will show the greatest selectivity towards the separation of Na^+ and K^+:

a 4% cross-linked resin,

an 8% cross-linked resin,

or a 12% cross-linked resin?

(*iii*) For concentrations < 0.1 mol dm^{-3}, will the selectivity coefficients for the following ions follow the order

$$K^+ < Ca^{2+} < Al^{3+}$$

or $Al^{3+} < Ca^{2+} < K^+$?

(*iv*) Will *increasing* the concentration of the ions in question (*iii*) increase or decrease the selectivity of the separation?

Response

(*i*)   The selectivity coefficients will follow the order:

$$Li^+ < Na^+ < K^+ < Rb^+ < Cs^+.$$

This is because, as the relative atomic mass of the ion increases so the degree of hydration of the ion decreases. This results in the effective radius of the *hydrated* (as opposed to free) ion actually *decreasing*. This in turn results in a higher charge density on the ion and, perhaps more importantly, greater ability by the smaller ion to penetrate the resin matrix, resulting in greater affinity of the resin for the ion. This, in its turn, leads to the higher selectivity coefficient.

(*ii*)   The 12% cross-linked resin will show the greatest selectivity towards the separation of Na^+ and K^+. This resin will have the tightest pore structure which, in turn will limit penetration of the resin matrix by the ions, this limitation being more pronounced the greater the size of the hydrated ion. The hydrated Na^+ ion is larger than the hydrated K^+ ion and hence the affinity of the more strongly cross-linked resin for Na^+ decreases more than it does for K^+. Hence the more strongly cross-linked resin is the more selective resin.

(*iii*) At low concentrations (generally below about 0.1 mol dm^{-3}) the affinity of the resin for ions of different charge will increase with increasing charge on the ion. Hence the correct order of selectivity is:

$$K^+ < Ca^{2+} < Al^{3+}.$$

(*iv*)   If the concentrations of the ions are increased, then activity effects will modify the situation in (*iii*) and the uptake of ions of low charge will increase in comparison with those of high charge. So we would expect the selectivity of the resin to decrease with increasing concentration. In fact in extreme cases the order of selectivity coefficients might actually alter.

****

SAQ 9.5a   Which of the following three ternary mixtures (as solutions in a suitable aqueous organic solvent) is the only one that can be easily resolved into its individual components by using simple ion-exchange methods (ie not ion-exchange chromatography)? Briefly list, in their correct order, the operations which would be necessary to achieve the separation.

(*i*)   Aniline ($C_6H_5NH_2$), phenylmethylamine (benzylamine) ($C_6H_5CH_2NH_2$), and phenyl methanol (benzyl alcohol) ($C_6H_5CH_2OH$).

(*ii*)   Benzoic acid ($C_6H_5CO_2H$), phenylethanoic acid (phenylacetic acid) ($C_6H_5CH_2CO_2H$), and aniline.

(*iii*)   Phenylmethanol (benzyl alcohol), phenylmethylamine (benzylamine) and phenylethanoic acid (phenylacetic acid).

Response

Mixture (*iii*) is the correct answer. This contains an acidic compound, a neutral compound, and a basic compound which all behave very differently with regard to conversion into ionic forms at different pH values.

Mixture (*i*) contains two basic and one neutral compound, whereas mixture (*ii*) has one basic and two acidic compounds. It would thus be possible only partially to separate the components of these mixtures.

Let us now see how to resolve mixture (*iii*) into its three components. There are two alternative approaches:

(*a*) Render the solution acidic and pass it through a cation-exchange column, when the phenylmethylamine will be retained as a cation and the other two components will pass through as neutral molecules. Recover the phenylmethylamine by passing an alkaline solution through the column.

Now render the remainder of the mixture alkaline and pass it through an anion-exchange column, when the phenylacetic acid will be retained as an anion and the benzyl alcohol will pass straight through. Final recovery of the phenylacetic acid simply involves passing an acidic solution through the column.

(*b*) The second method is exactly analogous to the first, except that you start with the anion-exchange column and the solution made alkaline to retain the phenylacetic acid in the anionic state. Subsequently you pass the remainder of the mixture in the presence of acid through a cation-exchange column to retain the phenylmethylamine as a protonated cation.

****

SAQ 9.5b

Six statements are given below about the extent of retention of ions on ion-exchange resins under specific conditions. Only *some* of these statements are true. Identify the true statements and then correct the false statements (correction when required will simply involve converting the statement into its *opposite*).

When you have done this, or as you do it, look at the list of explanations and select one that correctly explains each true and each corrected statement (each explanation may be used once, more than once, or not at all).   ⟶

SAQ 9.5b (cont.)

The statements

(*i*) Sulphate ions are retained more strongly than carbonate ions on an anion-exchange resin at neutral pH.

(*ii*) The order of retention of halide ions on an anion-exchange resin at neutral pH is:

$$Cl^- > Br^- > I^-.$$

(*iii*) Nickel ions are retained more strongly than cobalt ions in the presence of $9M$-HCl on a cation-exchange resin.

(*iv*) The order of retention of the following anions on an anion-exchange column at high pH and low concentration is:

$$ClO_4^- > SO_4^{2-} > PO_4^{3-}.$$

(*v*) At a mildly alkaline pH, leucine, an amino-acid, $RCH(NH_2)CO_2H$, with R = $-CH_2CH(CH_3)_2$, is less strongly retained than glutamic acid, an amino-acid with R = $-CH_2CH_2CO_2H$, on an anion-exchange resin.

(*vi*) Magnesium ions are more strongly retained than zinc ions on an anion-exchange resin in the presence of $9M$-HCl.

The explanations

A   *Because* differences in the acid dissociation constants lead to differences in the relative proportions of the species present in

$\longrightarrow$

SAQ 9.5b (cont.)

the anionic form, as opposed to the neutral molecular form.

B   *Because* differences in the size of the complex anions formed in the presence of the HCl lead to differences in the degree of penetration of the complex anions into the resin.

C   *Because* differences in the relative stabilities of the complex anions lead to differences in the ratio of cation to anion present.

D   *Because* the retention of the anion decreases in line with increasing radius of the hydrated anion because of decreased penetration of the ion into the resin.

E   *Because* retention of the anion increases with increasing charge on the anion under these conditions.

F   *Because* the radius of the free isolated anion increases along the series of increasing retention. The larger the free isolated anion then the smaller the hydrated anion, because of the reduced extent of hydration.

Response

There are three true statements and the other three become true when they are converted into their opposites.

Here is a quick-check table showing whether each statement is true or not (remember where a statement is false, then its opposite is true), together with the correct explanation for the statement or its opposite.

Statement	True/False	Correct Explanation
(*i*)	True	*A*
(*ii*)	False	*D*
(*iii*)	True	*C*
(*iv*)	False	*E*
(*v*)	True	*A*
(*vi*)	False	*C*

If you managed to match everything as above, well done! You are showing some understanding of the factors that govern retention on ion exchange resins. If you got the wrong match or had difficulty with the question, then study very carefully and think about the statements (corrected where necessary) together with the explanations.

Here they are again in their correct forms together with the appropriate explanations and additional comments in parentheses where appropriate.

(*i*)   Sulphate ions are retained more strongly than carbonate ions on an anion-exchange resin at neutral pH *because* differences in the acid dissociation constants lead to differences in the relative proportions of the species present in the anionic form as opposed to a non-ionic form (remember that sulphuric acid is a much stronger acid than carbonic acid).

(*ii*) The order of retention of halide ions on an anion exchange resin at neutral pH is:

$$Cl^- > Br^- > I^-$$

because the retention of the anion decreases in line with increasing radius of the hydrated anion because of decreased penetration of the ion into the resin.

(*iii*) Nickel ions are retained more strongly than cobalt ions in the presence of $9M$-HCl on a cation-exchange resin *because* differences in the relative stabilities of the complex anions lead to differences in the ratio of cation to anion present (in the presence of $9M$-HCl, cobalt forms a stable complex chloro-anion, whereas nickel does not).

(*iv*) The order of retention of the following anions on an anion-exchange column at high pH and low concentration is:

$$ClO_4^- > SO_4^{2-} > PO_4^{3-}$$

because retention of the anion increases with increasing charge on the anion under these conditions (note that at lower pH values a different order may apply because the acids corresponding to the anions vary in strength).

(*v*) At a mildly alkaline pH, leucine, an amino-acid, $RCH(NH_2)CO_2H$, with $R = -CH_2CH(CH_3)_2$ is less strongly retained than glutamic acid, an amino-acid with $R = -CH_2CH_2CO_2H$, on an anion-exchange resin *because* differences in the acid dissociation constants lead to differences in the relative proportions of the species present in the anionic form, as opposed to the neutral molecular form (in this case, glutamic acid, with two $-CO_2H$ groups and one $-NH_2$ group forms an anion at lower pH than does leucine with one $-CO_2H$ and one $-NH_2$ group, ie it has a lower isoelectric point).

(*vi*) Magnesium ions are more strongly retained than zinc ions on an anion-exchange resin in the presence of $9M$-HCl *because* differences in the relative stabilities of the complex anions lead to differences in the ratio of cation to anion present. (In practice, zinc forms a stable complex chloro-anion, whereas magnesium does not under these conditions. Hence zinc will be strongly retained, whereas magnesium will not be retained at all on the anion-exchange resin).

****

SAQ 9.6a

A sample of sea water is to be analysed for traces of a number of transition- and heavy-metal ions. The method of analysis is to involve a solution technique, and a 1000-fold preconcentration of the ions is required. Below are brief descriptions of *four* preconcentration procedures based on ion-exchange. Only *one* of these is likely to be successful. Which is it and what are the potential snags with the other three methods? For simplicity, assume that the sea-water matrix consists of high concentrations of sodium and magnesium ions and very low concentrations of the analyte ions.

(*i*)   The sample (10 dm^3) is passed through a small column packed with sulphonated PSDVB resin in the H$^+$ form. Subsequently the ions are recovered by passing 2M-HNO$_3$ (10 cm^3) through the column.

(*ii*) The sample (10 dm^3) is passed through a small column packed with sulphonated PSDVB resin in the H$^+$ form. The resin is then suitably ashed and the ash dissolved in 10 cm^3 of 2M-HNO$_3$.

(*iii*) The sample (10 dm^3) after adjustment to pH 6–8 is passed through a small column packed with PSDVB resin containing iminodiacetic acid groups in the cationic form. Subsequently the ions are recovered by passing 10 cm^3 of 2M-HNO$_3$ through the column. $\longrightarrow$

SAQ 9.6a (cont.)

> (*iv*) The sample (10 dm³) after adjustment to render it molar with respect to hydrochloric acid is passed through a small column packed with PSDVB resin containing $-CH_2N(CH_3)_3^+$ groups in the Cl^- form. Subsequently the ions are recovered by passing 10 cm³ of $2M\text{-}HNO_3$ through the column.

Response

Method (*iii*) is the only method without any readily identifiable snags, and is the correct answer. PSDVB resins containing iminodiacetic acid groups show a selectivity somewhat like EDTA. In other words, at neutral pH they will show a high affinity for a wide range of metal ions, including those of interest to us here, but not alkali metals or magnesium. Note, though, that calcium will be retained. If there is danger of the column being overloaded by calcium, then other, more selective chelating resins are available. The recovery of the ions is easily accomplished by using a solution of a strong acid because the ionogenic groups become protonated and lose their chelating ability.

Method (*i*) is unsuitable on two counts. First, as the resin chosen here is non-selective, the column will rapidly become overloaded with Mg^{2+} and Na^+ ions. Secondly, collection of the ions relies solely on displacement by H^+ which may not be quantitative under the conditions described. This is as opposed to the resin losing its exchange properties at low pH, as with the chelating resins.

Method (*ii*) represents a possible improvement on method (*i*) as we have addressed the problem of non-quantitative recovery of ions by ashing the resin. However, the problem of non-selectivity still remains.

Method (*iv*) is based on the formation of complex chloro-anions and would be potentially successful for those metal ions that form particularly stable complexes. However, it is unlikely to be wide-ranging enough for our purposes. For instance, if you look back to Fig. 9.5b, you can see that Ni^{2+}, Mn^{2+}, Co^{2+} and Cu^{2+} are all unlikely to be retained. You could increase the range of metals retained by increasing the HCl concentration, but this would present practical problems with a 10 dm³ sample!

****

SAQ 9.6b   Which of the following separations could be achieved by ion exchange or ion-exchange chromatography?

(*i*) Separation of U^{235} from U^{238} as gaseous UF_6.

(*ii*) Resolution of a mixture of rare-earth metal ions.

(*iii*) Resolution of a mixture of simple sugars.

(*iv*) Resolution of a mixture of halo alkanes.

Response

Separations (*ii*) and (*iii*) are possible by ion-exchange chromatography. The separation of rare-earth metal ions is a classic example of ion-exchange chromatography and is possible by using a cation-exchanger and citrate buffers. The separation of sugars is possible by first converting them into borate adducts, which are anionic, and then by using an anion-exchanger.

Separations (*i*) and (*iv*) are not possible by ion-exchange chromatography. We are perhaps stretching things a bit too far in trying to separate isotopes by ion-exchange, and anyway, gaseous UF_6 is not ionic. Likewise the simple halo alkanes are not ionic and it is not possible to derivatise them simply and reversibly into an ionic form.

****

SAQ 9.6c	Propose schemes based on ion exchange for carrying out the following: (*i*) separation of Fe^{2+} and Fe^{3+}, (*ii*) separation of Co^{2+}, and Pb^{2+} at low concentrations, (*iii*) preconcentration of all the metal ions present in a 1 dm^3 rain-water sample, by a factor of 1000, (*iv*) separation of a carboxylic acid, an aldehyde, and an alcohol, (*v*) preconcentration of common anions in a sample of tap-water.

Response

(*i*)   This separation might be possible by cation-exchange chromatography since it is reasonable to expect greater affinity of the resin for Fe^{3+} than for Fe^{2+}, at least at low concentration, because of the difference in charge. However, if you look at Fig. 9.5a you will see that there is a more certain possibility of separation by using an anion-exchange resin and a hydrochloric acid medium. If you look at the distribution coefficients for the two ions as a function of hydrochloric acid concentration, then you should see that a quantitative separation is possible by working in $7M$-HCl. Under these conditions, the Fe^{3+} will be retained by the resin, the Fe^{2+} passing straight through. So, to separate Fe^{2+} from Fe^{3+} place the mixture on an anion-exchange resin in a column in the presence of $7M$-HCl and pass $7M$-HCl through the column until all the Fe(II) has been collected. Then pass M-HCl through to recover the Fe(III). This separation is possible because of the large differences in the stabilities of the Fe(II) and Fe(III) chloro-complexes.

(*ii*) This is another separation best tackled by using an anion exchange resin and a hydrochloric acid medium. Once again, if we consult Fig. 9.5a, we can see that the three ions, in the presence of HCl, show different behaviour on an anion-exchange resin due to differences in the stabilities of their chloro complexes.

One possible procedure would therefore be to pass the sample through an anion-exchange column in the presence of $10M$-HCl, when the Co(II) will be retained as a chloro-complex whereas the Ni(II) and Pb(II) will pass straight through the column, leading to the separation of Co(II) from Ni(II) and Pb(II). The Co(II) can be subsequently recovered by running water through the column.

Now to separate the Ni(II) and Pb(II) we can again see from Fig. 9.5a that there are possibilities by working in $2M$-HCl. Under these conditions the Pb(II) is retained on the column, whereas the Ni(II) once more passes straight through. The Pb(II) could then be subsequently recovered with either water or concentrated acid.

(*NB* The lead concentration in this separation should be low because of the low solubility of lead chloride in water.

(*iii*) In this preconcentration problem, we have no difficulty with selectivity, because rain-water is unlikely to contain high concentrations of any ions, and anyway our brief is to collect all the ions. So perhaps here we would be best advised to

choose the most unselective cation-exchange resin available
to us, which would be, say, a sulphonated PSDVB resin in the
H^+ form. So we pass the 1 dm^3 sample through a small column packed with the above resin. Subsequent recovery of the
preconcentrated ions might be troublesome. It is unlikely that
we could achieve a quantitative recovery directly by passing
only 1 cm^3 of acid through the column. So we must either
pass sufficient acid through to collect all the ions and then reduce the final volume of the acid solution by evaporation, or
alternatively we could consider ashing the resin and dissolving
the small ash residue in 1 cm^3 of a suitable strong acid. A fair
amount of practical skill may be called for in this procedure.

(*iv*) In this separation, we have two non-ionic compounds (the alcohol and the aldehyde) and a carboxylic acid that forms an
anion in the presence of alkali. So we can remove the acid first
by making the sample alkaline (don't heat – we don't want to
polymerise the aldehyde!) and passing the solution through an
anion-exchange column. The aldehyde and alcohol will pass
straight through whereas the ionised acid will be retained and
can subsequently be recovered by passing dilute acid through
the column.

Now to separate the aldehyde and the alcohol. First neutralise
the alkali and add an excess of concentrated aqueous sodium
hydrogen sulphite to the sample. Do you remember the reaction between aldehydes and sodium hydrogen sulphite? Here
it is to remind you:

$$R-C\overset{\displaystyle O}{\underset{\displaystyle H}{\Big\langle}} + NaHSO_3 \rightleftharpoons R-\overset{\displaystyle OH}{\underset{\displaystyle H}{\underset{|}{\overset{|}{C}}}}-SO_3^- + Na^+$$

In other words, we have converted our aldehyde into an anionic addition compound which can be retained on our anion-
exchange column, while the alcohol passes through. The above
reaction is reversible and the aldehyde can be recovered from
the column by washing it with a sodium chloride solution.

(*v*)   Ion exchange is used to preconcentrate cations in water. Preconcentration of anions is more unusual, but in theory should be possible by ion exchange. It is, however, liable to be more difficult, and the purpose of this question was to get you to think about the possibilities and difficulties.

The general principles would be the same as for the preconcentration of cations except that the water sample would be passed through a strong, non-selective anion-exchange resin in the OH^- form. The anions could then be recovered by passing a small volume of concentrated NaOH solution through the column.

Resins containing $-CH_2N(CH_3)_3^+$ groups tend to retain OH^- ions only rather feebly, which will make for problems in recovering the ions. We can't expect to be able to ash the resins and recover the anions unaltered in the same way as we could with metal cations. It would be preferable to use a resin containing $-CH_2N(CH_3)_2C_2H_4OH^+$ groups (commercially available), which has a greater affinity for OH^- ions compared with the more conventional anion-exchange resins.

****

SAQ 10.2a

Which of the following processes can correctly be described as an adsorption?

(*i*)   Collection of trace-metal ions in water by passing the water through a column packed with a sulphonated polystyrene-divinylbenzene resin.

(*ii*)   Collection of traces of an organic vapour in the atmosphere by passing air through a glass trap maintained at liquid nitrogen temperatures. $\longrightarrow$

**SAQ 10.2a
(cont.)**

(*iii*) Collection of traces of a toxic gas in the atmosphere by passing air through a solution
with which the gas reacts.

(*iv*) Collection of traces of organic compounds
in a river water sample by passing the water
through a column packed with charcoal.

Response

Only *process* (*iv*) is a true *adsorption* process. Charcoal is the adsorbent, the organic compounds being reversibly adsorbed onto the
surface of the charcoal particles.

Process (*i*) is *ion-exchange*. Sulphonated polystyrene-divinylbenzene polymers are cation-exchange resins. Here we have preconcentration by cation-exchange, hydrogen ions replacing the cations
in solution.

Process (*ii*) is simply *cold-trapping*. At the low temperature of liquid
nitrogen, the organic vapours are simply frozen and retained in the
trap. No adsorption occurs here.

Process (*iii*) is *absorption*. The gas is retained in the bulk of the
solution. No surface effect is involved here.

****

SAQ 10.2b Offer explanations for the following observations.

(*i*) The more polar an organic compound is, the less efficiently is it adsorbed onto a non-polar absorbent from an aqueous medium.

(*ii*) Adsorption of a volatile compound from the vapour phase is more pronounced at lower temperatures.

(*iii*) Silica, which in a finely divided form adsorbs water very readily normally does not strongly adsorb compounds in general until after it has been heated at 100–150 °C for at least 30 min.

(*iv*) The adsorption capacity (the maximum possible amount of material that can be adsorbed per gram of adsorbent) of an adsorbent increases as the particle size of the adsorbent decreases.

Response

Here are my explanations. Your explanations should correspond to these in essence rather than in detail.

(*i*) The more polar an organic compound is the more soluble it is in water, ie the better solvent water is for the compound. Molecules are desorbed efficiently by good solvents. Therefore water will *desorb* polar compounds very efficiently, limiting the efficiency of adsorption of polar compounds from an aqueous medium. You might have argued that a non-polar adsorbent would show preference to non-polar compounds, which is true, but the solvent factor is likely to predominate.

(*ii*) Adsorption from the vapour phase involves a vapour–solid equilibrium and the position of this equilibrium will be governed by the vapour pressure of the compound, the higher the vapour pressure the greater the tendency towards desorption of the compound. At low temperatures vapour pressures will be correspondingly low. Hence adsorption will be more pronounced. So 'adsorb at low temperatures and desorb at high temperatures' is the rule for adsorption from the vapour phase.

(*iii*) Because the atmosphere usually contains a large amount of water vapour, particulate silica 'off the shelf' will have its adsorption sites already largely blocked by water adsorbed from the atmosphere. Remember, water being a polar molecule is likely to be attracted strongly to the polar Si–OH groups. The heating (known as 'activation') is in essence a thermal desorption, removing the adsorbed water, and thus making the adsorption sites available for use.

(*iv*) The adsorption capacity depends on the number of adsorption sites available per gram of adsorbent. This in turn depends on the surface area per gram of adsorbent, which in turn is greater for small-volume particles than for large-volume particles. (Two spheres will have a greater surface area in total than a single sphere if the total volumes are equal).

☆☆☆☆☆☆☆☆☆☆☆☆☆☆☆☆☆☆☆☆☆☆☆☆☆☆☆☆☆☆☆☆☆☆☆

SAQ 10.2c   Here is a description of an attempted sampling/preconcentration procedure, but unfortunately the analyst who carried this out made a number of mistakes. Can you spot four mistakes and suggest in each case what he should have done? $\longrightarrow$

SAQ 10.2c
(cont.)

'Water, which was to be analysed chromato-graphically for a wide range of organic compounds of various properties, was passed slowly through a short column in two sections, each packed with granulated charcoal. The flow-rate and sampling time were not recorded, although the total sample volume was measured as 10 dm³. The charcoal from the two sections of the tube was combined for desorption. The wet charcoal was dried by heating it at 150 °C for 30 min and the compounds were then desorbed by shaking the charcoal with 10 cm³ of hexane and filtering off the charcoal.'

Response

(*i*) The first mistake was to combine the charcoal from the two sections in the tube. The charcoal was purposely packed in two sections, so that the analyst could check for 'breakthrough' by analysing each separately.

(*ii*) He shouldn't have heated the charcoal like that to dry it. Any volatile organic compounds present (remember, you don't know much about the compounds present) would be lost by thermal desorption. Instead, he should have desorbed the compounds with a solvent and dried the solution with a drying agent.

(*iii*) The choice of hexane as solvent is surprising. This is a very non-polar solvent and many polar compounds may not be desorbed. Better to use a solvent of greater polarity such as diethyl ether (ethoxyethane).

(*iv*) The simple single-stage batch technique will not assure quantitative desorption. Better either to pass the solvent through the charcoal in a column, or to treat the charcoal with successive portions of solvent until analysis shows desorption to be complete.

Note that his failure to record flow rates and sampling times was not a critical factor. You might, however, have questioned his use of charcoal as adsorbent because of problems with desorption, and recommended a porous polymer adsorbent. If you had, this would have been a possible alternative answer, say, perhaps to (*iv*).

****

SAQ 10.3a

> Propose *two* outline schemes for preconcentrating the metal ions Cu^{2+}, Zn^{2+}, Cd^{2+} and Pb^{2+} present at trace levels in 1 dm^3 of tap-water. The main preconcentration step should be by coprecipitation. The final sample should consist of 5 cm^3 of aqueous solution, free of carrier.
>
> Make a few critical comments about the relative merits of your two schemes.

Response

The two schemes that I have in mind are based on adsorption onto a hydrous oxide carrier-ion precipitate, and solid solution formation with an indifferent organic compound, respectively.

Here are some more details.

Scheme A

Add to our 1 dm^3 water sample sufficient potassium aluminium sulphate *of the purest grade available* to make the concentration of the solution, say, 1000 mg dm^{-3} in Al^{3+}. Now make the solution alkaline to precipitate the Al^{3+} ions. Do not make the solution too alkaline, or the Al^{3+} will redissolve as an aluminate. Now, filter off the precipitate and redissolve it in a small volume of acid. I recommend hydrochloric acid for this purpose, since this will help us in the next stage of the process, which is to separate the carrier ion and trace-metal ions. We can achieve this by anion-exchange if we adjust the HCl concentration to $4M$ (see Fig. 9.5a). Under these conditions, all the trace-metal ions form chloro complexes and will be retained on the resin, whereas the Al^{3+} will not be retained. The stabilities of these anions suggest that recovery from the resin might sometimes be difficult, and therefore ashing of the resin may be necessary. Dry-ashing might result in losses of volatile metal chlorides, and so wet-ashing with hot oxidising mineral acids may have to be adopted. Finally evaporate the solution to a small volume and make it up to 5 cm^3.

Scheme B

Add a buffer to the 1 dm^3 water sample to ensure a neutral or slightly alkaline pH. Add sufficient 8-hydroxyquinoline to ensure all the trace-metal ions are complexed. Then add about 1 g of β-naphthol dissolved in a minimum of acetone and mix well. Filter off the precipitate. Then either redissolve the precipitate in trichloromethane and back extract the trace metal ions into 5 cm^3 of acid or ash the precipitate and redissolve the residue in 5 cm^3 of acid. Dry ashing or wet ashing could be used but watch for loss of volatiles if you use the former.

Brief comparison of the schemes

If you think about what is involved practically with both schemes for a moment, you will, I believe, agree that the operations involved in scheme B are quicker and easier to carry out than those in scheme A. The precipitate will probably be easier to filter off in scheme B, since hydrous oxide precipitates tend to be gelatinous (precipitation from homogeneous solution may help here). We need an ion-exchange separation in scheme A *plus* an ashing step. Scheme B requires a simple solvent extraction *or* an ashing step. Also, it would be easier to obtain the organic reagents free of trace-metal ions than it would an aluminium salt. So it appears that scheme B is the one to be preferred.

****

SAQ 10.3b
> Suggest how the following separations could be achieved by means of precipitation:
>
> (*i*)   copper(II) and silver(I) from cobalt(II) and nickel(II),
>
> (*ii*)   iron(II), chromium(III), and aluminium(III) from calcium(II), copper(II) and zinc(II),
>
> (*iii*) iron(II), cobalt(II), and nickel(II) from calcium(II), magnesium(II) and aluminium(III),
>
> (*iv*)   nickel(II) from iron(III), chromium(III) and zinc(II).

Response

Clues to at least some of these separations are to be found in Fig. 10.3a, 10.3b and 10.3c. Did you consult these? If not, and you got stuck with the question, then have another go and this time consult the figures.

(*i*)   Fig. 10.3b offers some possibilities for this separation, since the sulphides of silver and copper are much *less* soluble than those of cobalt and nickel. By consulting the figure you can see that, in the presence of $0.3M$ to $3M$ HCl, silver and copper are precipitated as their sulphides whereas cobalt and nickel are not.

(*ii*) This separation can be based on the precipitation of the metal hydroxides. (Fig. 10.3a). In fact, an ammonia/ammonium chloride buffer (pH *ca* 9) will cause the trivalent ions [Fe^{3+}, Cr^{3+} and Al^{3+}] to be precipitated, while the divalent ions [Ca^{2+}, Cu^{2+} and Zn^{2+}] are left in solution.

(*iii*) Nothing obvious emerges from the figures for this separation. However, we are trying to separate transition-metal ions [Fe^{2+}, Co^{2+} and Ni^{2+}] from main-groups metal ions [Ca^{2+}, Mg^{2+} and Al^{3+}] and this might be done by electrolytic precipitation by using a controlled potential mercury cathode, which is the approach that I recommend. There may also be possibilities by using a suitable organic precipitant together with pH control and possibly a masking agent.

(*iv*) Here we have a fairly selective separation. Fig. 10.3c holds the clue in which we see that dimethylglyoxime, if used under appropriate conditions, can be a fairly selective organic precipitant for nickel. If you consult the figure, you can observe possibilities for this separation if it is carried out in a slightly alkaline ammoniacal solution containing either tartaric acid or citric acid as a masking agent for the other ions.

****

SAQ 10.3c   In our discussions of separation of metal ions by precipitation we have proposed a number of separations, largely based on considerations of relative magnitudes of solubility products and on the possible existence of other chemical equilibria affecting the concentrations of one or other of the precipitating ions. However, if these

$\longrightarrow$

| SAQ 10.3c (cont.) | separations are attempted in practice you may come face to face with two or more problems which mean that these separations must be evaluated pratically before being finally accepted as feasible. Suggest *two* of these practical problems. |

Response

The two practical problems that I might have expected you to think of first are as follows.

(*i*) Problems associated with *coprecipitation*. For example, the metal hydroxides, and the metal sulphides for that matter, can act as efficient carriers for coprecipitation, especially in the presence of excess of OH^- or S^{2-} ions, resulting in the production of contaminated precipitates.

(*ii*) Problems associated with *the rate of precipitation*. Do you remember that at the beginning of Section 10.3.2 we noted that K_{sp} refers only to a system in thermodynamic equilibrium and that it told us nothing about the rate of precipitation or whether we might come across formation of supersaturated solutions? These are both real possibilities. Certain metal sulphides are precipitated only very slowly. Calcium and magnesium oxalate are both insoluble, but whereas the calcium salt is precipitated immediately, the magnesium salt is precipitated only slowly. Indeed there is the possibility of a separation based on kinetic considerations.

Another practical problem that might arise is the filtrability of the precipitates. For example, many hydroxide precipitates are gelatinous, and are difficult to filter. Precipitation from homogeneous solution may help here.

∗∗∗∗∗∗∗∗∗∗∗∗∗∗∗∗∗∗∗∗∗∗∗∗∗∗∗∗∗∗∗∗∗∗∗∗∗∗∗

SAQ 10.4a

Indicate which of the following statements are *true* and which are *false*. Where the statements are false, explain clearly what the true statement should be.

(*i*) Distillation is a technique for separating species on the basis of differences in boiling-point. All species that can be volatilised are theoretically separable by distillation as long as their boiling-points differ, the ease of separability depending on, and solely on, differences in boiling-point.

(*ii*) Fractional distillation is a variant of normal distillation in which the condensed vapours are collected in separate fractions of different boiling-range.

(*iii*) Fractional distillation offers the only feasible means of separating a very complex mixture of volatile liquids on an analytical scale.

(*iv*) If you attempt to purify a high-boiling liquid by distillation in a conventional still but find that it starts to decompose below its boiling-point, then you will need to seek some technique other than distillation for purifying it.

(*v*) Steam distillation is a special technique whereby the liquid being distilled is held in a vessel over a steam bath. This technique is designed to ensure that distillation takes place exclusively at 100 °C, thus permitting the separation of those species that are

$\longrightarrow$

**SAQ 10.4a
(cont.)**

> more volatile than water from those that
> are less.
>
> (*vi*) Azeotropic distillation means distillation
> under an atmosphere of nitrogen. This
> technique is used for the distillation of
> species sensitive to oxidation at high
> temperatures (from the French 'l'azote',
> meaning nitrogen).

Response

Did you, by chance, come to the conclusion that all the above statements were, in one way or another, *false*? If you did then congratulations because, indeed, *all six statements are false*! Let us now look
in turn at each statement and the reasons why it is false.

(*i*)   Distillation is indeed a means of separating species by differences in boiling-point, and if two species have the same
boiling-point then their separation by distillation is not possible.

However, that said, the full situation is more complex than we
have implied. For a start, the separability of species, although
often being eased by large differences in boiling-point, will
often depend on other factors in addition to differences in
boiling-point, mainly interactions between the species in the
liquid phase. In particular, the relationship between boiling-
point and composition of a binary mixture is linear only for
an ideal mixture. It is also possible and quite common to get
constant-boiling mixtures of substances with different boiling-
points. Here, the composition of the vapour is the same as that
of the liquid phase and no separation is possible. (Perhaps I
should remind you that when a mixture of liquids is brought
into equilibrium with its vapour, ie, boiled, then, in general,
we expect the vapour phase to have a different composition
from that of the liquid, ie to have a higher concentration of

the more volatile species. Only if this is so is separation by distillation possible).

(*ii*) Fractional distillation implies the use of a fractionating column, in which rising vapour is continually being equilibrated with falling liquid resulting in increased concentration of the more volatile component in the vapour phase, and the less volatile in the liquid phase. It is a more efficient means of separation than conventional distillation and may be likened to a continuous countercurrent separation, whereas conventional distillation is a batch technique. Collecting fractions of different boiling-point is a normal distillation practice whether or not a fractionating column is being used.

(*iii*) Gas chromatography is an alternative, and almost always a better way of separating the components of complex mixtures of volatile liquids on an analytical scale, and even on a small preparative scale.

(*iv*) Before abandoning the idea of distillation, you should try a low-pressure or vacuum distillation, whereby the distillation is conducted at a pressure less than atmospheric. Do you remember that the boiling-point of a liquid is a function of the pressure in the gas phase applied to the liquid and decreases as this is lowered? So, if you decrease the pressure at which the distillation is carried out, you will lower the boiling-point of the liquid phase, thus eliminating, you hope, the problems of decomposition.

(*v*) This is not the basis of steam distillation. It appears that many substances that are difficult to distil on their own show increased volatility in the presence of steam. Therefore, if steam is passed *through* the solution and then condensed in the manner of a distillation, these species will be concentrated in the distillate. The statement is therefore true to the extent that the distillation is conducted at 100 °C but this is quite incidental. The species to which this technique is applied are usually less volatile than water. They must, though, be volatile in steam.

(*vi*) This idea of distillation of oxygen-sensitive compounds under an atmosphere of nitrogen seems sensible enough and could well be useful. It is not, however, the basis of an azeotropic distillation. An azeotropic distillation is the distillation of a constant-boiling mixture in which the liquid and vapour phases have the same composition.

****

SAQ 10.4b

> If you study Fig. 10.4a, you should be able to spot possibilities for the sequential separation of two groups of elements by volatilisation. What are these two groups? There are also two possibilities for volatilising elements in a suitable form for atomic absorption spectroscopy, a technique applicable to metals and metalloidal elements and requiring the element either in the atomic state or in a form readily convertible into the atomic state. What are these two possibilities?

Response

The two groups of elements that can be separated area as follows.

(*i*)   The halogens: chlorine, bromine, and iodine (as halides).

First of all, remove iodide as iodine by distilling the mixture in the presence of nitrous acid. Then KIO_3 or telluric acid is added and the bromide is distilled out as bromine. The chloride ion is left in solution.

(*ii*)   Arsenic, antimony and tin.

The three separations shown in the figure may be carried out in the sequence As → Sb → Sn, to give a sequential separation of the three elements as AsCl$_3$, SbCl$_3$, and SnBr$_4$.

The two possibilities for linking a volatilisation step with atomic absorption are:

(*a*) The production of mercury vapour by reduction with NaBH$_4$, since this is the only quoted procedure for producing a metallic element in the vapour state.

(*b*) The production, also by reduction with NaBH$_4$, of metallic and metalloidal hydrides, which are thermally very unstable and readily dissociate to produce the element in the atomic state on being heated.

Both these procedures are, in fact, regularly used in atomic absorption spectroscopy.

****

SAQ 10.4c

In the course of this Unit, we have studied *four* separate techniques suitable for preconcentrating trace-metal ions in water. Name these four techniques and then briefly compare them under the following headings, highlighting those techniques that you think are best, or worst, under each heading.

(*i*) Preconcentration factors obtainable.

(*ii*) Metal ions that can be preconcentrated.

(*iii*) Suitability of the form of the preconcentrated sample for subsequent analysis.

⟶

SAQ 10.4c
(cont.)

> (*iv*) Speed, ease, and simplicity of the precon-centration procedure.
>
> (*v*) The effect of other ions present at high concentrations, which are not to be pre-concentrated.
>
> (*vi*) Actual number of applications in practice.

Response

The *four* techniques that we have studied are:

(*a*) Solvent evaporation (Section 10.4.4),

(*b*) Solvent extraction (Section 8.2 and elsewhere in Section 8),

(*c*) Ion-exchange (Section 9.6.3),

(*d*) Coprecipitation (Section 10.3).

Here now, are some of the points that I would have expected you to make in your comparisons of these four preconcentration techniques. Wherever possible, I have tried to indicate the 'best' and the 'worst' technique under each heading.

(*i*) *Preconcentration factors obtainable*

Solvent extraction, we found, was limited to a preconcentration factor of not more than about 50 for a one-stage extraction. With co-precipitation and evaporation, we cannot very realistically deal with samples of more than a few dm^3 in volume because of handling problems or the time taken for evaporation. No such restrictions apply to ion exchange, since any volume of water may theoretically be passed through the column. Bearing in mind that for all techniques, the preconcentrated sample will probably have a volume of a few cm^3, it appears that ion exchange has the greatest potential, whereas solvent extraction (as a one-stage process) is the most limited.

(ii) Metal ions that can be preconcentrated

Since solvent evaporation does not involve the metal ion directly, it is by definition totally unselective and therefore universal. Solvent-extraction and ion-exchange methods exist for nearly all metal ions, although no universal solvent-extraction scheme is known. Universal ion-exchange schemes are known, but are not necessarily the best for some metal ions. Coprecipitation, on the other hand, is restricted in its applicability to certain types of metal ion only. So, under this heading, solvent evaporation is the best, coprecipitation the worst, with ion exchange somewhat better than solvent extraction.

(iii) Suitability of the form of the preconcentrated sample for subsequent analysis

A one-stage solvent extraction produces a solution of the metal ions in complexed form in an organic solvent. Ion exchange results in the metal ions being preconcentrated onto a solid resin; coprecipitation results in the ions being preconcentrated into a precipitate consisting largely of a carrier material. With solvent evaporation, the original matrix is retained. If it is important that the preconcentrated sample be in the aqueous phase, then a back-extraction into acid is normally possible with solvent extraction. In ion exchange, the ions may be eluted from the resin with acid, or the resin may be ashed and the residue dissolved in acid. With coprecipitation, we should be able to redissolve the precipitate but we may have to separate the carrier. I should point out that it is often very convenient to have the preconcentrated metals in an organic solvent (eg for flame spectroscopy or solution spectrophotometry). I would nominate solvent extraction as the best under this heading and coprecipitation the worst.

(iv) Speed, ease and simplicity of the preconcentration procedure

Solvent extraction is no doubt the quickest of these operations. Solvent evaporation, on the other hand is perhaps the slowest, although for most of the time evaporation may proceed unattended. Assuming that our objective is a solution of the preconcentrated ions, then both ion-exchange and coprecipitation involve extra steps in the separation, thus lengthening and complicating the procedure. Arguably, solvent extraction is probably best and coprecipitation worst (because of the need to remove the carrier) under this heading.

(v) *The effect of other ions present at high concentration which are not to be preconcentrated*

What we want to know here is whether such ions would interfere with the preconcentration. Straight away, we notice that in ion exchange, unless a selective resin is being used that does not retain these ions, we might have problems with the resin becoming saturated. Solvent evaporation will not be affected by this factor, whereas sometimes such an ion in coprecipitation can actually be used as the carrier. In solvent extraction it depends on what ions are involved. You might encounter problems, for instance, if you were trying to preconcentrate one transition-metal ion in the presence of a large amount of another. 'Best' and 'worst' are less clear-cut here but I would suggest that evaporation was best and ion exchange was worst.

You might perhaps have looked at this slightly differently and considered the technique's ability to preconcentrate one trace-ion, leaving the other major ion behind. Selective preconcentrations of certain metal ions are possible for solvent extraction and ion exchange and to a less extent, coprecipitation.

(vi) *Actual number of applications in practice*

Looking at the above as a whole, you might be inclined to conclude that solvent extraction is generally the most useful procedure, provided that you do not require a very high preconcentration factor, whereas coprecipitation is the least. If you did, then in practice, you would be just about right. Because of mere convenience, solvent extraction is probably the most often used, whereas ion exchange would be used where a high preconcentration factor was needed. Evaporation is quite often used on its own in practice, since it requires the fewest additions of reagents, but it is also useful when used in combination with one of the other techniques. Coprecipitation is a rather specialised technique for which there are fewer applications.

I have produced a rather lengthy answer to this SAQ. Don't be dismayed if yours is not quite so lengthy, although I hope you will have mentioned a reasonable proportion of these points. However, you could well have come to different conclusions about 'best' and 'worst' techniques. I stress that this is a subjective judgement and it is really not very important if we find we have *some* differences in opinion here, as long as you have argued logically and convincingly towards your own conclusion.

Units of Measurement

For historic reasons a number of different units of measurement have evolved to express quantity of the same thing. In the 1960s, many international scientific bodies recommended the standardisation of names and symbols and the adoption universally of a coherent set of units—the SI units (Système Internationale d'Unités)—based on the definition of five basic units: metre (m); kilogram (kg); second (s); ampere (A); mole (mol); and candela (cd).

The earlier literature references and some of the older text books, naturally use the older units. Even now many practicing scientists have not adopted the SI unit as their working unit. It is therefore necessary to know of the older units and be able to interconvert with SI units.

In this series of texts SI units are used as standard practice. However in areas of activity where their use has not become general practice, eg biologically based laboratories, the earlier defined units are used. This is explained in the study guide to each unit.

Table 1 shows some symbols and abbreviations commonly used in analytical chemistry; Table 2 shows some of the alternative methods for expressing the values of physical quantities and the relationship to the value in SI units.

More details and definition of other units may be found in the *Manual of Symbols and Terminology for Physicochemical Quantities and Units*, Whiffen, 1979, Pergamon Press.

Table 1 *Symbols and Abbreviations Commonly used in Analytical Chemistry*

Å	Angstrom
$A_r(X)$	relative atomic mass of X
A	ampere
E or U	energy
G	Gibbs free energy (function)
H	enthalpy
J	joule
K	kelvin ($273.15 + t\,°C$)
K	equilibrium constant (with subscripts p, c, therm etc.)
K_a, K_b	acid and base ionisation constants
$M_r(X)$	relative molecular mass of X
N	newton (SI unit of force)
P	total pressure
s	standard deviation
T	temperature/K
V	volume
V	volt ($J\ A^{-1}\ s^{-1}$)
$a, a(A)$	activity, activity of A
c	concentration/ mol dm^{-3}
e	electron
g	gramme
i	current
s	second
t	temperature / °C
bp	boiling point
fp	freezing point
mp	melting point
$\approx$	approximately equal to
$<$	less than
$>$	greater than
e, $\exp(x)$	exponential of x
ln x	natural logarithm of x; ln $x = 2.303 \log x$
log x	common logarithm of x to base 10

Table 2 *Alternative Methods of Expressing Various Physical Quantities*

1. **Mass (SI unit : kg)**

$$g = 10^{-3} \text{ kg}$$
$$mg = 10^{-3} \text{ g} = 10^{-6} \text{ kg}$$
$$\mu g = 10^{-6} \text{ g} = 10^{-9} \text{ kg}$$

2. **Length (SI unit : m)**

$$cm = 10^{-2} \text{ m}$$
$$\text{Å} = 10^{-10} \text{ m}$$
$$nm = 10^{-9} \text{ m} = 10\text{Å}$$
$$pm = 10^{-12} \text{ m} = 10^{-2} \text{ Å}$$

3. **Volume (SI unit : m³)**

$$l = dm^3 = 10^{-3} \text{ m}^3$$
$$ml = cm^3 = 10^{-6} \text{ m}^3$$
$$\mu l = 10^{-3} \text{ cm}^3$$

4. **Concentration (SI units : mol m⁻³)**

$$M = \text{mol } l^{-1} = \text{mol dm}^{-3} = 10^3 \text{ mol m}^{-3}$$
$$\text{mg } l^{-1} = \mu g \text{ cm}^{-3} = \text{ppm} = 10^{-3} \text{ g dm}^{-3}$$
$$\mu g \text{ g}^{-1} = \text{ppm} = 10^{-6} \text{ g g}^{-1}$$
$$\text{ng cm}^{-3} = 10^{-6} \text{ g dm}^{-3}$$
$$\text{ng dm}^{-3} = \text{pg cm}^{-3}$$
$$\text{pg g}^{-1} = \text{ppb} = 10^{-12} \text{ g g}^{-1}$$
$$\text{mg\%} = 10^{-2} \text{ g dm}^{-3}$$
$$\mu g\% = 10^{-5} \text{ g dm}^{-3}$$

5. **Pressure (SI unit : N m⁻² = kg m⁻¹ s⁻²)**

$$Pa = Nm^{-2}$$
$$\text{atmos} = 101\ 325 \text{ N m}^{-2}$$
$$\text{bar} = 10^5 \text{ N m}^{-2}$$
$$\text{torr} = \text{mmHg} = 133.322 \text{ N m}^{-2}$$

6. **Energy (SI unit : J = kg m² s⁻²)**

$$cal = 4.184 \text{ J}$$
$$erg = 10^{-7} \text{ J}$$
$$eV = 1.602 \times 10^{-19} \text{ J}$$

Table 3 *Prefixes for SI Units*

Fraction	Prefix	Symbol
10^{-1}	deci	d
10^{-2}	centi	c
10^{-3}	milli	m
10^{-6}	micro	μ
10^{-9}	nano	n
10^{-12}	pico	p
10^{-15}	femto	f
10^{-18}	atto	a

Multiple	Prefix	Symbol
10	deka	da
10^2	hecto	h
10^3	kilo	k
10^6	mega	M
10^9	giga	G
10^{12}	tera	T
10^{15}	peta	P
10^{18}	exa	E

Table 4 *Recommended Values of Physical Constants*

Physical constant	Symbol	Value
acceleration due to gravity	g	9.81 m s^{-2}
Avogadro constant	N_A	$6.022\ 05 \times 10^{23} \text{ mol}^{-1}$
Boltzmann constant	k	$1.380\ 66 \times 10^{-23} \text{ J K}^{-1}$
charge to mass ratio	e/m	$1.758\ 796 \times 10^{11} \text{ C kg}^{-1}$
electronic charge	e	$1.602\ 19 \times 10^{-19} \text{ C}$
Faraday constant	F	$9.648\ 46 \times 10^{4} \text{ C mol}^{-1}$
gas constant	R	$8.314 \text{ J K}^{-1} \text{ mol}^{-1}$
'ice-point' temperature	T_{ice}	$273.150 \text{ K exactly}$
molar volume of ideal gas (stp)	V_m	$2.241\ 38 \times 10^{-2} \text{ m}^3 \text{ mol}^{-1}$
permittivity of a vacuum	ϵ_0	$8.854\ 188 \times 10^{-12} \text{ kg}^{-1} \text{ m}^{-3} \text{ s}^4 \text{ A}^2 \text{ (F m}^{-1})$
Planck constant	h	$6.626\ 2 \times 10^{-34} \text{ J s}$
standard atmosphere pressure	p	$101\ 325 \text{ N m}^{-2} \text{ exactly}$
atomic mass unit	m_u	$1.660\ 566 \times 10^{-27} \text{ kg}$
speed of light in a vacuum	c	$2.997\ 925 \times 10^{8} \text{ m s}^{-1}$